The Antidepressant Era

DAVID HEALY

The Antidepressant Era

Harvard University Press
Cambridge, Massachusetts, and London, England

First Harvard University Press paperback edition, 1999

Library of Congress Cataloging-in-Publication Data

Healy, David, MRC Psych.
  The antidepressant era / David Healy.
      p.   cm.
  Includes bibliographical references and index.
  ISBN 0-674-03957-2 (cloth)
  ISBN 0-674-03958-0 (pbk.)
    1. Antidepressants—History.   I. Title.
RM332.H42   1997
616.85´27061—dc21   97-23118

For Helen
Ever skeptical—occasionally puzzled

Contents

Acknowledgments

In early 1995 a white stretch limousine belonging to a Gulf sheikh was ordered off the road in a suburb of San Diego. It had thirty-five seats and was sixty-six feet long, which was one foot over the maximum allowed in California. There's stretching and stretching and this was just too stretched, a highway patrol spokesman said.

I would risk illegality if I attempted to stretch limo all of those to whom I owe a debt in the construction of this book. A great deal of my professional life and interests have come together here, and therefore I am indebted to many of the teachers I have had at various stages, a larger number of colleagues—both clinicians and those working within the pharmaceutical industry—a great number of friends, and a growing number of trainees. I also owe a clear debt to those I have interviewed in the course of my research, as will be apparent from the notes. I need also to particularly mention Brian Leonard, Ned Shorter, Tom Ban, Tony Roberts, and Brian Williams, as well as two former editors, Angela von der Lippe and Roger Osborne. I need to thank my colleagues who covered my absences while I was interviewing and writing and those who have had to listen to my attempts in seminars and lectures to put some of the material into narrative format.

The poem "The Building of the Skyscraper" is from George Oppen, *Collected Poems,* copyright © 1965 by George Oppen, and is reprinted by permission of New Directions Publishing Corporation.

Assuming all those to whom I am heavily indebted could squeeze into a very stretched limo, there is a further problem—my driving, which in the ordinary course of events is only marginally better than my ability to park anything larger than a mini. Other users of this road should therefore beware. Any crash is of course my responsibility alone. I expect those in the back are quietly keeping their fingers crossed and probably their eyes closed—they may even be hoping we get stopped for stretching things too far.

The Building of the Skyscraper

The steel worker on the girder
Learned not to look down, and does his work
And there are words we have learned
Not to look at,
Not to look for substance
Below them. But we are on the verge
Of vertigo.

There are words that mean nothing
But there is something to mean.
Not a declaration which is truth
But a thing
Which is . . .

O, the tree, growing from the sidewalk—
It has a little life, sprouting
Little green buds
Into the culture of the streets.
We look back
Three hundred years and see bare land
And suffer vertigo.

—*George Oppen*

Introduction

Antidepressants were introduced along with the first antibiotics, the first antihypertensives, and a range of other drugs in a therapeutic revolution that took place in the years just after World War II. For the first time, an armamentarium of specific treatments for specific diseases became available, an arsenal of magic bullets, as they were called. This development inaugurated a revolution that has transformed our ideas of disease, of health, and of treatment. It is a revolution that has brought health to the center of Western politics and culture. It may make health the primary focus for global politics in the twenty-first century, if links between human health and the state of the environment continue to converge. Where race, creed, and class divide us, health offers the possibility of a common language.

While modern pharmacotherapy has offered answers to health needs on one level, however, the revolution appears to be a force for alienation at other levels. For many, the idea that unhappiness can be categorized as a disease and should be treated with drugs rather than psychotherapy stands as a potent symbol of that alienation. And yet much of the evidence seems to point in just this direction.

The political, social, and medical issues involved in the treatment of depression came together in the 1980s in a set-piece drama, a court case about the adequacy of the treatment afforded Rafael Osheroff. As we shall see in Chapter 7, the arguments on Osheroff's behalf were that there now exist proven treatments for the management of the depression he clearly had and that an unwillingness to accept and prescribe accordingly had cost him dear and would cost both the psychiatric profession and society dear if the lessons of the case were not taken to heart.

Notwithstanding the Osheroff case, the following example is all too typical of a consultation for depression in primary care in the 1990s. The subject is liable to be a woman in her late twenties or early thirties. She may have been abused in childhood or have other significant events in her past. After the birth of her first child she becomes depressed. She is concerned that her state will rub off on her baby and destroy her relationship with her husband. She sees her mood as stemming from the events in her past. She has already been to see her local clergyman and talked to one or two other people who have felt "down in the dumps," but because she has continued to feel as bad if not worse, she has gone to her primary care physician. After a relatively brief consultation, he offers the opinion that she is one of those women who needs antidepressants. Neither of them takes up the postpartum option, which might have provided a convenient therapeutic myth.

He prescribes antidepressants but she doesn't take them and doesn't improve. Sensing treatment resistance, he offers to refer her to a psychiatrist, an offer she takes because she doesn't have any confidence in her physician's ability to know what is really going on. The expert, however, does little more than to reiterate the recommendation to take antidepressants. She can't see how they can help since she sees her problems as something that stems from her childhood. She believes she needs to talk about it. She had expected to talk or be offered hypnotherapy rather than a "biological" solution.

She is also skeptical of drugs in general. She is not particularly concerned about the immediate side effects that seemed to worry the doctor, but she is afraid of problems like addiction or side effects from long-term treatment—"they didn't know about thalidomide at the time did they?—or about Valium?" Furthermore she regards taking antidepressants as relying on a solution outside of herself and this adds to her sense of failure. She is feeling so bad and is so concerned about the effect on her husband and the development of her baby, however, that she tries the pills despite her fears. After six or nine months with little relief from a variety of antidepressants, including Prozac, she leaves the "system" and seeks help from a hypnotherapist. This works.

This woman has a number of problems. She has a problem with the experience of depression. But she also has a problem because her predicament is being reduced to the contours of a strict clinical target, while her situation is much more than this. Like many other people she is forced to take antidepressants and does so feeling that they are addressing the symptoms and not the cause. She is not offered a meaning, a narrative, or a myth to move forward with. There has been a "take it or leave it" offer to handle what she perceives as her psychic pain with psychoanalgesics, but she hasn't been told what the pain is. She hasn't been told that she is not going mad, and being given an antidepressant, far from being seen as helpful, in her eyes seems to confirm her as a failure.

At present health systems appear to be gearing themselves up to make sure that this woman is given the treatment she doesn't want. Guidelines are being constructed to specify which antidepressants clinicians might use in the event that a first course of treatment fails; these will not include hypnosis as an option. In some places clinicians are no longer reimbursed for doing anything other than prescribe an antidepressant in cases like this. Clearly something is going wrong here, but what? This book will attempt to pin down the problems by looking at the impact on psychiatric practice of one group of drugs—the antidepressants.

In looking at where we have come from and where we might be going, I am probably coming from much the same position as many people who will read this book from outside of psychopharmacology. Although I have been the secretary of the British Association for Psychopharmacology, I have also been a grant holder for a cognitive therapy of delusions project and so am clearly not committed exclusively to pharmacotherapy. I remain uncertain as to whether the drugs work as well as they are claimed to, or whether their apparent superiority to psychotherapy stems in part from the fact that current methods of assessment favor drug treatments, or indeed whether drug treatments could work even better than often appears to be the case in clinical practice. It is a valid question whether the therapists using these treatments are capable of pharmacotherapy or whether therapy is reduced in their hands to a lifeless rather than a life-giving act.

The tone of this story is in many ways skeptical. This came as something of a surprise to me, because by nature I am an enthusiast. There are, however, aspects of the antidepressant story that have forced skepticism on me and I think will do so for many readers. I am not certain that this is the right position to be in and my defense has been to be universally skeptical; skeptical of the motives of clinicians as well as the pharmaceutical industry, skeptical of both pharmacotherapists and psychotherapists. Universal skepticism, however, is a tense position to maintain. It risks toppling over into cynicism, which certainly would be a mistake. If the reader is left with a sense of unresolved tension, this will perhaps best reflect my own position.

This is not only a story about psychiatry; it brings in many areas of modern medicine. It does so because the antidepressants occupy a place somewhere near the fulcrum of the health care system. They do so because the common experience of depression lies somewhere midway between illness and disease, and the drugs themselves, the antidepressants, lie midway between magic bullets and snake oil. The same story could not have been told of schizophrenia and the antipsychotics. Because of this there are two stories interwoven here, a dramatic story about the antidepressants themselves, their discoverers, and those who worked on their mode of action, but there is also a broader story as the very ambiguity of the antidepressants acts as an attractor for many of the forces at play in the rest of medicine. As a consequence there is a sense in which the antidepressant story offers an opportunity to examine the medical grin without having the medical cat in the way.

The foundation of this book is that depression as it is known to or understood by the public in the 1990s was all but unknown as recently as thirty-five years ago. This book therefore necessarily begins with an appeal to the historical imagination. Can you imagine a world in which, effectively, there was no such thing as depression? This seems almost inconceivable now, and so Chapter 1 seeks to convey something of the cultural and social landscape of the predepression era. It attempts to set the scene on the medical and psychiatric stage, just as the antidepressants were introduced. The key players are the public, the pharmaceutical industry, the Food and Drug Administration (FDA), and the psychiatric profession. The key issues are the disease concept and whether depression is a disease as well as the appropriate level of regulation of the pharmaceutical industry. It is against this background that the antidepressant drama has unfolded. Many drugs follow a pattern of being hailed as a panacea at first, only to be later dismissed as poisons before achieving pedestrian status years or decades later. This was not the case with the antidepressants, which began life as the poor cousins of both the major and minor tranquilizers; their history has had a slow beginning. Recently, however, with the emergence of the selective serotonin reuptake inhibitors, it has appeared to visibly gather pace, just as the sand falling through an hourglass appears to do.

When the "antidepressants" were first "discovered" or "invented" depression was perceived as relatively rare. The first reports of new drug treatments were greeted with interest as much for the light they might shed on the nature of the disorder and on the question of how depressions might best be classified as for any benefits they might produce for depressed individuals. Indeed the idea that there might be a depression that drugs could treat had in one sense to be invented as had the idea of an antidepressant—and part of the problem to this day is that neither of these inventions has fully worked, at least not with the public at large. As Chapter 2 will show, five different investigators or groups could claim to have made the discovery—there were five different inventions. The conventional story assigns priority to only one of them, in some respects the one that now seems the least likely or the most mythical.[1]

Four of the five discoveries were made simply by clinical observations; one involved a randomized double-blind trial. One of the characteristics of the modern era is that nothing can be said to be discovered until it has been through the process of a randomized trial. The discovery of such trials has arguably done just as much to shape the parameters of health debates today as the combined discoveries of all our specific treatments. The path down which trials evolved is traced in Chapter 3 and a first sketch is offered of how the evaluative, economic, and regulatory forces that shaped the Osheroff case began to come together.

The interactions among economic, evaluative, and regulatory forces are traced through the 1960s, 1970s, and 1980s in Chapter 4, in what were a series of highly personalized battles about the place of the monoamine oxidase inhibitors in the treatment of depression, the place of lithium in depression's long-term treatment, and the proper place of regulation in drug development. The stakes being wagered by all parties grew progressively during this period, as we shall see.

The advent of the psychotropic drugs has also given rise to a new biological language in psychiatry. The extent to which this has come to be part of popular culture is in many ways astonishing. Shortly after the discovery of the antidepressants, the idea of deliberately tampering with anyone's brain chemistry was seen as all but godless, while now details of refinements in the designer specificity of compounds like Prozac are the stuff of fashionable coffee-table talk. The growth of this language, the early resistance to it, and its subsequent triumph are charted in Chapter 5. This triumph, however, is not without its ambiguities. It can reasonably be asked whether biological language offers more in the line of marketing copy than it offers in terms of clinical meaning.

In Chapter 6 a further perspective on the marketing of antidepressants opens up. In many respects the discovery of antidepressants has been the invention of and marketing of depression. In the last decade, the pharmaceutical companies have struck out from the beachhead of depression into the heartlands of the neuroses, marketing obsessive-compulsive disorder, panic disorder, and social phobia as they have gone. Ironically, a good case can be made that these developments have depended heavily on the regulatory mechanisms put in place in the 1960s, which were aimed at funneling pharmaceutical developments down "ethical" channels.

This is a story of ambiguities and ideologies. The promise of a pharmacological scalpel that the antidepressants brought to psychiatry and the rise of a biological language that they gave rise to, looked at from a different angle, can appear little more than a cloak of artistic verisimilitude for quite other processes. It is a story of ideologies that clashed head on in the Osheroff case. The development of these ambiguities and ideologies are outlined throughout the book; the current balance of the opposing forces is assessed in Chapter 7. The significance of the sound and fury of battle will be for the reader to decide.

My own views on depression and regulation are in the Postscript. These lie sufficiently outside the mainstream, outside the currently politically dominant views, I hope, to persuade readers that most of the extraordinary events outlined in the following pages were actually as extraordinary as I have portrayed them and that the tension and drama generated have not been part of a plot to land an unsuspecting reader at some predetermined landing place.

1 Of Illness, Disease, and Remedies

For most of human history, the first question in the face of ill health has been the question *Why?—Why Me? Why Now?* This was the question that older premodern humoral models of health and illness sought to address. In contrast, modern clinicians refuse to engage with the why of illness and focus instead on *How* a particular disease has evolved and *How* it may be corrected. This chapter will trace that transformation in medical approaches, for the interplay between these two questions is of crucial importance when it comes to the management of any disease but arguably for the management of depression in particular. Depression also raises questions about the nature of the disease concept, the extent of its application, and the differences between the idea of a disease and the experience of illness. The different ways depression has been seen over the past forty years and the way its management appears to be going offer, as we shall see in succeeding chapters, some feel for the interplay among the medical profession, the pharmaceutical industry, the regulatory apparatus, and the larger culture in which we live. Depression thus offers an insight on the forces shaping the conceptual landscape within which illnesses are experienced and diseases defined.

The story begins with the humoral models of health and disease, of which the best known are the Hippocratic system, which involved four humors, the Ayurvedic system of three Dhosas, and the Chinese Yin and Yang system. Even better known today, although not classified as a humoral system, is the alternative health movement, the many tributaries of which have a common commitment to what essentially are humoral principles. According to the Hippocratic system there were four humors: blood, which was hot and moist and came from the heart; phlegm, which was cold and moist and came from the brain; black bile (melancholy), which was cold and dry and came from the spleen and bowel; and yellow bile, which was hot and dry and came from the liver. Bile, phlegm, and the other humors could bring about disease when they got too dry or wet or hot or cold or out of balance. "If the brain is corrupted by phlegm the patients are quiet and silent, if by bile they are vociferous, malignant and act improperly. If the brain is heated, terrors, fears and terrifying dreams occur; if it is too cool the patients are grieved and troubled."[1]

In general in humoral systems, health is seen as a state of harmony of the humors and ill health or disease a state of dysharmony, imbalance, or dyscrasia. Imbalances could be brought about by eating the wrong proportion of hot or dry foods, for instance, by the wrong exercise regimes, by exposure to the wrong elements—too much of a hot or dry wind—by irregularity of bowel movements or by the inappropriate loss of fluids as in masturbation or menstruation. The treatment of a dyscrasia involved efforts to readjust the humoral balance. The preeminent treatment was bloodletting, which could be employed in the case of anyone who was flushed or whose circulation appeared speeded up. Alternatives included a range of enemas, purgatives, or emetics.[2] A cure of this sort was enshrined in myth when Melamphus cured the daughters of Argos of madness by giving them milk from goats that had eaten black Hellebore—a drastic purgative.

Following the eclipse of Greece by Rome, Galen of Pergamon, who lived from A.D. 129 to c. 200, became the preeminent medical authority for almost 1,500 years. Galen's commentaries on Hippocrates became better known than the Hippocratic corpus itself. But where Hippocratic thinking had been a loose collection of aphorisms put together by practitioners with very different viewpoints, Galen made a system of the Hippocratic writings by marrying humoral theory with Aristotelian logic. He took the theory of humors as a set of given first principles and argued that the job of the physician was to reason logically from those principles to the current state of the patient. The system was all but irrefutable in that superficially discrepant observations could be reconciled in an overall synthesis of humoral, personality, and environmental factors.[3]

With the collapse of Rome, Galen's works were taken up by Arab physicians and from there were incorporated into the medical thinking of Europe after 1250. His works were seen as so important that they were among the first books to be printed with the arrival of the printing press in the late 1400s. But whereas Renaissance discoveries in physics led to the overthrow of Aristotelian physics and the Ptolemaic cosmology, the emerging anatomical knowledge of the seventeenth century on the circulation of the blood, for instance, or demonstrations of the existence of nerves were incorporated into updated versions of Galen's models. A variety of systems emerged which substituted ideas of tension of the blood vessels or the nerves or obstructions in the flow of the lymph or even aberrant chemical reactions in one or another set of bodily tissues for humoral imbalances. But all such systems, in one way or another, remained true to Galen in that they attempted to relate local problems to disorder in the body as a whole.

When it came to treatment, the more dramatic the effects the more useful the medicine. The physician's aim, where possible, was to mimic the body's methods of healing. Diseases produced sweating, urination, phlegm, the eruption of blisters or skin lesions, or diarrhea, so the physician also aimed to produce diuresis (the passage of water), evacuation of the bowels, blistering, the discharge of blood, or even the production of abscesses elsewhere in the body to act as counterirritants. Benjamin Rush, who is commonly invoked as the father of American psychiatry, for example, saw all disease as involving an excess of vascular excitability. Rush's prescription was heroic depletion of blood or the use of calomel, jalap, or the emetic ipecac. During one epidemic, he bled his entire family of twelve children twice a week. He even bled his six-week-old child.[4] Such practices could be justified on the grounds that they were acting in imitation of nature. Given that much healing is spontaneous anyway, their use was almost certainly associated with some recoveries. Moreover, they produced effects—even clearly desirable effects, as when a sickly flush was cleared by bloodletting or the agitated patient was apparently calmer after venesection.[5] The distance from modern medicine is well illustrated by this point. Today the production of "side"-effects is more likely to lead to a lawsuit than congratulations on a job well done. But for the premodern physician the evidence that a humoral regimen was working was there for all to see—neither randomization nor statistical methods were needed to prove the point.

The job of the physician was to know the theoretical framework that would predict how a local procedure would affect and be shaped by the economy of the whole body. Even when inoculation for smallpox gained acceptance, for instance, its use was embedded in regimes of cathartics, rest, and diet aimed at treating the whole person. In practicing this way, the physician could depend on the fact that he shared the same world view as the patient he was treating. He also brought to the encounter a view of illness that was attested to by all of the significant authorities from antiquity through to the present day, thereby embedding the therapeutic act in a framework that made sense to the sufferer and perhaps contributed accordingly to successful outcomes and compliance with treatment.

A modern form of humoral ideas clearly remains in the health food industry, which capitalizes on our seemingly innate desire to regulate our internal balance with a range of vitamins, minerals, tonics, herbals, and other compounds. There are a few points worth making about this. One is that clearly such ideas seem intuitively right to many people. Second, these approaches are not specific to particular diseases; for the most part they aim at correcting imbalances affecting the body as a whole rather than the semimechanical defects that modern medicine sees as its targets. Third, they remain available to people to administer as and when they see fit. This element of leaving control in the hands of the sufferer, or one's destiny in one's own hands, will come to center stage at the end of the book.

In the humoral framework, the very word *disease* implies illness—the presence of symptoms. In contrast, many modern diseases such as hypertension or cancer are asymptomatic for the greater part of their courses. Does a modern presymptomatic diagnosis help? Arguably not. Being told that one is HIV positive or that one's blood pressure is elevated may extend the experience of illness. Once diagnosed, some people become symptomatic even if they were not symptomatic at the time of diagnosis. To diagnose or not to diagnose—there is no escaping this dilemma in the case of depression, because one cannot be depressed without being symptomatic. One can be symptomatic, however, without being diagnosed and treated with Prozac, for instance. Do diagnosis and treatment help in such circumstances? The evidence on this point is less than clear.

Specific Diseases and Specific Treatments

Until the last decades of the nineteenth century, the idea that there might be a specific remedy for a specific ill was tantamount to quackery. By the end of century, however, medicine was facing a crisis. It seemed as though a choice would have to be made between therapeutic visions—a choice between the treatment of the whole person or the rectification of specific semimechanical defects. A number of factors contributed to the crisis. One was the idea of disease specificity. Another was the emergence of ideas of therapeutic specificity, which crystallized at the turn of the century in the concept of the magic bullet and in the establishment of an industry geared up to produce the bullets. Finally there was the emergence of an interest in outcomes, to be explored in Chapter 3.

The first radical break from the idea of whole body diseases came from Giovanni Baptista di Morgagni in 1761 who, on the basis of efforts to correlate disease pictures with autopsy findings, came to the conclusion that some diseases at least could be caused by a lesion in one organ system only. This was a profound rupture with the past. Morgagni's proposal quickly led to proposals that inflammatory conditions of the kidney identified at post mortem, for instance, might constitute the disease of nephritis, and that inflammatory conditions of the nerves might be a completely different disease—a neuritis. In 1800 Xavier Bichat proposed that the pathological disturbances behind a disease might not even involve all of one organ, such as the kidney or the nervous system, but might rather be limited to one tissue type in an organ.[6]

These pathological findings did not lead immediately to the idea of specific diseases. It was always possible for those of a humoral bent to fit the findings into a picture of imbalances by claiming that there were probably counterbalancing disturbances elsewhere in the body or that nervous or vascular transmission, for instance, might transmit the disturbances throughout the body. It must be remembered that the investigative techniques that might have detected and been used to demonstrate localized pathologies in living subjects did not exist. Neither did either the medical or the surgical techniques to remedy any defect.[7] Nor did, as we shall see in Chapter 3, the statistical techniques that now allow us to look into the social body to track the course that specific diseases run. In a real sense, therefore, there was little to be gained by choosing the new views of disease, while there was much to lose. One American practitioner in mid-century commented: "They are tearing down the temple of medicine to lay its foundations anew . . . they lose more in therapeutics than they gain in morbid anatomy—they are explaining how men die but not how to cure them."[8]

It was the emergence of bacteriology in the 1880s that made the difference. Initially there was intense resistance to the new bacterial theories. When Joseph Lister attempted to introduce antiseptic techniques to surgical theaters in Glasgow in 1860 he made little progress at first, despite being able to demonstrate an improvement in outcome. It was thirty years before the new practice won acceptance. Louis Pasteur's pleas for sterilization went similarly unheeded, until he circumvented medical resistance by appealing to the public at large. Events were on his side. The isolation of the tubercle bacillus by Robert Koch, in 1882, and the organism responsible for cholera, in 1884, had a considerable impact. All of a sudden it became fashionable for medical men to be seen looking down microscopes. But even then many argued that these organisms would only cause disease against the background of a susceptible constitution. Given that disease was something that self-evidently involved the whole person, many physicians simply could not understand why a medical man would want to have a microscope.[9]

Some flavor of the way things were seen can be gauged from a famous self-experiment by Max von Pettenkoffer. While accepting that Koch's new organism might be involved in cholera, he pointed out that there were places where the organism could be demonstrated in the drinking water but cholera did not follow. Accordingly he suggested that the disease needed not only a specific germ but also the action on that germ of specific local factors, such as soil conditions, seasonal factors, and a susceptible individual constitution. To prove the point, on October 7, 1892, he swallowed a broth of cholera organisms, isolated from a patient who had died from it. Lest his stomach acids neutralize the organism, he drank some alkali as well. He had abdominal cramps and diarrhea for a week but was not otherwise affected by the disease. Fortunately, it would seem, because the general view now would be that the majority of subjects participating in such an experiment would be at serious risk.

Unlike Morgagni, however, bacteriologists such as Koch did more than identify pathogens. They established techniques for isolating and culturing them. The skeptics could literally be invited to come and put their hand in the wound. Furthermore, the isolation and culturing of microorganisms permitted the testing of a range of disinfectants for sensitivity to particular microorganisms and also the development of vaccines or antitoxins. This complex of theory, demonstration, and treatment finally overcame two thousand years of tradition.

One of the key episodes centered on the treatment of diphtheria, which at the end of the nineteenth century accounted for up to 5 percent of childhood deaths in urban settings in most of the Western world. It killed by secreting a toxin that was responsible for the formation of a membrane across the throat, leading to suffocation or bleeding to death because of efforts to tear the membrane. Even though the signs of the disease were distinctive, there was a confusion between diphtheria and other throat problems, which was only conclusively resolved when Edwin Klebs in 1883, working from Koch's laboratory, identified a specific diphtheria organism. This led to the production of an antitoxin, and by 1893 the first reports of success in children emerged. Large-scale production of the antitoxin began in 1895, and deaths from the disease began to fall dramatically from 1896.

The new bacteriology forced the creation of a new model of disease—a specific disease model. Where previously fevers had been manifestations of a humoral imbalance or nervous debility, all to be treated similarly with purgings or bleedings or tonics, it now seemed that each of the eighteen bacteria that had been identified by 1887 might be associated with its own disease and that each might need specific management. The new ideas attracted worldwide interest. It is estimated that upward of three thousand American medical students or clinicians went to Germany during the 1880s to attend lectures.

It will not come, therefore, as a surprise to the reader to hear that in 1896, the year diphtheria antitoxin was launched, the first proposals regarding specific psychiatric diseases emerged. In 1896 Emil Kraepelin proposed a division of mental disorders into two specific diseases, manic-depressive insanity and dementia praecox. At almost the same time, Sigmund Freud put forward ideas about a specific cause and specific cure for hysteria.

While Kraepelin's ideas only really took hold half a century later, following the emergence of the first psychiatric magic bullets (illustrating the point perhaps that specific diseases are of no use without specific cures), Freud's views had much greater impact. Ever one to be abreast of the fashion, Freud decided that if hysteria was a specific illness, then it must have a single specific cause and it should respond to a specific intervention. Because some of the hysterics he was seeing appeared to have been almost certainly sexually abused in childhood and this abuse seemed causally related to their condition, he argued that all patients suffering from hysteria must have been similarly abused. The idea that trauma might lead to hysteria was not unique. At much the same time in Paris, Pierre Janet was moving to a view that many hysterics had been traumatized but, unlike Freud, Janet appeared happy with a generic form of trauma rather than the very specific psychic germ Freud was suggesting.[10]

Part of Freud's claim stemmed from what appeared to be a specific therapeutic response. He had become involved in abreacting his patients—a therapy that supposedly involved reuniting memories with emotions that had been split off from them. This appeared to lead to more than a remembrance of abuse—it seemed in a number of cases to produce such a vivid reliving of the original trauma that he felt there was little room to doubt that the event had actually happened. All the ambiguities of what might have been happening are of course now being played out in acrimonious court cases about the validity of repressed memories, but he lacked the benefits of a century of debate on these issues. In addition, his specific therapy initially appeared to produce a cure. His later admission that the cures were not sustained was a crushing blow to the validity of his claim to have discovered the specific cause of hysteria.[11]

In the face of these difficulties, in 1897, he abandoned his claims for a specific etiology for hysteria in childhood sexual abuse, with immense consequences for psychopathology and twentieth-century culture generally. There has been considerable debate about the role that social opprobrium may have had in causing him to change his mind. It is assumed that people were shocked at his proposal, although there is little evidence that there was much overt hostility. What is rarely recognized is that the form of his hypothesis was cut from the cloth of his age and that just as the germ theory could be challenged by Pettenkoffer's apparent demonstration that a preexisting constitution was even more important than the supposed germ, Freud was able to take a similar escape route when the one-to-one link between sexual abuse and hysteria began to break down. He spent the rest of his career investigating the factors that made for a neurotic personality or constitution, possibly significantly extending the experience of illness for all of us in the process.

FROM BLACK HELLEBORE TO MAGIC BULLETS

What sense would it make for a physician if he discovered the origin of diseases but could not cure or alleviate them? . . . I am directing you, physicians, to alchemy for the preparation of the arcana, for the separation of the pure from the impure, to the end that you may obtain a flawless, pure remedy, God-given, perfect and of certain efficacy, achieving the highest degree of virtue and power. For it is not God's design that remedies should exist for us ready-made, boiled and salted, but that we should boil them ourselves . . . and learn in the process.

—Paracelsus[12]

Paracelsus was born in Switzerland in 1493. He qualified as a physician in 1515. In 1526 he was appointed professor of medicine at the University of Basel. Less than two years later, on February 15, 1528, he was stripped of his title and a warrant for his arrest was issued. Why? There were a number of factors. One was that he rejected the doctrines of Galen, Avicenna, and Aretaeus in favor of a new doctrine of signatures, which was to be built up from contemporary observations of nature and man's responses to her remedies. "My proofs derive from my own experience and my own reasoning and not from reference to authorities." A second reason was his willingness to judge remedies by their outcomes, which we shall consider further in Chapter 3. A third reason was his advocacy of chemical treatments.

The chemistry of the early Renaissance was entirely different from modern chemistry. There were no periodic tables and no understanding of the principles that might lead to predictable interactions between compounds or to their rational conjunction. The branch of alchemy that was later to develop into chemistry was called the Spagyric Art. This aimed at finding methods to separate pure from impure compounds and to correct the imperfections that might exist in natural compounds.

It was this that Paracelsus turned to in an effort to seek out poisons to counteract the poisons that might be causing disease. One of his famous maxims was that "everything is a poison; it is but the dose that determines whether it kills or cures."[13] He championed a belief, possibly widely held by country healers at the time though rejected in the temples of orthodoxy, that in addition to herbs, chemicals and metals of various sorts could be used in healing, as a benign God would not have put them there otherwise. They were there so "that by a little industry of ours they might be changed into pledges of His love, for the use of mortals against the rage of future diseases."[14]

The discovery of medicines was to be made in accordance with the doctrine of signatures. Nature, according to Paracelsus, had made liverwort and kidneywort with leaves in the shape of the parts of the body each of these plants could cure. She had made the leaves of thistles prickly like needles and thanks to this sign the therapeutic art discovered that there is no better herb against internal pricklings. Syphilis was associated with the marketplace, which comes under the sign of the planet mercury, and therefore mercury should be used for syphilis.[15]

Adopting a related approach in 1763, Edward Stone had stumbled on the fever-quelling properties of the bark of the willow tree. He reasoned that Providence would place cures close to the diseases they might remedy. Aware that some tree barks could help fevers, he felt that the bark of the willow tree might be a good candidate for a remedy, given that it grew best by swamps and people seemed more likely to get fevers near swamps. Paracelsus would almost certainly have been pleased with both this and the subsequent isolation of an active principle, salicin, by German chemists in 1828. This was improved by French chemists in 1838 by conversion to salicylic acid and finally transformed in 1896 into acetylsalicylic acid, a compound which was more effective and had much fewer side effects than the original. As Aspirin, it became one of the most successful compounds ever launched.[16]

Although some distilled oils and metals had been used before Paracelsus, he inaugurated an era in which large numbers of chemical and metallic compounds were tried on patients for the first time—calomel, ether, mercury, antimony, and lead—leading John Donne to complain a century later that Parcelsus must have been in league with Satan, as he had introduced a whole new family of death-dealing drugs. By the time Donne wrote, the use of "chemicals" had made considerable inroads on medicine and a new breed of physicians, the chemical doctors, had appeared and achieved influence in the courts of Henry IV in Paris and James I in England.[17]

THE MARKET IN MEDICINES

A large market in drug therapies emerged in the seventeenth century, serviced by the druggist or dispensing chemist. The chemist could synthesize and sell one of two types of compound—either a concoction, later to be termed a patent medicine, or more specific and purer ingredients, whether herbal or chemical. The pure ingredients were in general sold to doctors, who might compound concoctions from them in turn. The sale of pure ingredients was later to form the basis for distinctions between ethical, or prescription, medicines and patent medicines. Ethical drugs were advertised only to physicians and not to the public at large, whereas proprietary or patent medicines were sold directly to the public and their ingredients were a secret.[18]

The various patent concoctions sought to capitalize on an increasing public demand for medicinal products and an apparent willingness of people to self-medicate. The norm was almost certainly self-treatment for almost all ailments, with recourse to a physician only in extremis. The customers were Galenists in their approach. They were more likely to think the medicine was efficacious if it produced visible reactions, and so the various medicines contained a range of tonic, purgative, stimulant, or sedative factors. The market grew dramatically. A variety of nostrums became household names. In the main these were aimed at universal complaints, such as halitosis, indigestion, nervous fatigue, and debility—at illnesses rather than at what we now think of as diseases.

Many important marketing principles were established in the process. The first U.S. patent went to Samuel Lee, Jr., of Connecticut for "Bilious Pills." Another Samuel Lee from Connecticut also obtained a remedy for Bilious Pills and a war was launched in the newspapers between the two proprietors. Far from leading to a collapse in the market, the indications are that this dispute helped to make the market for Bilious Pills, judging by the number of other proprietors across the country who joined in with their own preparations shortly thereafter. The increase in sales for a particular commodity by competition among producers is a story that has been repeated over and over again from the marketing of aspirin to that of Prozac, Zoloft, and Paxil.[19]

In 1804 there were some 90 patent medicines listed in New York. By 1857, this had swollen to over 1,500 nationwide. The growth in business paralleled the growth of the press and literacy. Where there were 200 papers in 1800, there were 4,000 by 1860. The proprietors of patent medicines were among the first in the United States to market nationally rather than just locally. They also marketed heavily. According to congressional documents, in 1849 one producer was spending $100,000 per year advertising his purgative. The proprietary medicines industry spent more than any other industry on advertising in the second half of the nineteenth century—more than clothiers, haulers, or any other group. By the end of the century up to $1,000,000 was being spent annually on telling the American people of the benefits of Scott's Emulsion, which was just one of an estimated 50,000 compounds, in a trade that had a retail value of several hundred million dollars.[20]

Many of the principles of modern advertising were created around efforts to market such compounds. One of the leading early lights in advertising, Claude Hopkins, recalled that "the greatest advertising men of my day were schooled in the medicine field."[21] The patent medicines industry was among the first to market lifestyles rather than the compound per se. As the marketing of aspirin was to show, there appears to be almost no limit to the amount of money that can be put into an advertising campaign—whatever is put in always seems to yield a return in terms of increased consumption of the product, even when most observers feel that prior consumption has been at maximal levels.

The first steps toward what later became known as the ethical pharmaceutical industry came with the isolation of morphine and quinine. Although opium had been used for at least three millennia, and Paracelsus among others had used it in the treatment of depression in the form of a laudanum, it was not until the early years of the nineteenth century that efforts began to isolate its active principle. The breakthrough came in 1806 when Friedrich Sertürner isolated morphium. He had been stimulated by observations that some samples of opium relieved pain much more effectively than others—some were ineffective and some produced overdoses. This suggested an unevenly distributed active principle, which if it could be isolated might be standardized so that exactly the same amount could be given each time. Sertürner's discovery and his methods opened the way for the subsequent isolation of emetine, strychnine, codeine, caffeine, atropine, and other salts.[22]

The second important salt came from the bark of the cinchona tree. This was first mentioned in Europe in the Chronicle of St. Augustine, which was written sometime around 1633 by an Augustinian monk, Father Antonio de la Calancha, who had lived in Peru. He and others had noted that the bark of the Cinchona (fever-bark) tree could alleviate the fever of malaria. Because Europe at the time was subject to malarial plagues (ague), the finding was of some significance. The bark was shipped to Seville, Rome, and other European cities, where it was known as Jesuit's bark. Efforts to isolate an active principle failed until in 1819 Carl Runge in Jena and in 1820 Pierre-Joseph Pelletier and Joseph-Bien-aimé Caventou in Paris isolated China base or quinine from it. Pelletier and Caventou published their results and urged clinicians to make use of pure quinine. An epidemic of malaria in Barcelona led Pelletier to establish a factory for its production.[23]

Another development took place in the United States in mid-century. A number of firms that had set up in the Philadelphia area, such as the later Smith, Kline & French, began to look to sales across the country. This necessitated large-scale production and mechanization, advertising, and discounting. The variation in products from manufacturer to manufacturer and from batch to batch led to pleas for standardization of the dose and form of preparations. Against this background, some firms established reputations for quality on the basis of the production of accurate amounts of medicines such as quinine, for which Smith, Kline & French won the contract with the U.S. army during the war against Mexico. E. R. Squibb and Wyeth similarly secured contracts during the Civil War. These developments led to the creation of pharmacopoeias, the establishment of pharmacy courses and colleges, the publication of pharmacy journals and the foundation of pharmaceutical associations, which pressed for the enforcement of the licensing of practitioners.[24]

Both quinine and morphine to the modern eye are demonstrably specific active principles. It was later claimed that their isolation did for medicine what gunpowder had done for war, but this is a judgment colored by retrospection. Their isolation did not, initially, lead to any great change in the concept of therapeutics. For the better part of half a century, quinine was thought to be active by virtue of being a tonic and morphine was regarded as either a stimulant or a sedative, depending on the case being argued. The real change came as a consequence of the development of organic chemistry. This for the first time enabled entirely new compounds—not otherwise found in nature—to be synthesized.

Atomic physics and evolutionary biology are commonly seen as the pillars on which modern culture rests, but organic chemistry has arguably been every bit as important as these in shaping the modern world. Its development and the growth of the chemical industries that stemmed from it spanned the same years that saw Darwin and others formulating the theory of evolution. The opening discovery in what was initially called coal tar chemistry is usually cited as the formation of aniline purple by William Perkins in 1856. When coal is heated in the absence of oxygen, a range of volatile products known as coal tars are produced. Working in the laboratory of August von Hofmann at the Royal College of Science in London, Perkins was engaged in just such a process in an attempt to synthesize quinine, when he produced aniline purple—a mauve substance that he quickly realized could act as a dye. He set up his own company to exploit this possibility, and a number of other companies were soon established as it became clear that variations on this process could give rise to benzene, toluene, xylene, naphthalene, creosote, and a range of additional products for which there was a substantial market.[25]

The exploitation of Perkins's finding required a new breed of chemist— the organic chemist. Organic chemistry is often regarded as having been clearly established as a separate field with the publication of Marcellin Berthelot's *Chimie Organique,* in 1860.[26] This work was made possible by developments in the 1850s that saw the establishment of the principles of organic structure—isomerism and valency—by August Kékulé, Archibald Scott Couper, and others. It was now possible to offer models of how organic molecules (carbon-based molecules) hung together. The availability of these models made it possible to manipulate structures systematically to yield a range of new compounds—soaps, detergents, plastics, resins, adhesives, preservatives, pesticides, cosmetics, and textiles.[27]

It was soon to become clear that these compounds could also interact profoundly with biological processes.

In Basel, in 1859, J. R. Geigy established a dye company and Alexander Clavel set up the company that was later to be Ciba. In 1862 Jean Gaspar Dolfus established a company that would later be Sandoz. All were dye producers. Ciba only produced its first pharmaceutical preparation in 1889; Sandoz produced a medical remedy for the first time in 1921 and Geigy in 1940 (although the key compounds from which the psychotropic drugs were later to come had all been synthesized by the turn of the century).[28] After World War I, all three of these companies branched out into textiles and in the 1930s into plastics and insecticides. In addition to setting up a home base, all three quickly moved to set up branches outside Switzerland, establishing the basis for later multinational developments. There were a number of reasons for doing this; one was to circumvent patent laws, another to avoid import duties or export tax.[29]

In Germany, a comparable industry was developing. Its turn to pharmaceuticals in the 1890s underpinned the interest Ciba and the other Swiss companies had in this sector. The vagaries of German patent law were such that the process of producing a discovery was patented rather than the end product. When another coal-tar derivative, acetanilid, was accidentally found in 1886 to have fever-reducing properties, Kalle & Company, unable to find a patentable way to produce it, came up with the idea instead of registering a memorable brand name, Antifebrin, under which it sold the product. Kalle's sales boomed.[30]

Another German company pursued this route with even more dramatic results. In 1862 Friedrich Bayer had established the forerunner of the Bayer Company, again to make dyes from coal tar. In order to survive in what had become a very competitive business by the 1880s, Bayer began to build a research base. A number of scientists were appointed and Carl Duisberg, an accomplished chemist, was put in charge of a research department. Aware of Kalle's example, he looked through the company's stock of coal tar derivatives and tested a series of variations on one, para-nitrophenol, for antipyretic (fever-reducing) properties. This produced acetophenetidin, which Bayer sold under the brand name Phenacetin. It was the first time that a drug had been conceived, developed and tested in-house, and subsequently marketed by the same company.

Duisberg hired more chemists on the basis of Phenacetin's success. By 1890 the company had ninety scientists on staff. In 1896 an entirely separate pharmaceutical division was set up, which in 1897 began to sell diacetylmorphine, under the brand name Heroin, a product that has had considerable consumer loyalty ever since. In 1899 Bayer brought out acetylsalicylic acid under the brand name Aspirin. Following the model established first in Europe by Kalle but already widespread in the United States, Aspirin went on sale worldwide at virtually the same time, trading under the same name wherever it was sold—in countries as diverse as Argentina and Thailand. Sales of Aspirin and of other compounds, branded and sold in a similar manner, saw to it that the stock of pharmaceutical companies uniquely rode out the Great Depression.

Many of what are now the best-known American companies became heavily involved in these markets as a consequence of World War I. Compounds such as aspirin had to be supplied to the American market during the war, when supplies from Germany were interrupted. At the end of the war, the U.S. government intervened to confiscate German properties and patents in the United States, selling them off to American producers. Sterling Health acquired Bayer US and Aspirin, for instance. The United States took this step in large part because of the German patent on an apparently specific treatment for syphilis, Salvarsan, whose supply had been disrupted by the war.

Salvarsan had been synthesized by the chemist Paul Ehrlich. He had been impressed by the specificity of antibodies produced by vaccines, which zeroed in on and destroyed target cells without damaging anything else. He reasoned that developments in organic chemistry were such that it should be possible to produce comparably specific compounds industrially by manipulating the side-chains of the various coal-tar derivatives to produce what he termed *magic bullets*. The finding that a dye, trypan red, was effective in the treatment of trypanosomiasis (sleeping sickness) appeared to support his argument. Salvarsan was a derivative of arsenic. It killed syphilitic spirochetes in the test tube. It was introduced in 1910 as the first magic bullet. In practice, Salvarsan's results were disappointing, but this only became apparent after the war was over. During the war, it seemed that American troops were being denied an important treatment.

Among the American companies to benefit from the moves of the federal government were Abbott Laboratories, Smith, Kline & French, and Parke-Davis, who along with Eli Lilly and a few others already had a research base as a consequence of their involvement in the production of antitoxins and vaccines. This group of companies developed further in the 1920s, following the isolation of and synthesis of vitamins. In 1929, for instance, Abbott set up a facility for the production of cod liver oil, followed by a variety of other fish liver oils. This was followed in the 1930s by multivitamin preparations, containing vitamins A, B, C, and D. Eli Lilly had become involved in the production of standardized preparations of insulin following its discovery by Frederick Banting and Charles Best.[31]

Then, in 1931, the first true magic bullet was found. Gerard Domagk of I. G. Farben (formerly Bayer), working with dyes with a sulfur-nitrogen chain (hence the name sulfonamides), synthesized a red dye, streptozon. This turned out to be antibiotic in laboratory mice. Two years later it was tried on a ten-month-old child apparently dying of blood poisoning and the child recovered. Domagk's own child developed blood poisoning and was saved. The first sulfa drug was launched worldwide. The Pasteur Institute, in 1935, found that one of the metabolites of streptozon, sulfanilamide, was also an active antibiotic, which the institute then patented and launched. In addition to other antibiotics, the first diuretics, antihypertensives, and oral hypoglycemic agents were developed from the sulfa drugs during the 1950s. A new era had begun. But whereas consumers had up till then been free to purchase whatever compounds they wanted for whatever they wanted, this liberty was taken from them just as the first safe and relatively effective medicines came into existence. The new era was to be a regulated one.

Regulation and the Medico-Pharmaceutical Complex

Roper: So now you'd give the Devil the benefit of the law!

More: Yes. What would you do? Cut a great road through the law to get after the Devil?

Roper: I'd cut down every law in England to do that.

More: Oh? And when the last law was down, and the Devil turned round on you, where would you hide, Roper, the laws all being flat? This country is planted thick with laws from coast to coast—man's laws, not God's—and if you cut them down . . . d'you really think you could stand upright in the winds that would blow then?

—R. Bolt, *A Man for All Seasons*[32]

In 1901 there were outbreaks of tetanus in Camden, New Jersey, and St. Louis that appeared to be connected to supplies of smallpox vaccine. There was an immediate concern that the problem might have stemmed from unsanitary practices in the production of the vaccine by one or more of the large producers. Another possible culprit was the novel type of vaccine preparation, which most of the major producers had switched to, in which case the market as a whole could be seen as responsible. The outcome of the ensuing debate in Congress was a 1902 act that left the production of medicines in the private sector but aimed to regulate it.[33]

In 1905 a bill to regulate the production of food came before Congress. Current wisdom is that the bill would probably have failed but for the publication of Upton Sinclair's novel *The Jungle,* which painted an extremely disquieting picture of the meat packing industry. The public clamor in response to this book may have added that something that pushed the 1906 Food and Drug Act into being. The inclusion of medicines in the act owed a great deal to the efforts of Samuel Hopkins Adams, who in 1905 had run a series of articles in *Collier's* magazine under the headline "The Great American Fraud." These were aimed at bringing to public attention the fact that the better part of a $100,000,000 would be spent that year on mixtures containing alcohol, opiates, narcotics of various sorts, and any number of unspecified ingredients, which were liable to be harmful to one's health. Adams exposed Radam's Microbe Killer as containing over 99 percent water. He inveighed against catarrh powders that contained cocaine and against pain killers with toxic concentrations of acetanilid. In response, the 1906 Food and Drug Act specified that the main ingredients of foods and drugs had to be identified on package labels and that the labels could not be misleading.

Along with the 1902 act, this act had implications for the companies serving the market. The effect of regulation was to debar the field to smaller companies and to push the larger companies in the direction of having a scientific capability. Links were developing between academia, industry and the government. As Willard Graham of Smith, Kline & French put it, the act meant that only competent and reliable firms could be engaged in the manufacture of pharmaceuticals. Carl Alsberg of the Bureau of Chemistry (later to become the Food and Drug Administration), wrote that "the Bureau is not infrequently appealed to by industry to compel the cessation of unfair practices and to encourage the standardization of products when the industry is incapable by itself of bringing about these rules." He continued: "Since the Bureau of Chemistry has always regarded it as its duty not merely to report violations of the law but also to prevent violations by constructive work intended to improve methods of manufacture, it cooperates actively with such associations of manufacturers. Such cooperation by the various Government agencies is bound to exert the profoundest influence on the country's industrial and social development."[34] This influence has been branded a conspiracy by critics such as Peter Breggin and groups like the Church of Scientology. Whether a conspiracy, necessary evil or force for the good, this setting of the rules of the game has a pervasive influence that is commonly unrecognized or underestimated.

The next tightening of the regulatory net involved sulfanilamide, which had been introduced, as mentioned, in 1935. In 1938 reports emerged of deaths in individuals taking the drug (subsequent estimates suggest there were at least 107 deaths, a number of whom were children). The problem was traced to Tennessee, where a chemist at a pharmaceutical plant had dissolved the sulfanilamide in diethylene glycol, which produced an acceptably fragrant and passably tasting mixture that the company sold as an elixir. The Bureau of Chemistry intervened but was only able to do so on a technicality—an elixir must contain alcohol and therefore the preparation was misbranded.[35]

This disaster led to a Food, Drug and Cosmetics Act in 1938, which banned false or misleading statements in the labeling of a remedy and prohibited the marketing of any preparation of a compound until it had been shown to be safe. In practice, this meant that in addition to a list of ingredients, the producer of a medicine had to specify its use and that specification had to be acceptable to a newly created agency, the Food and Drug Administration (FDA).

In 1951 the Humphrey-Durham amendment to the 1938 act gave the FDA the powers to decide which drugs should be made available by prescription only. The category of prescription-only drugs had been introduced before World War I to control the consumption of narcotics and psychoactive drugs such as cocaine, but apart from these the market in drugs was essentially a free one. Addicts might need to be controlled but free citizens didn't. The creation following World War II of entire categories of drugs that were available by prescription only radically changed the health marketplace.[36]

The first drugs affected were the antibiotics. In 1928, Alexander Fleming had made his famous observation that the *Penicillium* mold killed the staphylococcus bacterium responsible for a great number of general infections. In something of a testimony to the inability of pure science to effect any change in the world simply by virtue of scientific discovery, the penicillin story went nowhere until 1938, when Ernst Chain began to administer penicillin to infected animals and found that it cured them. There was a problem, however. A liter of penicillium culture produced 1,000 units of penicillin but the treatment of pneumonia, for instance, commonly takes 100,000 units per day. Clearly the discovery's effects would remain limited until someone found a way to produce penicillin more efficiently.

World War II inhibited European companies from doing much about it. In the United States, the Office of Scientific Research and Development took on the task of coordinating the efforts of several of the larger American companies, Merck, Squibb, and Pfizer (all relatively small still). Different penicillium molds were discovered with higher yields, different media were found to promote their growth, and finally Pfizer took a considerable financial risk on deep tank fermentation, which paid off. Commercial penicillin production began. Other companies followed their lead. Yields jumped from 1,000 to 1,000,000 units per liter and production capacity from 400 million units in the first half of 1943 to 650 billion units per month by August 1945. Deaths from pneumonia, influenza, syphilis, rheumatic fever, and other infectious diseases dropped dramatically.[37]

A framework for cooperation among scientists, companies, and the government had been outlined. After quinine had become difficult to obtain, owing to Japanese expansion in the Far East, a similar cooperation between the government and research laboratories at Goldwater Memorial Hospital, headed by James Shannon, had led to the development of synthetic antimalarials. This led to the production of Atabrine, the virtual founding of the science of pharmacokinetics, and the launching of the careers of Bernard Brodie and Julius Axelrod, who were to be key players in the development of biological psychopharmacology (see Chapter 5). The war also launched the administrative career of James Shannon, who later, as director of the National Institutes of Health (NIH), helped forge the basis for the medico-pharmaceutical complex that is one of the driving forces behind the modern health industry.[38]

During Shannon's directorship, the budget of the NIH increased fifteenfold. In part this was due to Shannon. In part it was due to the climate of the times. The Cold War and the Russian launch of Sputnik dramatically symbolized the importance of science. Science found further funds easy to come by. As we shall see, in 1955 Nate Kline, Frank Ayd, and others lobbied Congress for funding to support the evaluation of the newly introduced psychotropic drugs. The resulting funding was so generous that the man charged with the responsibility of allocating it, Jonathan Cole, had difficulty giving it all away.[39] Health research, including mental health research, was being capitalized. The first academic research positions in medicine and psychiatry were set up in the United States at this time, a few years ahead of comparable establishments in Britain and the rest of Europe. It was a period in which a new kind of corporate knowledge was being generated.

Exactly the same development of an interlinking network of basic science research, private industry, and government funding, in the case of physics and chemistry, had been termed by Eisenhower in 1958 the military-industrial complex. In a speech recognizing the benefits and dangers of the new arrangements, he noted: "In holding scientific research and discovery in respect, as we should, we must be alert to the equal and possibly opposite danger that public policy itself could become the captive of a scientific-technological elite." Scientific knowledge in some new sense was becoming a commodity in its own right. The new university professors in health would be a different breed from the ivory-towered dons of yesteryear.[40]

The Dark Side

Even before the 1950s were out, the new magic bullets, which had appeared to promise specific effectiveness with a minimum of other effects, had begun to appear to some as a poisoned chalice. The first indications that the ground was shifting came from academic physicians, who began to note increasing problems such as idiosyncratic responses and treatment resistance, whether to antibiotics or to the new psychotropic agents, as well as cumulative toxicity and other iatrogenic problems. In 1952 Leo Meyler, who had himself suffered from a severe adverse reaction to a drug, published his *Side Effects of Drugs,* which was the first volume of its kind.[41]

There was a concern that prescribing was exceeding reasonable limits. If, as defenders of drug development argued, there was no such thing as a safe drug, then clinicians ought to prescribe responsibly—but there seemed to be a default toward ever more prescribing. From the middle of the nineteenth century, clinicians had been distinguished from quacks by their tendency to prescribe fewer medications, grounded in their belief that there was little that was specifically effective. After the War there appeared to be a change of culture, with an ever-increasing level of prescribing.

Clinicians complained that often they had little to go on other than promotional literature from the companies. Concerns were voiced that have been repeated over the ensuing decades that the greater part of the much vaunted research budgets of ethical companies went toward the marketing of compounds rather than toward the development of new compounds. When research monies went toward new compound development, moreover, that development was often said to be a matter of a minor modification of an existing molecule—sufficient to get around the patent laws, rather than a development that constituted a substantial scientific advance. There was a concern that the advertising of compounds did not give a balanced indication of the good and bad points of a compound. Pharmaceutical companies were guilty, it was claimed, of placing research papers in medical journals that fell short of acceptable standards. There were claims that some journals were in the pocket of certain companies—that there was an effective threat of withdrawal of advertising revenue, for example, if articles inimical to a company's interests were published.[42] Questions were raised about the scientific competence of FDA staff and about whether the FDA didn't help businesses rather more than it should at the expense of the drug-consuming public.[43]

There were a variety of responses to the emerging situation. In the case of psychopharmacology, Jonathan Cole established the Early Clinical Drug Evaluation Unit (ECDEU) in order to keep a closer account of what was going on. ECDEU served to provide clinicians involved in testing new compounds with a forum in which they could meet with company clinicians and scientists as well as staff from the FDA, in order to track the issues.[44] On a wider front, however, the growing concerns led to hearings presided over by Senator Estes Kefauver, which started in December 1959 and ran until September 1960.

Meanwhile, in 1957, a compound called Contergan had been launched in Germany as a sedative and tranquilizer. It was available over the counter. Contergan was the trade name for thalidomide, which had been discovered by a company named Chemie Grünenthal in the early 1950s. In early testing and subsequent clinical use, it appeared safe. It was claimed that it was both safe in overdose and nonaddictive. This safety profile led to suggestions that it might be useful during pregnancy, and it was rapidly adopted. But in line with the practices of the day, impressions of both efficacy and safety hinged on the testimonials of a few clinicians rather than on demonstrable effects from multicenter studies and a systematic cataloguing of adverse events. All too often the clinicians were everyday practitioners rather than anyone with recognized expertise in clinical pharmacology. A further problem was that Chemie Grünenthal had not tested thalidomide in pregnant animals prior to marketing.[45]

By 1961 reports had begun to emerge from Australia and Germany of possible birth defects occurring as a consequence of thalidomide. On November 26 the German newspaper *Welt am Sontag* published the suspicions of a Professor Lenz from Hamburg under the headline "Malformations from Tablets." Grünenthal attacked the article as sensationalist but nevertheless withdrew the drug from the German market "because press reports have undermined the basis for scientific discussion." America was largely spared the problem because of bureaucratic red tape within the FDA and because the official handling the case, Dr. Frances Kelsey, had concerns about the drug's safety.

It is estimated that six thousand children were born with thalidomide-induced deformities. The disaster raised the problem of the adverse effects of drugs on the developing fetus for the first time. It called into question the procedures companies had for testing against such an eventuality as well as the procedures by which both the FDA and the companies generally decided that a drug was safe and efficacious. Why were nonexperts being used to test out new compounds? What kind of study would establish whether a compound was effective or not? And what kind of compounds were wanted in the first place?

The thalidomide disaster led to the passage of the Kefauver-Harris drug amendments, which passed both houses of Congress unanimously on October 10, 1962, although arguably none of the provisions of these amendments would have made any difference in the case of thalidomide.[46] This legislation put in place more rigorous requirements on manufacturers to satisfy the FDA that a new compound was both safe and efficacious for the ailment for which it was designed. On the question of safety, systematic testing for possible injury to the fetus was made mandatory. The FDA was also required to approve all plans for clinical testing of compounds and to ensure that such testing be done by investigators competent to conduct them. For the first time companies were required to notify the agency of any adverse events that occurred after launching a new medication.

The 1962 amendments charged the FDA with establishing the efficacy not only of prescription-only compounds but also of over-the-counter products (OTCs). It was estimated that there might be up to half a million OTC products on the market—there were eight thousand antacids alone. A preliminary investigation of five hundred suggested that anywhere between half and three-quarters were ineffective. The FDA established the recommended dosages and labeling for the approximately five hundred active ingredients there were thought to be in OTC compounds. As a result a large number of "antidementia" drugs and "antidepressants," for instance, vanished.

The 1962 amendments had three major consequences, all critical to the plot of this book. One was the institutionalization of the view that randomized, placebo-controlled, double-blind trials are the appropriate means, indeed almost the only scientific means, to establish the efficacy of a treatment. This was a development as significant as the development of the magic bullets themselves. Trials of this sort, ideally suited to the evaluation of drug therapies, were introduced early into psychiatry and in recent years have extended to all areas of mental health work, including psychotherapy (see Chapter 7). The claim that the efficacy of treatments can be determined objectively by this means has also been central to the current health care reforms, as it appears to offer the promise that costly and inefficient treatments can be eliminated.

The second concerned the fate of the prescription drugs. In 1959 there was considerable hostility to continuing the 1951 arrangements. "A unique characteristic of [prescription drugs] is the difference between the buyer and the orderer; in the words of Chairman Kefauver, 'He who orders does not buy; and he who buys does not order' . . . Hence in ethical drugs the ability of the ordinary consumer to protect himself against the monopoly element inherent in trademarks is nonexistent. The consumer is captive to a degree not present in any other industry."[47] But although the antiprescription lobby had almost won the day in 1959, by the early 1960s, as part of the response to thalidomide, which had been available over the counter in many countries, the prescription status of the new generation of treatments was maintained; the contemporary forms of captivity that this has given rise to will be outlined in some detail in Chapters 6 and 7.

Finally, a view that had been present since the 1951 amendments came more to the fore, which was that the FDA was in the business of licensing compounds for indications that medical experts agreed were the indications for which compounds were needed—diseases. This latter point arose in response to the continuing indications from the market that what consumers, in contrast, wanted were drugs for halitosis and nervous fatigue—illnesses. Consumers wanted to manage their ailments, the experts wanted to slay dragons. In proceeding down this route, the FDA was putting a premium on what has been termed the medical model of disease but is perhaps more appropriately termed a bacteriological model. It was one thing to do this for mainstream medicine but for psychiatry and for depression?

Depressive Illness and Depressive Diseases

THE ARCHEOLOGY OF MOOD DISORDERS

> Months of Delusion I have assigned for me
> Nothing for my own but nights of grief
> Lying in my bed, I wonder when will it be day?
> Risen I think how slowly evening comes.
> Restlessly I fret till twilight falls.
> Swifter than a weaver's shuttle my days have passed
> And vanished leaving no hope behind.
> Remember that my life is but a breath
> And that my eyes will never see joy again . . .
> In the anguish of my spirit I must speak
> lament in the bitterness of my soul.
> My days run hurrying by
> Seeking no happiness in their flight . . .
> If I resolve to stifle my moans
> Change countenance and wear a smiling face
> Fear comes over me at the thought of all I suffer.
>
> —Book of Job

The terms *mania* and *melancholia* date from the Greeks. There are recognizable pictures of both manic and melancholic subjects in the works of Hippocrates, Galen, Homer, and other Greek and Roman authors as well as in the Bible. However, while some of the descriptions, such as that in Job, speak powerfully across the millennia and in so doing attest to the stability of certain basic human reaction patterns, and while some of the terms used to designate these states might even be the same as have been used in previous centuries, it would be a mistake to think that there is a continuity of understanding spanning the three thousand years from Homer and Job to ourselves. The modern meaning of the terms *mania* and *melancholia* barely goes back one hundred years.[48]

For the Greeks and through subsequent history until after 1800, the mad person was a maniac—he or she was insane. The typically mad person was thought to be wholly insane, in the sense of someone who had lost complete possession of all his or her faculties. Madness was seen as a delirious and raving state. It came in underactive and overactive forms. Overactive states were designated as mania; states of insanity involving underactivity were designated as melancholia. This contrast between over- and underactive insanities fitted Greek humoral models of health, in which what counted was the balance between hot and cold as well as moist and dry humors in the body. Too much heat, it was thought, led to frenzy and overactivity and too much cold to underactivity.

Some echo of what was involved can be heard in the terms *maniac* and *maniacal,* which derive from mania, and even today are applied to someone who is wholly and completely out of control. The madman, before 1800, was a maniac, with all that that entailed in terms of compromised judgment and frenzied or overactive behavior. This is a very different picture from the one we now have of someone who has a diagnosis of mania, in which the affected individual is more likely on balance to be pleasantly grandiose and perhaps somewhat overactive but not insane or maniacal.[49]

One reason maniacs were seen as raving mad may have been because many of them actually were, in modern terms, delirious.[50] For a great deal of the nineteenth century and the early twentieth, it now seems clear that a substantial proportion of people committed to asylums died within weeks or months of being admitted, which strongly suggests that underlying physical illnesses played a part in the production of their madness. Many states of mania, up to 1900, may have stemmed from infections of one sort or another: either direct infections of the brain, such as syphilis or tuberculosis, or the consequences of febrile states secondary to any other infection. Indeed the first drugs to bring about a calming of the asylums were the antibiotics, penicillin and streptomycin, when they were introduced in the 1940s.[51]

But the problem is not simply one of a difference in the frequency of infections. Looking at admissions to asylums around the turn of the century, it turns out that upward of 50 percent of the diagnoses made were for mania. Since World War I, in contrast, a diagnosis of mania has only been likely to occur as part of a diagnosis of manic-depression, and at any one time, probably no more than 1–2 percent of the old asylum population, or of the population of general hospital psychiatric units today, would attract such a diagnosis. If we can assume that what was a very common disorder did not simply vanish around 1900, such findings point to a substantially different use of the term mania before and after 1900. In fact as we shall see, virtually every other term, including the terms *neurosis, psychosis,* and *melancholia,* have similarly changed in meaning, and the diagnosis of depression did not exist before 1900. Writings before this date, therefore, were describing a landscape very different from the one we now inhabit.

This change may be brought home by considering descriptions of melancholia. Richard Napier described Alice Davies in 1630: "Extreme melancholy, possessing her for a long time, with fear, and sorely tempted not to touch anything for fear that then she shall be tempted to wash her clothes, even upon her back. Is tortured until that she be forced to wash her clothes, be them never so good and new. Will not suffer her husband, child nor any of the household to have any new clothes until they wash them for fear that the dust of them will fall upon her. Dareth not go to the church for treading on the ground, fearing lest any dust should fall upon them."[52]

Now consider also this quote from Henry Maudsley in the 1897 edition of *The Pathology of Mind.* "In another group of cases of simple melancholy, the cause of affliction is a morbid impulse to utter a bad word or to do an ill deed. The impulse is bad enough, but the essence of the misery is not always so much the fear of actually yielding to it as the haunting fear of the fear; it is that which is the perpetual torture, an acute agony when active, a quivering apprehension of recurrence when quiescent. Sometimes the impulse is of a dangerous character, as when it prompts a father to kill himself or urges a mother to kill her children; not unfrequently it is an impulse to utter aloud a profane or obscene word or to do an indecent act; now and then it is an impulse to do some meaningless and absurd act which has taken hold of the fancy and will not let it go, and to repeat the act over and over again."[53]

To the modern eye these are clearly descriptions of obsessive-compulsive disorder. To both Napier and Maudsley, however, they were descriptions of melancholia. Any state that could lead to underactivity or inactivity, for one reason or the other, was diagnosed as melancholia. Conditions that we now recognize as negative schizophrenia, obsessive-compulsive disorder, social phobia, panic disorder and others would all do this and hence would attract a diagnosis of melancholia, or underactive insanity, in contrast to the manias of florid psychosis, delirium, and what would now be called mania.

While the propensity of infections to cause delirium almost certainly influenced how people saw madness, another influence was important as well. The mad person was necessarily seen as wholly mad because insanity was somehow dependent on the soul, and as the soul was a single indivisible entity, insanity therefore also had to be indivisible. If a person's mind was deranged it had to be totally deranged, because the soul could not have parts; a derangement must necessarily affect the entire set of functions of the mind.

Several factors came together from the turn of the nineteenth century to transform this perception. One was the development of faculty psychology. While the mind might be a single spiritual entity, Thomas Reid, a

Scottish philosopher, argued that from the practical point of view of those wishing to investigate it, it made sense to conceive of it as having separate faculties. Initial speculation settled on three broad sets of faculties, the faculties of cognition, emotion, and the will. In due course, these were further subdivided, with a number of different schemes, some of which proposed over forty different faculties. The development of faculty psychology created the possibility that mental disorders could be conceived of as affecting one faculty only rather than the mind as a whole.

Such a conception was all but forced on philosophers by the development of neurophysiology, which was effectively established by the discovery of the reflex arc in 1823. With this development it became clear that certain behaviors could happen reflexly—without an input from the soul. Thereafter progressive investigation of the nervous system led to the description of reflexes at higher and higher levels of the spinal cord. In 1862 Ivan Sechenov described inhibitory reflexes in the brain stem. By 1880, Hughlings Jackson and others were to conceive the frontal lobes of the brain as the ultimate inhibitors of nervous system functioning and accordingly the seat of executive control. Jackson saw madness as something that occurred when the inhibitory control of the frontal lobes was loosened for whatever reason. With this progressive retreat of the soul into the inner fastnesses of the brain and a growing acceptance that areas of the brain might be acting outside the control of the soul, a great deal of the underpinning for the idea that insanity had to be a single homogenous entity was undercut.[54] But in order to realize the possibilities of the new concepts being offered by neurophysiology and faculty psychology, what was needed was a natural experiment that would bring the insane together in large numbers. If there were differing species of insanity, these might be more readily distinguished in settings where large numbers of the insane were gathered together. As it happened, such an experiment had just begun.

FROM ESQUIROL TO KRAEPELIN

In both France and Britain in the eighteenth century, there was a rise of what is now thought of as the statistical movement (see Chapter 3). One of the first efforts of this movement was the establishment of a census of the population. A dominant French concern was to determine whether the size of the French population was rising or falling. There was alarm over the possibility that it might be falling in relation to an expanding German population and that the centuries-old rivalry between these people might be tipping in favor of Germany, simply by virtue of a failure of the French population to reproduce itself. Reassuringly, the findings of the first censuses suggested that far from falling, the French population was actually increasing.

What also emerged, however, was that death rates varied across the country, with greater longevity being found in Alpine villages and much shorter life spans in villages that were low-lying and closer to marshes. These findings brought home an awareness that environmental influences, about which something could be done, might contribute to disease and mortality. It was out of such impressions that the moral movement in medicine was born. If poor personal habits and unsanitary living conditions could cause illness, as the French surveys appeared to show, then correcting these should help. The moral movement aimed at improving health by regulating the environment, through influencing what people ate and the regularity with which they ate it, the regularity of their work habits, the cleanliness of the air they were breathing, the amount of rest and exercise they were getting, and other such factors.[55] Hospitals were transformed from hostels where the sick went to die to institutions dedicated to providing a therapeutic milieu.

The development of the hospital was not confined to general medicine. The same impulses to provide an environment that might be conducive to recovery led to the establishment of the first asylums. While institutions had existed before 1800 to enforce the isolation of the dangerous, the emergence of the Retreat at York, with its policy of nonrestraint, where Samuel Tuke, a Quaker, and his family lived in with their patients, was an example of the kind of asylum that moral reformers wished to see. The impulses that created the Retreat were involved in part, at least, in the creation of most of the other asylums erected during the first half of the nineteenth century. This is not to say that there were not also interests at stake that were happy to see the roving and raving insane confined out of public view, but confinement of the insane as an oppressive policy was not initially at least the intention of any of the European states or of the nascent American democracies. Where older asylums had existed before the advent of the moral movement, we have the famous instance of Philippe Pinel unshackling the insane. Jean-Ètienne-Dominique Esquirol, his pupil, described an insane asylum as a "therapeutic instrument in the hands of an able physician, and our most powerful weapon against mental diseases."[56] These intentions were clearly corrupted later in the century, when overcrowding in the asylums meant that efforts to provide a suitable environment broke down and oppressive regimes and bureaucratic ineptitude conspired to aggravate or in some cases even create the problems they were supposed to mitigate.[57]

Whatever one's beliefs about the purposes for which the asylums were created, one of the consequences of their creation was that a large number of "lunatics" were collected together in the one place for the first time. In the presence of three hundred or four hundred such individuals (or even greater numbers later in the century), it became increasingly difficult to maintain the view that there was essentially only one form of insanity—that all insanities were the same. It was clear there were a number of insanities and that these must be classified in some way if progress in their understanding was to be made. This impulse to classify mental disorders should be set against a background of the successful classification of plant species by Linnaeus in the 1720s and the first attempts by Thomas Sydenham, François Boissier de Sauvages, and others in the course of the eighteenth century to classify illnesses.

While there had been a number of relatively theoretical classifications of illnesses that included mental illness put forward by Pinel, William Cullen of Edinburgh, and almost all the prominent medical thinkers of the time, the first practical asylum-based classification of mental illnesses was adumbrated by Esquirol in a series of pamphlets that were collected in book form in 1838.[58] Esquirol distinguished between insanity proper and the partial insanities, among which he listed monomania. Individuals with monomania, he believed, could not be termed insane because for the greater part their conversation and behavior was entirely reasonable. They had lively and accurate memories on issues other than those to do with the issue on which they were deluded. They reasoned in a logical and reasonable way, and even on issues on which they were deluded, they were liable to appeal to evidence in the same kinds of way that individuals ordinarily defending a strongly held position would appeal. Such individuals, Esquirol argued, could only be partially insane, an argument he was able to make because the idea of partial insanity was then in the process of becoming comprehensible. Karl Ludwig Kahlbaum, in Germany, taking a very similar approach to Esquirol, in 1863 described the partial insanities of dysthymia and paranoia. The old humoral polarity of mania and melancholia was breaking down.

Esquirol began by accepting that a partial insanity could selectively affect thinking and judgment or mood or the will or subdivisions of these. Among the disorders of the will that he delineated, for instance, were conditions such as dipsomania (alcoholism), kleptomania, pyromania, nymphomania, graphomania, and others. In essence, the use of mania as a suffix like this translated into an individual's having a mania for a specific behavior but otherwise being normal. While the concept of monomania did not survive, the survival to this day of this use of the mania suffix indicates where one of the corpses on the road to modernity is buried.

In proposing the disorder of lypemania (*lupos* = sadness), a selective disorder of mood, Esquirol all but created modernity. This was the cornerstone on which the modern edifice of depression has since been erected. A number of implications of this move are important here. One was that Esquirol was proposing the existence of a disorder that was not a disorder of the whole person. Previous concepts of a melancholic disorder did not remove disturbances of a specific faculty from the envelope of temperamental and constitutional considerations—it had been assumed that melancholia involved some input from a propensity to a melancholic constitution. In separating out these possibilities Esquirol set in train a debate about the merits of categorical as opposed to dimensional concepts of mental illness that, for instance, is at the heart of the debate stimulated by Peter Kramer's 1993 *Listening to Prozac* (see Chapter 5).[59] At first blush the discovery of antidepressants appeared to confirm Esquirol's idea, but as we shall see in Chapters 6 and 7, there is a considerable amount of evidence that suggests that the antidepressants may act on something that is an antecedent for mood and other neurotic disorders. Hence the idea that there is a specific disorder of the mood of the kind that was introduced first by Esquirol may yet need to be revised.

The more immediate problem posed by Esquirol's approach was that it could potentially be applied to any symptom, yielding a vast number of monomanias. As pointed out earlier, Morgagni and Bichat had introduced a new disease concept whereby various clinical presentations were tied down to tissue pathology. In 1822 Antoine Bayle made the first discoveries which suggested that this disease concept might apply to psychiatry also, when he found particular changes at post mortem in the brains of individuals who were suspected of having general paresis of the insane (GPI, or tertiary syphilis). An important aspect of Bayle's findings was that as a result GPI became recognized as a disease in which individuals could pass through a range of different clinical states (different monomanias) from grandiosity to delirium and subsequently on to a dementia. The implication was that a range of clinical presentations might be tied together in one disease entity. Against this background, questions began to be asked as to whether each of Esquirol's monomanias represented specific disease entities. There did not appear to be enough different areas in the brain to accommodate so many disease entities. Longitudinal descriptions, furthermore, began to suggest that monomanias, far from remaining constant, over the course of time might transmute into one another, suggesting that there might be a common disease underlying both.[60]

The monomania concept began to fall apart, but from its shell a number of disorders emerged. Two of these are particularly important to this story. One was *démence précoce* and the other *folie à double forme* or *folie circulaire*. Démence précoce was outlined first in 1853 by Benedict Morel. He described "young mental patients who appear to the observer to have every chance of recovery but after careful examination, one becomes convinced that idiotism and dementia are the sad fate that will terminate the course."[61] In 1860 he gave the name démence précoce to this "sudden immobilisation of all the faculties." This was a disorder that affected young people in their late teens or early twenties who previously might have shown considerable intellectual and social promise, but who thereafter displayed a marked deterioration in social and personal functioning from which they never recovered.

Between 1851 and 1854 there were also independent publications by Jean-Pierre Falret on folie circulaire and Jules Baillarger on folie à double forme. Both described a picture in which two monomanias, a particular kind of melancholia and of mania, were linked together. While mania and melancholia had been linked before by observers going all the way back to Greek and Roman physicians, the new idea was that an individual might cycle systematically between one pole and the other of a specific mood-regulating faculty, while all else remained normal. This is now what would be termed bipolar affective disorder or manic-depressive illness. There were bitter disputes between Falret and Baillarger as to who had recorded the new state first, disputes that have never been conclusively resolved.[62]

Between 1893 and 1899 the ideas of Falret, Baillarger, Morel, and others were taken up by Emil Kraepelin. He proposed that there were essentially two forms of mental illness, manic-depressive insanity and dementia praecox. Dementia praecox was the démence précoce of Morel, but Kraepelin enlarged the concept. He argued that the illness could present in the classical way that Morel had outlined, but that there were also a number of individuals who might appear to have a partial insanity or paranoia at the start, but whose illnesses, if tracked over time, evolved into the typical picture of dementia praecox—that is, an individual who had formerly shown considerable promise but who went on to show a marked deterioration of functioning from which he or she essentially never recovered.[63]

It was on this disease that Eugene Bleuler wrote an important monograph entitled *Dementia Praecox or the Group of Schizophrenias*. He proposed that the basic pathophysiology of the illness involved a loosening of associations but perhaps even more important, in the light of the splitting of the mind that resulted from the loosening of associations, in 1908 he proposed that the disease might be more appropriately termed schizophrenia. The term schizophrenia and the idea of a loosening of associations caught on, probably in part because people thought they knew what was meant by a loosening of associations but also in part because an adjective could be made of the term schizophrenia in a way that it couldn't be made of dementia praecox. Once coined, words like schizophrenia, neurosis, and psychosis can be like harpoons; if they go in they can be very difficult to get out. On such details can important aspects of the history of medicine turn.[64]

In just the same way that he had collapsed a number of apparently dissimilar disorders into the one disease concept of dementia praecox, based

on a consideration of their longitudinal course, so also Kraepelin subsumed a number of mood disorders under the general heading of manic-
depressive insanity. Again his disease concept was a much broader term
than the originally limited description of a bipolar affective disorder given
by Falret and Baillarger. For Kraepelin, any disease of the mood qualified
for a diagnosis of manic-depressive insanity. Individuals with recurrent
episodes oscillating between mania and depression clearly qualified, but
so did individuals with current bouts of mania only, or with the more
common recurrent bouts of depression only. This latter syndrome is what
is now termed as unipolar depression. Kraepelin, however, argued that
distinctions between bipolar and unipolar depression were not essential.
They both formed part of the same illness in much the way that whether
one has tuberculosis of the brain or of the lung is fundamentally unimportant, tuberculosis is still tuberculosis wherever in the body it takes
root. There may be something about an individual's constitution that
makes one person bipolar, another unipolar, yet another dysthymic and
that would make some others cycle in recurrent brief episodes, but for
Kraepelin the disease was still all one:

> Manic-depressive insanity . . . includes on the one hand the whole
> domain of so-called periodic and circular insanity, on the other hand
> simple mania, the greater part of the morbid states termed melancho-
> lia . . . Lastly, we include here certain slight and slightest colourings
> of mood, some of them periodic, some of them continuously morbid,
> which on the one hand are to be regarded as the rudiment of more
> severe disorders and on the other hand pass over without sharp
> boundary into the domain of personal predisposition. In the course
> of the years I have become more and more convinced that all the
> above-mentioned states only represent manifestations of a single
> morbid process.[65]

In contrast to schizophrenia, for Kraepelin, manic-depressive disorders
showed a remitting course. As regards symptomatology, his description
of the depression found in manic-depression was very close to current descriptions of what is now called major depressive disorder or what until
quite recently was called endogenous or vital depression. This is a disorder characterized by apathy, loss of energy, retardation of thinking and
activity, as well as profound feelings of gloominess, despair, and suicidal
ideation. He described a characteristic diurnal variation of mood,
whereby the individual was usually worse in the morning but improved as
the day wore on. In addition, he pointed to vegetative symptoms, such as
poor sleep and loss of appetite. The earlier picture from Job fits
Kraepelin's portrait of depression very well. The idea that this was a dis-

ease was something that European clinicians in 1900 were prepared to adopt on the basis that many of the symptoms such as diurnal variation of mood and the characteristic early morning wakening of depressed patients are difficult to explain plausibly without invoking a biological disturbance.[66]

The fact that Kraepelin initially called his illness manic-depressive *insanity* betrays its roots in older ideas of mental illness, but one of the consequences of the new formulation was an almost immediate drop in the diagnosis of mania. From being the diagnosis given to almost any agitated or raving person admitted to the hospital, the frequency of its use rapidly dropped. Indeed, the terms *mania, manic,* and *maniac* were dropped completely in many circles during the middle years of the century, owing in part to their pejorative connotations. Patients were described as hypomanic rather than manic. It is only in the past decade or two, as these connotations have finally evaporated, that mania and manic have begun to be used again with any degree of frequency—but they now mean something quite different from what they once meant.

It is difficult now to overestimate Kraepelin's contribution. For many the emergence, in the 1950s, of what were basically two classes of psychotropic drug, the antipsychotic and the antidepressant, was the ultimate vindication of his point of view. In Europe, every view on the affective disorders since 1896 has defined itself with reference to him. In the United States, in contrast, until recently his work was downgraded, so that by 1960 Kraepelin was essentially seen as an obscure German from the prehistory of psychiatry. Now, however, to the surprise of many observers, American psychiatry is dominated by a neo-Kraepelinian school, whose influence did much to mold the classification of psychiatric disorders found in the third edition of the American Psychiatric Association's *Diagnostic and Statistical Manual of Mental Disorders, or DSM-III.*[67] Part of my story will lie in attempting to indicate how work done at the turn of the century by someone who knew nothing of modern treatments could in the course of twenty years move from the backwaters of psychiatry to take over the mainstream.

DEPRESSIVE DISEASE AND DEPRESSIVE NEUROSIS

Following Kraepelin, two broad trends emerged, one in Germany and the other in the United States. The German contribution came from Kurt Schneider, in Heidelberg, who was concerned to isolate the key clinical features within the manic-depressive and schizophrenic complexes. While Kraepelin had distinguished these disorders on the basis of their longitudinal course, Schneider attempted to identify typical features of the disorder that would allow a clinician to make a diagnosis based on one presentation. He described the first-rank symptoms of schizophrenia which are used to make a diagnosis of that condition to this day. As regards the mood disorders, he picked out a constellation of vital disturbances of drive and will as the hallmarks of a process that has been variously called endogenous, vital, biological, or melancholic depression, in contrast to those states of misery or unhappiness that were clearly present for a reason and that lacked such hallmarks. Schneider was concerned with affective diseases or affective psychoses, as he called them. He did not believe in affective neuroses, depressive reactions, or depressive illnesses.

Adolf Meyer, in contrast, who is cited by many as the dominant figure in American psychiatry from 1900 through World War II, felt that Kraepelin's disease models were wrong. He proposed the idea of reaction formations instead, and rather than a unitary disease concept he offered a spectrum of affective reactions. He also proposed jettisoning the term *melancholia* completely because, he argued, it implied a knowledge of something that clinicians at the time did not possess, whereas the term *depression* would designate exactly what was meant—a depression of mental energies, of an as yet unspecified nature.[68] Where Schneider sought to distinguish those who could be termed diseased from the merely chronically unhappy, Meyer's approach led in the opposite direction. He was interested in what were coming to be called the depressive neuroses.

The concept of a depressive neurosis found favor in American psychiatry which, unlike German psychiatry, was as much an outpatient and office psychiatry as an asylum-based discipline. Depressive psychoses, as conceived by Schneider, were extraordinarily rare conditions, leading to admissions to a mental hospital between the years 1900 and 1945 of the order of fifty patients per million per year[69]—not enough to build a clinical practice on or indeed, as we shall see, not much of a market for a new drug. In contrast, the United States had given birth in the 1870s and 1880s to the concept of neurasthenia, the prototype of what was later to become the depressive neurosis,[70] one of the striking features of which was the frequency with which it came to be diagnosed. Distinctions between depressive neuroses and psychoses, as we shall see in the next chapter, played a critical part in shaping the discovery of antidepressant drugs and, as will become clear in Chapter 7, they have been central to the politics of psychiatry ever since Meyer.[71]

But where did these terms neurosis and psychosis come from? The word *psychosis* had first been introduced by Ernst von Feuchtersleben in 1845 as an alternative to *insanity* and *mania*. The word was used to distinguish those disorders which by virtue of disordered judgment and higher-level psychological functions were thought to be primarily disorders of the mind as opposed to the neuroses which were behavioral disorders, thought to stem from faulty nervous functioning.[72]

The coining of the term *neurosis* has been variously ascribed to Robert Whytt and William Cullen. Both worked in Edinburgh. One or the other of them used the term first somewhere between the years 1780 and 1785. As used initially it referred to a presumed disorder of the nerves. It was used for conditions where individuals appeared to be lethargic and faint, where they fell for no obvious reason, where they had fits or convulsions or they appeared perhaps to be blind or anesthetic but where also, on postmortem, no clear disorder of their nerves could be demonstrated. The term was used to distinguish such disorders, which were presumed to be of nervous origin, from neuritis, where, on post mortem, individuals showed very clear abnormalities of their nervous system—inflamed nerves or other nervous destruction.[73]

The distinction was much the same as that drawn at the same time between the terms *nephritis* and *nephrosis*. When patients presented with signs such as the passage of large amounts of water, sometimes bloodied, clinicians assumed that there was a disorder of the kidneys. In some cases, on postmortem, an inflammatory condition was plainly visible and the disorder was termed a nephritis but in others there were no obvious abnormalities of the kidney. There was a reluctance, however, to give up the idea that there must be something wrong with the kidney and accordingly the term nephrosis was coined to cover those states in which it was expected that in due course scientific advances would lead to a demonstration of renal pathology of some sort. As events transpired, these expectations were fulfilled and the nephroses have been shown to involve microscopic disturbances of the kidney that are responsible for the symptoms presented by the patient.

There was a similar expectation in the case of the neuroses that in due course scientific advances would locate the invisible pathology. The classic neurosis where this has happened, or is still expected to happen, is epilepsy. This involves very clear disturbances of behavior of the kind that led nineteenth-century observers to postulate nervous dysfunction. To call epilepsy a neurosis now would cause considerable confusion, however, which indicates how much the meaning of the term has changed.[74]

As efforts to isolate a clear structural pathology of the nervous system, which might underpin the neuroses, failed, attention turned to the possibility of a functional pathology. The discovery of the reflex arc opened up the possibility that some reflexes were aberrant. The emergence of such an idea as early as the 1860s attests to readiness of psychiatry to jump on the bandwagon of the latest research—a phenomenon that will be found repeatedly in succeeding chapters. From the 1880s onward, the work of Wilhelm Wundt pointed to the possibility that disordered nervous reflexes might produce a loosening of mental associations among an individual's thoughts, images and memories.[75]

These emerging concepts of the neuroses as involving a functional change in the operations of a nervous system laid the basis for the breakthroughs made by Freud and Janet at the turn of the century. Both, at much the same time, came to the conclusion that hysteria involved a disorder of remembrance, a splitting of associations, rather than a primary disorder of nervous function. In coming to this view they independently created the modern notion of the psyche—a realm of higher-level psychological function, which is not as dependent on nervous system functioning as the vegetative operations of eating and sleeping and appetite, but equally a realm that appears to be somewhat different from the realm of moral behavior, which might be more properly designated as mental rather than psychological. Janet left open the possibility that psychological disorders might be more likely in a weakened nervous system. Freud initially made a more complete break and suggested that the psychoneuroses were precipitated by environmental factors alone, most famously in the case of hysteria, which he proposed was caused by sexual abuse during childhood.

For a time, Freud used the word *psychoneurosis* to differentiate the types of disorder that he felt were psychological from the group of actual neuroses that he felt were those disturbances of behavior, involving fatigue, weakness, or convulsions, which stemmed from an actual disorder of the nerves. These latter he termed the actual neuroses. The primary candidates for the causes of actual neuroses, he believed, were masturbation and coitus interruptus, which could be expected to give rise to byproducts within the body that were not supposed to be produced. These toxic products would affect the *actual* functioning of nerves.

While Freud's idea about the actual neuroses may now sound rather strange, it indicates how the term neurosis was viewed around 1896 and also how much it has changed in meaning. His concept of a psychoneurosis quickly came to be accepted, so much so that there was a general acceptance that what were considered the neuroses were more likely to be disorders of psychological functioning rather than gross disturbances of nervous functioning. The term neurosis had changed meaning completely and had come to mean almost exactly the opposite of what it had originally meant.

This change in meaning, however, led to something of a problem in that both neurosis and psychosis now referred in some sense or other to psychological dysfunctions—how were they to be distinguished? An attempt was made to restrict the use of the term psychosis to mean something closer to what the term neurosis had originally meant—a disorder of nervous system functioning. This usage was championed by Karl

Jaspers, Kurt Schneider, and others. According to this usage, epilepsy would, for instance, be a psychosis on the basis that it was a disorder of nervous functioning that led to a disturbance of behaviour. Jaspers and Schneider categorized manic-depressive illness and schizophrenia as psychoses on the same basis.[76]

For Schneider, a depression was psychotic if it showed evidence of *vital* disturbance, whether or not there were delusions or hallucinations. The advent of electroconvulsive therapy (ECT) and demonstrations that it seemed preferentially useful for the more severe forms of depression only was a strong argument in favor of the view that there was a radical disjunction between psychotic and neurotic depression. They could not be seen simply as mild and severe forms of the same disorder, as one would not expect a treatment to work for the severe form of a disorder but not for its milder forms. This view gets strong support from common sense, in that very few people would accept that all cases of chronic misery or unhappiness should be seen as a disease. Intuitively it seems reasonable to suppose that there must be some such states that are states of behavioral maladaptation, neuroses, in other words, rather than diseases.

In the United States, under the influence of Meyer and Freud, where the neuroses were thought to involve a partial loss of insight, the psychoses were seen as involving a more complete loss of insight or a loss of contact with reality. Meyer's concept of a spectrum of affective reactions has built into it an implicit continuity between the milder forms of depressive reaction seen in an outpatient office and the more severe forms that require admission to hospital. The Freudian concept of depression, in practice, reverted to the state of affairs pre-Esquirol when rigid distinctions between personalities or constitutions and diseases were not drawn. Neurotic depression, in due course, became the bread and butter of American office practice.

Under Meyer's and later Freud's influence, American psychiatry drifted toward an extreme diagnostic nihilism. The schizophrenia concept expanded hugely, probably owing partly to an acceptance of Bleuler's view that schizophrenia involved a loosening of associations—loosening of associations being something that can be rather readily detected in anyone with nervous problems. Human nature being human nature, moreover, loosening of associations seems even easier to detect in members of the opposite sex, different social classes, or different ethnic groups. In the process even the boundaries between depression and schizophrenia were blurred and there was increasing recourse to the concept of a schizoaffective disorder, which had been first put forward by J. Kasanin in 1933.[77]

The French and the English approaches lay midway between the German and American views. Anyone who was chronically anxious and stressed, it was argued, would end up showing the signs of so-called vital depression—poor sleep, poor appetite, and loss of energy—while primary or vital depressions, supposedly characterized by apathy, anhedonia, poor sleep and appetite, and general retardation, would in many cases be expected to have a secondary anxiety overlay. Some argued that they could distinguish a primary depressive illness on the basis of the presence or absence of a "vital" feel, but authorities such as Edward Mapother and Aubrey Lewis, in the United Kingdom, felt that making distinctions among the depressions was futile and unwarranted as there were no clear indications for treatment of one kind of depression rather than the other.[78]

The advent of ECT made little impression on British views, even though after some years of trial and error it became clear that ECT was a treatment for affective disorders rather than for schizophrenia and, indeed, for affective disorders with melancholic features and that "vital feel" rather than treatment for the minor neurotic depressive disorders. What retrospectively can be seen to have made some difference was World War II and the emigration to Britain of many prominent German psychiatrists, notably Willy Mayer-Gross and Erwin Stengel. Mayer-Gross introduced a distinction between endogenous and reactive depression, between depressive diseases and depressive responses, that was to be made famous in the 1960s by Martin Roth in Newcastle.

Does diagnosis count, then? Should the symptoms of unhappiness or hopelessness count for more in making a diagnosis than looking for a picture in which a larger number of vital disturbances hang together? Do antidepressants treat a disease or symptoms? If they treat symptoms of unhappiness or hopelessness, then all such symptoms presumably should respond regardless of diagnosis. If they treat a particular disease then many cases of unhappiness and hopelessness should not respond. Or did Esquirol get it wrong and do such drugs treat something that is an antecedent for mood and other disorders—a constitutional factor perhaps, in which case their effects might be termed cosmetic.

By the 1950s, the field awaited the cut of a pharmacological scalpel. The advent of the antipsychotic and antidepressant drugs, as we shall see, appeared to provide just such a scalpel. Diagnosis seemed to count. One group of drugs appeared to treat Kraepelin's dementia praecox; the other seemed to treat manic-depressive insanity and associated conditions. Twenty years after the emergence of imipramine, American psychiatry, which had been Freudian and analytic and had thought in dimensional terms, was to be transformed into a psychiatry which thought in categorical and Kraepelinian terms—terms that appealed to regulators faced with a crisis provoked by a German drug sold over the counter and promoted, among other things, for the management of neurotic depression.

2 The Discovery of Antidepressants

In 1952 the first antipsychotic drug, chlorpromazine (Thorazine/Largactil), was discovered.[1] Its discovery was the critical event in the foundation of psychopharmacology. It quickened the pace of all other developments, including the discovery of antidepressants. In one sense, therefore, the antidepressant story effectively starts in 1952. The mix of elements that went into making chlorpromazine, however, was essential to the discovery of the antidepressants as well, and this mix goes back to the development of coal-tar chemistry.

In 1868 Carl Graebe and Carl Liebermann synthesized yet another dye from coal tar—alizarin. This dye had been known from antiquity to occur naturally, but its synthesis allowed easier production. An arrangement was made with the Badische Anilin und Soda Fabrik (BASF) to produce it commercially. BASF then went on to produce a range of dyes, including methylene blue. Working on methylene blue in 1883, August Bernthsen produced the first phenothiazine compound, the parent molecule from which chlorpromazine was to come. There things rested for sixty years until a linking of the phenothiazines to the neurohormone histamine catalyzed further developments.[2]

Before World War II, the idea had emerged that histamine, a hormone now recognized to play a key part in allergic responses, might be involved in stress and shock reactions. By stress and shock in this context, however, is meant primarily the collapse that may be witnessed in cardiovascular systems in response to blood loss or infections, rather than the emotional state that the word suggests to most people. Daniel Bovet, of the

Pasteur Institute, began a search for antihistamine agents in the early 1930s. A number of compounds were isolated but were too toxic for use in humans.

In 1939 the Pasteur Institute linked up with the pharmaceutical laboratories of Rhône-Poulenc, and B. Halpern, of Rhône-Poulenc, found that dimethylamines were more powerful and less toxic antihistamines than Bovet's original compounds. These were subsequently marketed as phenbenzamine (Antergan), diphenhydramine (Benadryl), and others, from 1942 on. It quickly became clear that the antihistamines had some action on the brain and behavior—they were sedative at least. As early as 1943, French psychiatrists sought to exploit this property in the management of schizophrenia and manic-depression.

At the time Rhône-Poulenc's primary research focus was on antimalarial and antiparasitic compounds. Because methylene blue had shown some antimicrobial activity, one of their chemists, Paul Charpentier, set about the synthesis of a series of phenothiazines, in the hope that these might be useful as antimalarial agents. The new compounds were of no use for this purpose but, because Rhône-Poulenc investigators were also working with antihistamines, it was clear to them that these new compounds also had antihistaminic properties. This led to the marketing of promethazine (Phenergan), which is still in use as a mainstream antihistamine and as an aid for travel sickness. Another compound, diethazine, which was only weakly antihistaminic, appeared to be of some use for both the physical and the mental symptoms of Parkinson's disease and to have some ill-defined analgesic properties.

Thus the antihistamines generally appeared to have some central actions. The possibilities that this might offer had begun to attract attention within Rhône-Poulenc, in part owing to Henri Laborit, a French military surgeon. He had first used promethazine in 1949 as part of an anesthetic cocktail, aimed at anesthetizing in a manner that would minimize the risk of the circulatory shock with which anesthetic agents such as ether and chloroform were associated. He found promethazine's analgesic and sedative properties as well as the hypothermic response it appeared to produce useful. On this basis he encouraged Rhône-Poulenc to explore further this aspect of the antihistamine group.

In 1950, then, Rhône-Poulenc set out produce a compound with clearer central effects. It was unclear what would result. Pierre Koetschet, who was responsible for the development, thought there was a chance that any compound that might come out of the program could be of some use in anesthesia, analgesia, Parkinson's disease, and possibly psychiatric problems or epilepsy. He proposed screening for such a compound using a variety of behavioral tests in animals, which itself was an innovation.

Charpentier, therefore, synthesized a further series of phenothiazines. He picked one of the phenothiazines with the weakest peripheral effects and on December 11, 1950, produced RP4560, chlorpromazine.

Simone Courvoisier, who ran the behavioral tests, found that chlorpromazine had novel behavioral effects. Rats trained to climb a rope to a platform for their food, when given chlorpromazine, appeared to have little or no interest in climbing. As far as could be made out, this was not because they were sedated nor was it because their coordination had been interfered with. They had what seemed to be a loss of interest.

Between April and August 1951, chlorpromazine was given to a wide range of French and international physicians to test. Among others, it was given to Laborit. He and colleagues soon found that the drug had potent central effects. It led to a dropping of body temperature (thermolysis), analgesia, and sedation. The analgesia and sedation appeared to have a quality of indifference to them. They thought it would be useful of the management of medical stress, serious burns, cardiovascular disturbances such as Raynaud's disease, and as a potentiator of anesthesia. This profile was of great interest to physicians and to the industry—of much greater interest than a psychiatric indication would have been.

On the basis of his impressions that patients taking chlorpromazine seemed indifferent rather than truly sedated, however, Laborit thought it might also be of psychiatric interest. He persuaded a friend who was a psychiatrist, Cornelia Quarti, to take some. She did and reported a feeling of detachment. He then urged the psychiatrists at the Val-de-Grâce military hospital in Paris, where he was based, to give it to psychiatric patients. On his recommendation, it was tried in 1952, along with the barbiturates and other sedatives that these patients were routinely on. The impressions were that the new agent might enrich psychiatric therapy, but there were no clear indications that anything remarkable had been discovered.[3]

At much the same time, chlorpromazine was given by Pierre Deniker and Jean Delay to patients in the Hôpital Sainte-Anne.[4] Delay was one of the leading figures of the day in French psychiatry, along with Henri Ey. He had hosted the first World Congress of Psychiatry in Paris in 1950.[5] Even then he was "biologically" oriented. Delay and Deniker gave chlorpromazine on its own to a group of thirty-eight patients who were psychotically agitated—the predominant diagnosis was one of mania. This time the drug was given without the confounding effects of other sedatives. Deniker, Delay, and Jean Marie Harl described their findings in a series of publications that made it clear that the discovery of chlorpromazine represented a breakthrough for psychiatry. Delay, subsequently, organized a conference in Paris in 1955, to which the leading figures in European and world psychiatry were invited.

So who discovered chlorpromazine? Charpentier, who synthesized it? Courvoisier, who reported distinctive effects on animal behavior and neurophysiology? Laborit, who first noticed distinctive psychotropic effects in man? Or Delay and Deniker, who clearly outlined what has now become its accepted use in psychiatry and without whose endorsement and prestige Rhône-Poulenc might never have developed it further as an antipsychotic? There have been bitter disputes over this issue, as a result of which no Nobel Prize was ever awarded for what has been the single most important breakthrough in psychiatric treatment.

Despite the eminence of Delay and an increasing appreciation of chlorpromazine's striking effects, of central importance to the issues developed later in this chapter, there remained something of a bias against developing it as a psychiatric drug. When Smith, Kline & French, as part of its efforts to bring chlorpromazine to the United States in 1954–1955, brought French clinicians to America to meet American clinicians, Henri Laborit and his ideas on chlorpromazine's possible use in anesthesia were still being seen as the leading indication. Indeed, possible uses of the drug as an antiemetic or as an antipruritic agent seemed as likely as a future use as a major tranquilizer. In the course of the visit, however, Laborit's demonstrations using chlorpromazine as part of an anesthetic cocktail given to dogs more often than not failed to work, whereas a number of state hospital psychiatrists found themselves persuaded by what Pierre Deniker had to say.[6] The international meeting organized by Delay in 1955 ensured there was no going back.

Deniker's visit to America took place against a background of clinical studies conducted by a number of psychiatrists, following the publication of Delay and Deniker's reports. These had begun to generate an awareness that chlorpromazine was distinctly different from anything that had gone before. Heinz Lehmann, working at the Verdun Protestant Hospital[7] in Montreal, had read Delay and Deniker's report while soaking in the tub one Sunday. Having tried everything there was to try in those days, with little or no success, he was skeptical of claims for a breakthrough. Yet it seemed to him that Delay and Deniker's descriptions of the effects of chlorpromazine differed from the claims that were being made for anything else. He tried the drug out on seventy patients and found the effects dramatic. Patients who had been deluded and hallucinating for a decade emerged from their psychoses for the first time.[8] The use of chlorpromazine (Thorazine) spread like wildfire through American asylums. The state of Maryland, for instance had three state asylums with up to eight thousand patients per asylum. In 1955 almost all patients in each of these institutions were given Thorazine. It has been estimated that Smith, Kline & French made $75 million in 1955. Other companies, not surprisingly, decided that they wanted a piece of the action.

Findings such as these, replicated worldwide, led to a large turnout at the meeting organized by Delay at Val-de-Grâce in 1955. Most of those who had been using chlorpromazine in America, Britain, France, Germany, Switzerland, Italy, and the rest of Europe were present. There was a mood approaching euphoria, a feeling that for the first time a window might be opening up on the therapy of mental illness. The first reports that another drug, reserpine, was also useful in psychosis had been published in 1954, adding to the sense of excitement.[9]

Although there was a clear delay in accepting that chlorpromazine was a breakthrough, once it was established in 1954, the pharmaceutical industry was quick to explore the further implications of the new compounds. A number of meetings were convened by the large pharmaceutical houses—Geigy, Ciba, Roche, Sandoz, and Rhône-Poulenc—between 1953 and 1958 to explore the implications of developments. Indeed the industry played a key role in establishing the international organizational framework necessary for the development of psychopharmacology. They sponsored the attendance of both clinical and preclinical investigators, who were thereby enabled to meet and explore the shape of the new science. Such meetings led to the creation of the Collegium Internationale Neuropsychopharmacologium (CINP) and subsequently to the establishment of many national psychopharmacology societies.[10]

At some of these meetings, there was a view that while the neuroleptics, as they came to be called in Europe, or antipsychotics as they were more often called in the United States, represented a real development, and while there might be related developments, it was unlikely that there would ever be drugs for depression—because depression in some manner stemmed from object loss, and how could a drug be expected to reverse that? Minor tranquilizers, yes, safer barbiturates, yes, but antidepressants? There already were stimulants, but these had been of little use for depression. The amphetamines had been around since 1935, but the general consensus was that they had little to offer in cases of depression.

In contrast, there was the example of ECT, which, after being introduced for schizophrenia, was becoming in the opinion of many a treatment for depression. This issue was obscured by the fact that in many places, especially the United States, virtually all serious mental illness was being diagnosed as schizophrenia and there was accordingly a substantial response rate among "schizophrenics" to this treatment. There were many such as Heinz Lehmann who believed that because depression was a remitting disorder, finding a treatment for it should be easier than finding something helpful for schizophrenia.[11]

Lehmann, Delay, and others had tried a range of compounds in search of an antidepressant pill. Delay had experimented with dinitriles, now used for the treatment of angina and other cardiovascular conditions, but without obvious success. He had also used combinations of barbiturates and amphetamines, as part of an amphetamine shock treatment, in the belief that shock treatments were likely to be of benefit.[12] In some cases of agitated depression, chlorpromazine seemed to help. Most intriguingly, perhaps, both isoniazid and iproniazid had been tried by Delay. As early as 1952, he and colleagues had reported that isoniazid had an antidepressant effect. The findings were reported in a very brief paper at the end of 1952.[13] They were resurrected in 1958 after news of Nathan Kline's discovery of iproniazid. Perhaps because of the more dramatic effects of chlorpromazine, these first glimmers of what was to come were rather surprisingly overlooked.[14]

Kuhn, Geigy, and Imipramine

Although the discovery of antidepressants came after the discovery of chlorpromazine, their discovery was in many ways much more remarkable than that of chlorpromazine. Chlorpromazine has clearly apparent effects which have an onset within the hour. There were also dramatic effects on animal behavior. Imipramine, as will become clear, had none of these qualities. Giving it to depressed patients or to healthy volunteers produces nothing remarkable except for side effects. Far from being a stimulant, even slightly stimulant in the way Prozac is for some people, imipramine is for most people sedative.

Like chlorpromazine, the development of imipramine (Tofranil) came from an interest in antihistamines. This interest had not been confined to Rhône-Poulenc. Other companies, such as Merck, had already been active in the field, but now yet others moved in or became more active. One of these was Geigy. As early as 1948, their chief of pharmacology, Robert Domenjoz, had suggested that it might be worth looking for other compounds with structures similar to those of the phenothiazines.[15] They decided, therefore, to look at all molecules with a broadly similar structure. In the basement of the Geigy building there was, at the time, an archive of compounds that had been synthesized over previous years—stretching back, as it turned out, well into the nineteenth century. Attention focused on a compound that was to become the prototype of the tricyclic antidepressants—iminodibenzyl. This had been synthesized in 1898 by Thiele and Holzinger, who had described its chemical characteristics but had no ideas for a possible use for it.[16]

The central ring structure, a three-ring or tricyclic structure, resembled the phenothiazines, which also had a three-ring central structure. Laborit's reports of the use of promethazine in anesthesia quickened the pace and led Geigy to look further at this compound. As is usual, the chemists were asked to make a series of derivatives. A better idea of the properties of the class of compounds can be got in this way—having only one compound in the series risks missing a possible activity because that particular compound isn't absorbed well into the brain, for instance. A series of forty-two related compounds was put together by W. Schindler and F. Haeflinger, the Geigy chemists, from the original iminodibenzyl nucleus.[17]

These were tested on animals for sedative and thermolytic properties. All of the series were antihistaminic and anticholinergic.[18] Some were sedative and some were thermolytic, and these few were selected to be tried in humans—company employees. This was not a risk-free procedure, in that while a rough guide to the lethal dose would have been established by extrapolation from work in animals, the possible therapeutic dose would not have been known. As it happens, company scientists such as Alan Broadhurst ended up taking doses of these compounds considerably in excess of what are now recognized to be the therapeutic doses. The hunch, based on the animal and human tests, was that the series might be useful in anesthesia, or it might at least be hypnotic. The idea that there might be some other psychotropic action was not there—the antipsychotic effect of chlorpromazine had still not become apparent.

One compound was selected for clinical testing. This was G22150, which was the iminodibenzyl with the same side chain as promethazine (see Figure 2.1). As with the pattern of development of chlorpromazine, G22150 was given during 1950 to a wide range of clinicians to test. This group included some psychiatrists. Testing might involve little more than giving the drug on one or two occasions to a number of patients, often regardless of diagnosis. Among the psychiatrists asked to test the drug was Roland Kuhn, at the Münsterlingen Hospital near Konstanz.

Kuhn had been born at Bienne, a small town near Berne, on March 4, 1912. He studied medicine at Berne and Paris and trained in psychiatry with Jakob Klaesi, who had introduced prolonged sleep treatment, at the University Clinic in Berne. After Kuhn's training, he went to work at the psychiatric hospital in Münsterlingen, the state hospital for the canton of Thurgau, which had at that time up to seven hundred patients and—what was somewhat unusual—a growing outpatient service. His work before 1950 (and indeed subsequently) was largely conventional, with an interest in research of a psychotherapeutic bent and a leaning toward existential analysis. As such there could have been some concerns that he might have been hostile to drug treatment, but he wasn't. It would probably not have been possible to be trained by Klaesi and emerge believing that physical treatments could not be of some use, and of course ECT was being used at Münsterlingen—particularly for depression, as it turned out.[19]

Kuhn noticed little effect with G22150. It didn't even seem to be a useful sedative. Little more happened until 1953.[20] By then the use of chlor-

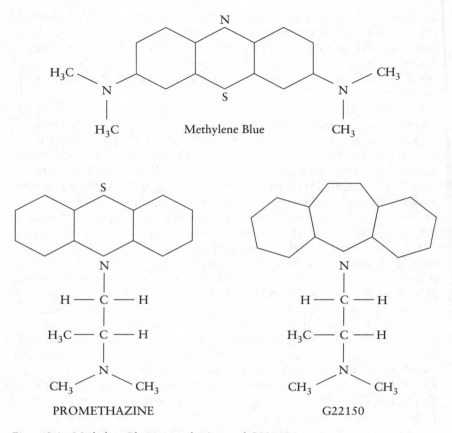

Figure 2.1   Methylene Blue, promethazine, and G22150

promazine had begun to spread beyond Paris. Paul Kielholz, professor of psychiatry at Basel, had been using it and like many others was convinced of its potential. He convened a meeting of fifty Swiss psychiatrists to discuss chlorpromazine. One of these was Kuhn, who began using the drug at Münsterlingen.

The published accounts diverge, somewhat, at this point.[21] When the effects of chlorpromazine became apparent to him, Kuhn claims he wrote to Domenjoz seeking to try out G22150 again, believing that he might have seen some sign of effects similar to those of chlorpromazine when he had tried it out in 1950. The alternative account maintains that Geigy reapproached clinicians. At this stage, in the light of chlorpromazine's growing success, they would have sought out psychiatrists, rather than clinicians generally. Given also that the beneficial effects of chlorpromazine were first appreciated by those working in large state hospitals, whether Delay and Deniker in the Sainte Anne, or Lehmann in the Verdun, company strategy would have been to contact doctors in these locations, because they were still likely to have substantial numbers of patients on minimal or no medication. Whichever way the direction of contact went between Kuhn and Geigy, Otto Kym and Paul Schmidlin from the company went back to visit him at Münsterlingen. G22150 was tried out again, but any psychotropic effects were indistinct—at least in the patients to whom it was given.

Geigy then sent Kuhn (and many other clinicians) G22355, which was the iminodibenzyl closest in structure to chlorpromazine (see Figure 2.2). It was given to a group of patients with schizophrenia. The study design was that those on chlorpromazine were taken off that drug and put on the new one while those newly admitted, who were drug free, also had G22355 given to them. It was administered to over three hundred patients with a variety of conditions. Many of the patients previously on chlorpromazine began to deteriorate, becoming increasingly agitated. Some appeared to become hypomanic. When one patient escaped from the hospital and rode into town on a bike in his nightshirt, singing at the top of his voice, the company decided to discontinue the study.

CHLORPROMAZINE   IMIPRAMINE

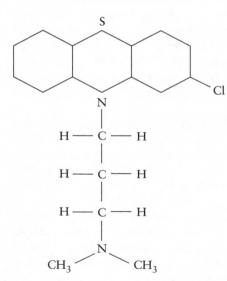

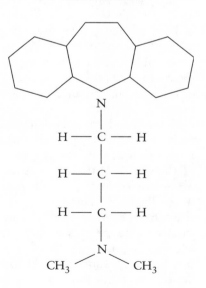

Figure 2.2   Chlorpromazine and imipramine

G22355, however, seemed to have some effects, even if not exactly the ones that were being looked for. Schmidlin, Kuhn, Broadhurst, and others spent time on several occasions trying to draw lessons from the events. What had happened was not simply consequent on the withdrawal of chlorpromazine, because the effects had occurred in both those previously on and those not on chlorpromazine. Because some of the patients had seemingly become "high," the idea that G22355 might have euphoriant effects was proposed. Kuhn has claimed that he later went on to test depressed patients with the drug for the sake of being comprehensive in his assessment. Alan Broadhurst, working at Geigy at the time, and others suggest that the idea that imipramine might be antidepressant was Paul Schmidlin's. The effects of the drug on mood in one sense were unfortunately public, so much so that when the idea was then put to Kuhn that he might try out G22355 again in depressed subjects he was rather unhappy at the prospect.

Whatever the input from the various parties, in 1955, a second study was set up to look at the effects of G22355 in patients who were depressed. Forty patients were ultimately studied, but the responses of the first three were so dramatic that company scientists, ward nursing staff, and Kuhn had little doubt that the treatment was effective. The first patient, Paula J.F., who was depressed and deluded, began treatment on January 12, 1956, and six days later, on January 18, was completely transformed. Kuhn sent a report on the study to Geigy on February 4, strongly endorsing the drug as a potential antidepressant. On August 31, 1957, he published his findings in a Swiss medical journal. On September 6, he presented his research at a poorly attended session of the second World Congress of Psychiatry, which was held in Zurich. Those in attendance suggest that there were probably no more than twelve listeners in the room. The presentation didn't electrify many in the audience. The rest of the meeting remained unaware that they had missed something important.[22]

Kuhn visited the United States in 1958 and on May 19 delivered essentially the same paper to an audience at Galesburg State Hospital in Illinois. This was published in the November issue of the *American Journal of Psychiatry* of that year. It has been compared to the Gettysburg Address—a piece that was neglected at the time but which on rereading has the hallmarks of a classic.[23]

By then he had treated over five hundred patients over a three-year period. Despite the observations that certain patients had become manic while taking the drug, Kuhn declared that imipramine was an antidepressant but not a euphoriant. He picked out the features of a syndrome that he felt was particularly likely to respond; a state that for some time had been called endogenous depression, vital depression, or melancholia. Patients typically had a general retardation in thinking and acting, reported that their oppressive mood was worse in the morning but improved in the afternoon, had a general loss of interest, slept poorly, had lost their appetite, and were preoccupied by fixed ideas of hopelessness, guilt, and despair.

Treatment produced an increase in vivacity and a restoration of interest in activities in general and in social interaction in particular. Sleep was restored and felt normal and refreshing, unlike the sleep that followed the then available hypnotics. Appetite was stimulated. Although improvement might be apparent after two to three days, Kuhn claimed that it could take up to four weeks to become established. He described all the side effects now associated with tricylic use—dry mouth, a tendency to sweat more profusely and sometimes paroxysmally, some constipation, possible drops in blood pressure, and possible confusional conditions in subjects with other brain disorders. He also proposed a dose range that remains the same today. Given modern concerns with the length of time that treatment needs to last, he argued that treatment with imipramine is symptomatic in the sense that if stopped while the underlying disease is active, it will lead to a reemergence of symptoms. Even as early as 1958, he claimed to have had patients in treatment with the drug for as long as two years (see Chapter 4).

Of interest in the light of Kramer's *Listening to Prozac,* Kuhn noted that "not infrequently . . . sufferers and their relatives confirm . . . the fact that they had not been so well for a long time."[24] He suggested, as does Kramer, that sometimes on looking back, patients realized that they had been mildly ill for much longer than they suspected. He then went on to claim that patients with obsessional and a range of phobic and psychosomatic conditions might respond to imipramine, noting the cure of a case of impotence of long standing, although the explanation that he gave for these responses differed from that of Kramer. For Kuhn vital depression was a state that did not necessarily reveal itself in overt sadness but involved a lowering of central vitality in a manner that might well contribute to the development of phobic, obsessive, or hysterical reactions. He explained the response of some of these states to imipramine on the basis that an underlying depression was being treated, the resolution of which could be expected to lead to an improvement in any accompanying neurotic condition. This point in many ways is absolutely central to the question of who discovered what, as will become more apparent.[25]

Finally he pursued the question of possible ethical and moral implications. He was aware that there might be concerns that treatment with a compound like imipramine might have an effect on conscience and resistance to criminal or immoral actions. He noted the case of a patient who had been charged with homosexual offenses, whose sexuality was transformed back to heterosexuality through treatment. This case has echoes of Kramer's Sam, an architect with an obsessive penchant for pornographic videos who was cured by Prozac, but Kuhn cautioned, "One must of course avoid drawing too far-reaching conclusions from a single case." Kuhn, then, ended with a comment that could have come straight from Peter Kramer, thirty-five years later: "An important field of research opens up here, rendered accessible for the first time by the recent development of psychopharmacology, and touching not only problems of psychiatry but also those of general psychology, religion and philosophy." Until the advent of Prozac, however, these aspects of the antidepressant story remained unexplored. Some of the reasons for this neglect may become more apparent in subsequent chapters.

G22355 was given the name imipramine. In November of 1957 it was launched in Switzerland, and in the spring of 1958 in a number of other European countries under the brand name Tofranil. It was 1960, going into 1961, however, before its use became established. In 1960 the journal *Nature,* convinced that the field of antidepressant drug treatments was an emerging one, commissioned Linford Rees to write an article on it. His article covered the effects of four stimulants, followed by three monoamine oxidase inhibitors, before touching on the only tricyclic— imipramine—which he suggested was not as effective as ECT but might have some use in mild depressive states.[26]

This chain of events raises a number of questions. One is the issue of who actually discovered the antidepressant effects of imipramine. A second is why it took so long for imipramine to have an impact. There had been some delay in recognizing the effects of chlorpromazine, but this might have been expected in that no previous drug had shown a specificity for psychiatric disturbances. Once the dike had been breached, however, the companies flooded the market with phenothiazine analogues. By 1955 many clinicians were trying out a number of possible other major tranquilizers. Not only that but different routes to a neuroleptic or antipsychotic effect had been discovered. The Lundbeck Company, for instance, had discovered the thioxanthenes in 1958, and Paul Janssen had found the butyrophenones, both of which were considerably more potent than the original chlorpromazine. Yet no other tricyclic emerged until the launch of amitriptyline (Elavil/Endep, Tryptizol/Lentizol) in 1961.[27]

On the question of who discovered imipramine, many "discoverers" of drugs or those who were around at the time recall discovery as something of a collaborative effort. Nathan Kline, who was later to become involved in a number of bitter priority disputes over iproniazid, when talking about the discovery of the tranquilizing effects of reserpine would allude to the observations of the hospital glazier, who wondered what patients were getting on one particular ward as he seemed to have many fewer windows to replace there.

Similarly Donald Klein, then of Hillside Hospital in New York and later of Columbia, credited the observations of nursing staff on one of the wards at Hillside regarding the response to imipramine of the patient in whom he first recognized the features of panic disorder. Hanns Hippius, later professor of psychiatry in Munich and the discoverer of clozapine, has suggested that in the case of new drugs it is almost always the nursing staff who make the initial observations.[28] In none of Kuhn's accounts of the discovery, however, is there any reference to input from nursing staff or to discussions with Geigy employees about the meaning of the effects that had been observed during the course of the first study with imipramine. Unlike Kline's papers and those of others, none of Kuhn's papers carried the names of any coauthors. This seems to have left a bad taste in the mouth of some.[29]

Indeed, somewhat curiously, in Kuhn's various accounts of the discovery, there are very few details about the actual chain of events. What one finds instead are two central points repeated consistently over thirty years. One was that imipramine treats a particular form of depression— vital or endogenous depression. The other was that discovery was made by a process of empathy. Without a deep, preverbal identification with the patient, he implies, the discovery would not have been made. He has contrasted this approach with the impersonal processes of modern research with its double-blind randomized methods and concern for quantifiable ratings.

For now the first point is the one that concerns us. In essence, what Kuhn claimed is not that he discovered imipramine but rather that he discovered the response of a particular kind of depression to a particular form of drug treatment. This response was moreover somewhat counterintuitive in that the state that benefited was an inhibited, retarded state and the treatment, far from being a stimulant, was something of a sedative. If it had been states of agitated depression or anxious unhappiness that responded, the effects would not have been so problematic.

In part the problems for anyone wanting to pinpoint exactly what happened lie in Kuhn's philosophy. He was one of a number of discoverers in biological psychiatry, who attended a meeting at Taylor Manor Hospital in Baltimore in 1970 convened by Frank Ayd.[30] His contribution to that meeting is quite different from those of any of the other participants. Where others deal in detail, describing events, dates, and personalities, Kuhn apologizes for the anecdotal quality of the few details he does offer, reluctantly conceding that they are part of the story. Coming from an existential background, he was interested in the dialectical evolution of ideas, which he appears to have assumed would inevitably evolve to a particular endpoint. Accordingly the lack of details about the route being taken can be interpreted as part of his general orientation rather than a specific defense against awareness of the part that others played in the process.

His concern, at Taylor Manor, was largely with the theoretical implications of the discovery—what it meant for psychopathological theory—rather than for the practical implications of having a treatment that works. He found himself acknowledging that in part the low-key reception his discovery received owed something to his failure to appreciate that not everyone would understand terms such as vital depression in the same way that he did—that not every psychiatrist would have the necessary familiarity with the work of Emil Kraepelin or Kurt Schneider. As such, from Kuhn's point of view ward nursing staff and even pharmaceutical company scientists could not have been expected to appreciate the significance of what they were seeing.

Indeed, from Kuhn's perspective in one sense they were not even seeing it, in that what was to be seen was not an antidepressant effect so much as the outlines of a disease—whose existence had been proposed before but which was now being revealed by a pharmacological scalpel. An acceptance of this proposal led to the use in Switzerland and Germany during the 1960s and 1970s of the term thymoleptic to distinguish what was involved from lay ideas of an "antidepressant." The term, perhaps ironically, given Kuhn's ideas of what a nonpsychopathologist should have been able to appreciate, was coined by Paul Schmidlin of Geigy. It never caught on. There were mixed reactions from other nonclinical investigators from Geigy. Alan Broadhurst for instance, a pharmacologist, found himself unsurprised by the improvements he saw, but he concedes that he might have been more surprised had he had psychiatric training at the time.[31]

Kuhn's attitude was very typical of German-speaking psychiatry during the 1950s. The interest in psychotropic drugs, since Kraepelin, lay more in what they might reveal about the nature of disease or the workings of the mind than in any use they might have to cure disease or help people. When LSD was synthesized, many philosophers were keen to try it in order to test out implications for theories of the mind. The attitude was very much the same with chlorpromazine and later imipramine. There was a feeling that one couldn't be scientific about treatment—that it was too essentially empirical. As late as 1976 Werner Janzarik, one of the successors of Kraepelin and Schneider, bemoaning the crisis in psychiatry, argued that the cause of the crisis lay in the decline of psychopathology and that, in part, this stemmed from the "triumph of therapeutic empiricism over theoretical examination and interpretation of the nature of psychiatric disorders."[32]

The World Psychiatric Association meeting at which Kuhn presented his findings was dominated by Europeans. Lehmann, who was there, re-members many people staying up till late at night discussing the finer points of existential psychiatry. When flying back on the plane to Canada with a professor of psychiatry from Toronto, the Englishman Aldwyn Stokes, Stokes asked him why there was so much fuss about existential-ism. Having a captive audience, Lehmann, who was himself German, de-cided to try and explain it, taking two hours to do so, but found Stokes unimpressed by what he was hearing. Lehmann concluded that German existentialism couldn't be understood by Americans or Anglo-Saxons.[33]

After Stokes fell asleep, Lehmann read Kuhn's paper. He had heard about the talk and made sure to get a copy. Having been the first to inves-tigate chlorpromazine in North America, he kept his eye out for new agents which might be worth trying. He was impressed enough to contact the Geigy offices in Canada when he got back and ask them for some imipramine. They knew nothing about it. Within a week, however, they had some for him and he started the first trial of imipramine in North America, reporting in 1958 on its use in eighty-four patients. Two-thirds of the sample responded.[34]

Any attempt to answer the question of why it took Geigy so long to market imipramine is also revealing. The company was faced with a num-ber of problems. One was what weight to give to the observations of Kuhn and the second had to do with the issue of how effective imipramine was and what size market it might command. Many other in-vestigators were skeptical. Some felt that the compound was so like chlor-promazine, it was hard to believe that it could be an entirely different agent.[35] A large number of psychiatrists from central Europe had tried it out, and some had pushed the dose up to 1000 mg and had seemed to find chlorpromazine-type effects at that level.[36] Compared with some of those who had tested the drug on patients, Kuhn appeared something of an Ichobod Crane figure.[37] When it came to convening the First Interna-tional Congress of Neuropsychopharmacology (CINP) in Rome in Sep-tember of 1958, Kuhn was not an invited participant. There were presen-tations on imipramine by W. Schmitt from Heidelberg, C. Fazio from Genes, Paul Kielholz from Basel, and Fritz Freyhan from America and on iproniazid from Nate Kline and others, but nothing from Kuhn.[38] Nor was he asked to the second or third CINP meetings.

Kuhn's interest in the nuances of psychopathological theory were of lit-tle interest to Geigy, who were more concerned with the question of whether there would be a market for the compound. He was also some-thing of a country doctor—in the sense of someone who was almost the opposite of the cosmopolitan Nate Kline. When Roche were uncertain about marketing iproniazid for depression, as we shall see, Kline forced the company's hand by personal interventions with senior company management and by announcing his discovery to the media—even before he presented the findings to a scientific audience. This and the fact that Münsterlingen was out of the way, whereas Rockland State Hospital, where Kline worked, was in New York and the Hôpital Sainte-Anne, where Delay worked, was in Paris, may have combined to cause something of a delay.

Geigy was faced, therefore, with the problem of what weight to put on the opinions of this man, when many of the leading lights in the field, even those committed to drug treatment, felt that there would never be such a compound and, beyond that, clearly did not view Kuhn as an emerging scientific star. Kuhn complicated the picture, moreover, by stressing that imipramine was not a stimulant, so company executives had to grapple with the idea of a drug that lifted mood without being a stimulant. Furthermore, he was proposing that it would only be of use in a limited number of depressive states and indeed that its benefits would be most marked in those depressions that were known to respond to ECT—a rather small market; there was no suggestion that there were large numbers of people out in the community who had a condition for which it was appropriate to think about giving ECT.

Was imipramine as effective as ECT?—most probably not. Was there a market for such an antidepressant?—who knew! Clearly there was a market for a stimulant; the widespread use of amphetamine in the 1940s and 1950s had shown that. But this compound was not a stimulant. Some impression of how the company must have viewed the problem can be gleaned from the fact that in 1958, when they produced chlorimipramine (clomipramine [Anafranil]), notwithstanding the example of imipramine, in a move that seems scarcely credible now, they chose to try it out first in a group of schizophrenic patients, before giving it to depressed patients and concluding that it too had its primary effects on mood (see Chapter 6). The impact of Thorazine's sales in the United States must be kept in mind here—this was what the other companies wanted to buy into.

Through 1956 and 1957, the company called in a number of experts to look at Kuhn's findings and to advise them whether there might be a market for the compound. This is very difficult to understand now, when depression is seen as such a widespread illness—almost the common cold of psychiatry. But at that time many of those affected suffered in silence in the hope that the condition would remit, as it very often did. It was often only when sleep and appetite were completely lost or the individuals were suicidal that their relatives committed them to the asylum, when ECT, after its introduction, often turned them around. Subsequently in the 1960s, when the use of chlorpromazine, Librium, and imipramine had created outpatient psychiatry, primary care physicians began to detect depressive disorders earlier and refer them for appointments. As recognition of the syndrome increased, the World Health Organization set up studies to establish the prevalence of depression and came to the conclusion that up to 100 million people worldwide were probably depressed on any one day. But in 1957 and 1958 this scenario was a long way off.

In the end Geigy proceeded with imipramine but only after two further developments. One of the most powerful shareholders in the company was Robert Böhringer, whose family also had a controlling interest in Böhringer-Ingelheim. A relative of his became depressed and Böhringer, aware of work within the company on an "antidepressant," asked for samples, brought them home, and treated her. She recovered within days. He threw his weight behind the development. Shortly afterward Paul Kielholz, professor of psychiatry at the University of Basel, who was emerging as the foremost psychopharmacotherapist in Switzerland, reported to the company that in his hands also imipramine appeared to produce antidepressant effects. But even more persuasively, by April of 1957 another "antidepressant" had hit both the market and the front pages of the *New York Times*. This was the monoamine oxidase inhibitor iproniazid, whose advent helped create the antidepressant market.

KLINE, ROCHE, AND IPRONIAZID

Multiple independent discovery happens often enough in science for it to be something of the norm rather than the exception. In the case of the antidepressants, however, as Kuhn's story indicates, the occurrence of multiple discovery is surprising in that no one was expecting something of the sort that Kuhn discovered. The fact that another group of antidepressants was discovered more or less simultaneously, in a different part of the world, by clinicians who had probably no idea what was happening in Münsterlingen adds to the surprise. Perhaps given the example of so many other discoveries we ought not to be so surprised but the story is, in many respects, far more complicated than one of simple independent discovery.

In terms of contrast, there could be no more dramatic a contrast between two individuals than that between Kuhn and Nathan Kline, who is generally credited with the discovery of the antidepressant effects of iproniazid. Where one was donnish, retiring, and rigidly consistent in his views, the other was flamboyant, cosmopolitan, and ever ready to adopt new ideas. The contrast between the modes of discovery was equally great. By a supreme irony, Kuhn the theoretician made his discovery as a result of observations on patients, while Kline, a confirmed empiric, came to his following a theoretical hunch—or at least he claimed he did. Where the imipramine story springs from one source, the iproniazid mainstream was formed by the confluence of many tributaries.

One part of the story goes back to 1877, when Oswald Schmiedeberg discovered that when a dog takes benzylamine it is excreted as hippuric acid.[39] In 1883 O. Minkowski demonstrated that benzylamine could be oxidized to benzoic acid, raising the possibility that the conversion of benzylamine in the body proceeded by an oxidative route.[40] In 1928 Molly Hare, working in Cambridge, found that a substance called tyramine was converted in the liver by a process of oxidation. She inferred the presence of an enzyme which she called tyramine oxidase.[41]

In 1932, shortly after Hare had left Cambridge, Derek Richter came to work there with Frederick Hopkins, one of the discoverers of vitamins and one of the founding fathers of biochemistry. Richter was given the task of finding out what happened to adrenaline in the body. It was already known that acetylcholine[42] was inactivated by choline esterase— could adrenaline similarly be inactivated by an enzyme? A co-worker in the department, David Green, who was working on the liver cytochrome system, noticed that it oxidized adrenaline to an unknown product, which Richter and Green worked on and characterized as adrenochrome. But metabolism to adrenochrome only seemed to account for a small amount of the available adrenaline. To find out more about where the rest might be destroyed in the body required the collaboration of a physiologist.[43]

This was at a time when physiologists and the early neurochemists were at war over the nature of neurotransmission—was it chemical or electrical? Unwittingly, Richter approached Adrian, professor of physiology at Cambridge and one of the protagonists in the disputes, to inquire whether anyone in his department might be interested in pursuing the issue. He was informed that the fate of adrenaline was not a matter of concern to physiologists. Richter, however, had been made aware of a German refugee in Adrian's department named Herman Blaschko, and asked whether he could approach him. Adrian was prepared to allow this on the grounds that a refugee might be unwise enough to be interested in such matters. Blaschko was interested, and he and Richter set to work on the problem.[44]

They weren't the only ones working on oxidases. In the Whitchurch Hospital in Cardiff, the resident medical superintendent, P. K. McCowan, had set up a laboratory and recruited an early refugee from Middle Europe, Juda Hirsch Quastel, to work there. McCowan and Quastel had looked at the metabolism of depressed and schizophrenic patients through the 1930s. Quastel had become interested in the issue of oxidases and found an enzyme that oxidized some amines, although he did not think adrenaline was one of these. Hearing of Richter's work, Quastel visited him in Cambridge. By this time Richter and Blaschko had discovered that there was an oxidizing enzyme that metabolized adrenaline, but they had also found that the same enzyme oxidized norepinephrine, dopamine, and 5HT (5-hydroxytryptamine, or serotonin),[45] which made it a vastly more significant entity—they called it amine oxidase. It was the same enzyme that Hare and Quastel had stumbled on.[46]

Quastel, shocked to find himself trumped, hurried back to Cardiff, ran some experiments, and submitted his findings.[47] Richter and Blaschko got wind of this and submitted theirs and were published first. Quastel's paper subsequently appeared with no mention of Richter and Blaschko's work.[48] Later that year Quastel and Pugh demonstrated that the new enzyme was to be found in the brain in addition to the liver—although the significance of this at the time remained uncertain. The following year, Al Zeller found a diamine oxidase and suggested renaming Richter and Blaschko's enzyme monoamine oxidase (MAO), a name that stuck.[49] Thus the monoamine oxidase story, appropriately, began with multiple independent discovery and disputes about priority.

A second strand in the story concerns the synthesis and use of isoniazid and iproniazid. Isoniazid had first been produced in 1912,[50] but its resynthesis in 1951 and the discovery that it was tuberculostatic was much more important. During World War II, the V2 and other rockets were initially propellant by liquid oxygen and ethanol. When stocks of this ran low toward the end of the war, German engineers resorting to using a compound called hydrazine. When the war ended large stocks of hydrazine were left over and these were taken over largely by the chemical companies, many of which were in the process of developing a pharmaceutical division. Hydrazine could be manipulated in a number of ways to produce a series of derivatives, and using isoniazid as their starter, Herbert Fox and John Gibas produced in 1951 at the Hoffman-la-Roche laboratories in Nutley, New Jersey, iproniazid. Both were tested for activity against tuberculosis and found to be potently tuberculostatic. Both were pressed into use.[51]

In 1952 Zeller and colleagues, who had previously found that antituberculous drugs tended to inhibit diamine oxidase, demonstrated that iproniazid was also a diamine oxidase inhibitor. They proposed that this might be sufficient to increase a variety of monoamine levels in the brain, perhaps leading to the mental side effects that were already being noted. Subsequently they found that iproniazid, although not isoniazid, inhibited monoamine oxidase. It was therefore a monoamine oxidase inhibitor—an MAOI.[52]

Within a few months of the first use of isoniazid and iproniazid for tuberculosis, mental side effects had been noted with both. Patients were reported to be doing well on them, particularly iproniazid. In addition to the healing of tubercular lesions, initial reports suggested that the general demeanor of those taking the drugs seemed to be favorably affected. One hospital in particular, the Sea View Hospital, and one orthopedic surgeon primarily, David Bosworth, were responsible for generating interest in iproniazid. This interest hit the headlines. The mythology of discovery has it that newspapers reported patients dancing in the wards even though they had holes in their lungs.[53]

This interest led to a study by Jackson Smith, of Houston, Texas, on the mental effects of iproniazid, which was published in the *American Practitioner* in 1953. It carried an unusual editorial note across the top of the first page to the effect that "the possibility of a favorable action of pharmaceuticals in mental disease, even by indirect action, warrants such a trial under the proper circumstances."[54] Smith gave iproniazid to a group of eleven patients, of whom five were depressed, on the basis of reports that it and isoniazid led to both weight gain and mental stimulation. This then was clearly an attempt to influence mental state, even if only as an indirect consequence of improving the general nutritional and metabolic state of the patient by promoting weight gain. Smith saw the drug as a possible tonic. Patients gained a considerable amount of weight in a short period of time and some of the depressed patients appeared to improve, as well, but he appears not to have made the connection to an antidepressant action.

As noted earlier, in Paris Jean Delay, Pierre Deniker, Jean-François Buisson, and others were trying both isoniazid and iproniazid in depressed patients. Their "antidepressant" effects were first discussed at the Société Médico-Psychologique in December of 1952. The effects were sufficient for Delay to commit Buisson to study them as part of a thesis. The date of this work is notable, but it is also notable that none of these authors ever claimed priority for the discovery of antidepressants—they appear happy to give the credit for this to Kline and Kuhn. The reports of Delay and his colleagues do not appear to have impinged on the American investigators of iproniazid despite the contact that existed between them—particularly with Deniker's visit to the United States in 1954.

A number of other reports appeared in America and were reviewed by George Crane in 1956.[55] In general, the mental side effects of iproniazid had become something of a problem, so that its use in tuberculosis was falling out of favor. Very large doses were being used and, at these doses, up to 10 percent of individuals were becoming confused or psychotic. Drawing on the latest developments in neurochemistry, Crane speculated that the mental side effects were due to its inhibiting effects on monoamine oxidase, either by an increase of 5HT, thereby mimicking the effects of LSD, or by an increase in one of the breakdown products of adrenaline, such as adrenochrome, discovered by Richter and Green. In 1955 Abram Hoffer, Humphrey Osmond, and James Smythies had proposed that abnormalities of adrenochrome might be the cause of schizophrenia, and this possibility was hotly debated topic at the time. In Baltimore, Frank Ayd, a psychiatrist in private practice and attached to Taylor Manor Hospital, gave iproniazid to depressed patients, on the recommendation of his local thoracic specialist, who commented on its mental side effects. Ayd too reported that it might be of some benefit but not that it was an antidepressant.[56] This was before Ayd heard Kuhn talk and before his role in the discovery of the antidepressant effects of amitriptyline.

Crane, however, took the issue further than anyone else. He felt that the problems associated with the use of iproniazid probably stemmed from the dose being used, and he decided to investigate whether a lower dose might produce beneficial effects without undue side effects. He was, according to himself, impressed with the possibility of using the drug as a stimulant for debilitated or depressed patients. But he found that very few patients could tolerate iproniazid indefinitely, even at the lower dose, without clear mental changes of one sort or another. Some had to discontinue treatment because of psychotic reactions; others benefited emotionally.

The problem in interpreting these finding is that Crane was looking at mental reactions to iproniazid in patients with tuberculosis. He went on to characterize it as both a stimulant and a tranquilizer. He claimed that many obsessive and phobic responses were diminished against a background of increased vitality. In some respects, he said, it resembled amphetamine but was potentially of greater usefulness than amphetamine in that amphetamine leads to a loss of appetite, whereas iproniazid caused marked weight gain. Crane had two substantial publications on what broadly speaking were the antidepressant side effects of iproniazid in 1956. Yet the honor for the discovery that this drug was an antidepressant went to Nate Kline—why and how?

At the time of his involvement with iproniazid, Kline was an assistant clinical professor of psychiatry at Columbia University and director of the research facility at Rockland State Hospital in New York. Born in 1923 in Atlantic City, he had begun as a psychologist before coming to psychiatry. Interested in research in mental illness, he had founded the research facility at Rockland in 1952, to which he recruited, over the years, a stream of others interested in psychiatric research.[57]

In 1952 there had been reports that Rauwolfia serpentina, a plant root used in India for the treatment of hypertension, snakebite, and insanity, could indeed be shown to be of benefit in the treatment of hypertension, even when given in the West. These reports led Kline to consider the possibility of trying it on psychiatric patients.[58] In 1953 Ciba, which had been investigating it, reported that it had isolated an active salt from the root, which it called reserpine.[59] Kline visited the American headquarters of Ciba at Summit, New Jersey, where he met and spent considerable time with one of the company pharmacologists, Jack Saunders. Another

Ciba pharmacologist, F. F. Yonkman, studying the effect of reserpine on animal behavior, had coined the term *tranquilizer* to describe these effects. Klein persuaded Ciba to support a study comparing rauwolfia, reserpine, and placebo in 710 psychiatric patients. Although there do not appear to have been the dramatic stories of patients waking up from psychosis that were reported by the pioneers of chlorpromazine, the sedative and tranquilizing effects of reserpine were quickly apparent to everyone from the hospital glazier through to the nursing staff.

The study was carried out in 1953 and reported by Kline on April 30, 1954, to the New York Academy of Science, making 1954 something of an annus mirabilis with the introduction of not just one but two major tranquilizers. Reserpine, however, never caught on clinically the way chlorpromazine did. Many hospitals tried both and opted for chlorpromazine. The beneficial effects of reserpine were of somewhat slower onset. In addition there were a number of reports of difficulties using it in the treatment of hypertension. However, where no one was honored for the discovery of chlorpromazine, Kline was awarded the Lasker Prize in 1957 for his role in demonstrating the effects of reserpine. (Much sought after, the Lasker Prize was awarded annually by the Lasker Foundation, set up by Albert and Mary Lasker in 1942 to support research on mental health, birth control, and cancer research. Mary Lasker was a vigorous campaigner for mental health, and she had considerable influence in the NIH and with Congress.)[60]

The reserpine story is quite unlike the chlorpromazine one in two respects. First, reserpine had no offspring. Unlike chlorpromazine the reserpine molecule could not be easily manipulated to yield a series. In contrast, reserpine almost immediately led to a fertile group of hypotheses—both biochemically and psychologically. Biochemically, observations made in Bernard Brodie's laboratory at NIH, correlating the sedative effects of reserpine with a lowering of brain 5HT, provided the first bridge between neurochemistry and behavior and are generally credited with establishing biochemical psychopharmacology. Subsequent observations of the effects of reserpine on catecholamines (epinephrine, norepinephrine, and dopamine) made by Arvid Carlsson extended the bridgehead in two important ways. First, it stimulated a very public debate between two schools of thought, a 5HT camp and a catecholamine camp, with competing experiments and fine-grained analyses of results that brought dynamism and a clear sense of direction to the field.[61]

The second effect was to provide the pharmaceutical industry with a principle to guide drug development. Thereafter, drugs could either be designed to produce similar depletions of 5HT or catecholamines with a view to seeing what they did in either animals or man or alternatively they could be designed to modify or block the reserpine effect. In this manner drug development began to become systematic. Indeed, a demonstration shortly afterward by Erminio Costa and Silvio Garattini that imipramine blocked the effects of reserpine produced the primary screening test for the development of two decades worth of antidepressant compounds.

Kline was invited to formally present his studies on reserpine to the annual American Psychiatric Association meeting in May of 1955. By that stage many clinicians had tried both it and chlorpromazine; there were papers on both agents at the meeting and considerable interest in both. But both were trumped by news of another compound—meprobamate. The first reports of meprobamate's efficacy as a barbituratelike tranquilizer were reported in the April issue of the *Journal of the American Medical Association*. Although there was no formal presentation on meprobamate to compare with the papers on chlorpromazine and reserpine, meprobamate was the news from the conference. Why? This is an important issue in terms of the later plot of this book. The answer almost certainly is because it promised to be a drug for office-based outpatient psychiatry—and where this might not have aroused much interest in Europe, in America it was of capital importance.[62]

Meprobamate was an anxiolytic rather than an antipsychotic or an antidepressant. The barbiturates had been anxiolytic but in the process had produced both sedation and muscle relaxation. They were, however, dependence producing and lethal in overdose, which limited their usefulness in the treatment of anxiety—an illness often seen in psychiatrist's offices. Meprobamate produced much less sedation. Its muscle-relaxing properties were therefore much more salient. This in turn stimulated thinking about the mechanisms involved in the anxiety disorders. Meprobamate also proved that a nonbarbiturate barbiturate could be produced, and at least initially it seemed likely to be safer than the barbiturates. Without it, the development of Librium or Valium would have been much less likely.[63] Meprobamate opened up the question of the mass treatment of nervous problems found in the community. In terms of American models of mental illness and of the later antidepressant story, it was possibly as important to the establishment of psychopharmacology as chlorpromazine and reserpine.[64]

Mike Gorman, an experienced press officer and member of the national executive for the National Committee against Mental Illness and an associate of Mary Lasker's, was at the meeting. In the light of the excitement, he thought it would be appropriate to capitalize on what was happening. He approached Kline, Frank Ayd, Henry Brill, and other prominent psychiatric advocates of drug therapy during the course of the meeting and suggested that they come to Washington to lobby Congress to put money into mental illness, given the success of the new treatments.

Kline and the others, therefore, testified before Senator Lister Hill's committee on the benefits of the new medications and the need for them to be properly evaluated. Congress had passed a Mental Health Study Act in July 1955, which provided the framework for large-scale donations to research. Two million dollars, annually, were subsequently allocated for research on psychopharmacology. By most accounts the figure who did most to extract this money was Kline. The amount involved in terms of what could be done with it in the 1950s was vast—so much so that those charged with administering it found it difficult to give it away. Biological psychiatry had been capitalized.[65] Although the money didn't go in the direction that Kline favored, his efforts at lobbying were probably the single most important input to the establishment of a new psychiatry.[66]

In 1956 a further project was taking shape for Kline. He was working with Mortimer Ostow within the psychoanalytic framework of American psychiatry that was as much de rigueur for him as classical psychopathology had been for Kuhn. Kline and Ostow speculated that as psychic conflicts all involved the binding of psychic energy in various different ways and as a great deal of ego energy went into binding instinctual (or id) energy down to produce a range of inhibited states, it was conceivable that a drug that took energy away from the ego might lead to a liberation of instinctual energy—it might be a psychic energizer. Alternatively, a drug that reduced libidinal energy might release ego energy. Such a drug, Kline suggested, in principle, would increase levels of energy and drive up appetite—both for food and for sex. It would increase responsiveness to stimuli and would leave available supplies of energy, the awareness of which would lead to a sense of joyousness and optimism. Getting the dose right would be important, because releasing too much id energy could be expected to tip the subject over into psychosis.[67]

It wasn't immediately clear, on this basis, why reserpine or indeed chlorpromazine shouldn't be antidepressants or psychic energizers at the right dose. They weren't thought to be—although most neuroleptics have been shown to be useful in depression (see Chapter 6). But two groups suddenly reported findings that seemed to fit with Kline's speculations. Charles Scott and M. Chessin, working at the Warner-Lambert Laboratories, and Bernard Brodie, Alfred Pletscher, and Park Shore at NIH found that animals, pretreated with iproniazid, when given reserpine, instead of becoming sedated, were activated.[68] One of the other interesting things about iproniazid was that, just as Kline's speculations suggested, people on too high a dose regularly became psychotic.

Kline approached Roche in April 1956 with a proposal to look at iproniazid in psychiatric patients. The company wasn't interested, even though the research director from its Basel division, Alfred Pletscher, had been involved in the work with Brodie, which had attracted Kline's attention. Iproniazid was slipping out of favor because of its side effects. In such situations, companies often try to kill off a compound quickly, and the last thing they want is some investigator to come along and propose a new indication—particularly when the size of the market for the proposed indication is uncertain.

Meanwhile, in May of 1956, Jack Saunders had left Ciba and joined Kline at Rockland State Research Facility. According to Kline later, Saunders was interested in the ideas about iproniazid. According to others, questions have to be asked about how many of these ideas and even ideas about reserpine stemmed in part from Saunders's input. He and Kline had been in contact ever since they first met over the reserpine studies in 1953. After joining Kline, he came in on the iproniazid project.

Saunders and Kline were then approached by Harry Loomer, who was a nonresearch member of staff at Rockland Hospital. Loomer wondered if they had anything that might be of help for regressed patients who had failed to respond to reserpine or chlorpromazine. This query provided a basis for the project. At the start of November, seventeen inpatients—all of them diagnosed as having schizophrenia—were put on iproniazid. Given the state of diagnosis in U.S. psychiatry at the time, this may have meant little more than that they were severely ill. A few weeks later Kline began enrolling some private practice patients with a diagnosis of depression. The original plan for the study involved pretreatment with iproniazid and subsequent administration of reserpine, but they opted for a lengthier period with iproniazid to determine what it did on its own. Like Crane, they also took care with the dose, using a regime substantially lower than that then being used for tuberculosis.

According to Kline's version of the events, by February they knew they were on to something. When they collated their data, they found that up to two-thirds of the patients showed some response. Part of the problem in attempting to evaluate what happened, however, is that the clinical features of the patient group remain unclear from the various publications—"most of the patients were withdrawn and deteriorated, with a heavy weighting of hebephrenics." Or as Kline had briefed the *New York Times:* "The drug . . . has produced 'remarkable' mood improvement and activity among long-term 'untouchable' psychotics of the 'burned-out' kind as well as among non-hospitalized neurotics." Kline had also given it to nine patients being seen in his clinic, who were more likely to have been depressed. Out of this came the proposal that iproniazid might be of value in influencing the course of individuals suffering from mild withdrawal or depressive symptoms.[69]

The enthusiasm of Kline and his colleagues, however, was not reciprocated at Roche, leading Kline to set up a clandestine meeting with David Barney, Roche's president, to alert him to the fact that his company might have something important on its hands. Yet Roche remained unenthusiastic and was finally only persuaded to cooperate further with the research by being told that the Rockland group was going to go ahead and present the findings anyway, as publicly as possible. Kline was not faced by the same difficulties as Kuhn in that iproniazid was already on the market, and this difference was crucial.

Kline at the time was the chairman of the Committee on Research of the American Psychiatric Association. In 1957 the regional meeting was scheduled for Syracuse, New York, and included a session on Research on Affects. Even though the patient numbers were modest, Kline decided to get his group's findings included in the program. George Crane, who had done more work than anyone in the area, was not invited as a speaker although subsequent arrangements were made to have him there as a discussant.[70]

A few days before the Syracuse meeting, Kline held a press conference at which he announced the findings. When Loomer subsequently presented the paper on April 6, the *New York Times* report on April 7 implied that the findings had been presented by Kline.[71] This led to some ill-feeling between Loomer and Kline and a subsequent correction by the *New York Times*. Kline also presented the findings at a congressional hearing in May and had them written into the record at the time—ensuring earlier publication than he would get with the Research Reports proceedings. Quite unlike the imipramine story, within a year of this paper large numbers of patients were treated with iproniazid. The fact that it was already on the market clearly made some difference—the company didn't have to agonize about whether there was a market that would make it worth its while to seek registration—it could sit back and see what happened. Kline, who was well connected to everyone who counted in American psychiatry, made it happen for Roche.

Kline was subsequently awarded a second Lasker Prize, in 1964, for his contributions to psychiatry—the only psychiatrist apart from Karl Menninger to be so honored. He wrote an acceptance paper in the *Journal of the American Medical Association (JAMA)* on the practical management of depression that also outlined the process of discovery. Saunders wrote a letter of protest to *JAMA* stating: "Neither the terms of the award nor the paper recognize that my role was primary in the development and administration of this drug for this purpose. There can be no doubt that intellectual dishonesty is involved here." Kline replied that accusations of lack of interest on his part until Saunders and Loomer had prepared a report for the Syracuse conference were unfounded and that furthermore "the hospitalized patients in this study did not have depressions but were primarily schizophrenics of the burned-out type . . . The specific effect of

Marsilid for the treatment of patients with . . . depression was begun at about the same time in my private practice in New York."[72]

Saunders and Loomer sued for recognition of their role. Loomer dropped out, but Saunders pursued the case and a lay jury initially found for him. His case was that at an APA regional meeting in Galveston, in 1955, he had suggested to a group of psychiatrists that it might be worth investigating amine oxidase inhibitors, on the basis that their actions might be expected to increase brain monoamines and that this might produce mental effects. He also suggested that his reason for leaving Ciba and joining Kline was because it "became necessary eventually to join the do-it-yourself camp in order to get things moving in this area." He claims he argued that there was a "probability that they (MAO inhibitors) would alleviate depressions." He also wrote that he had in "1956 . . . suggested and initiated studies with iproniazid." In all cases, he appealed to page 152 of the first volume of APA Research Reports, which appeared in 1955. Chasing up this lead some years ago, Merton Sandler established that the first volume had 152 pages but that there was no reference in it either to Saunders or to MAO inhibition.[73]

A judge restored the award to Kline after an appeal. There was a further series of court cases and eventually one-third of the award was given to Saunders. Saunders dropped out of medical and pharmacological circles. In 1970, a conference on Discoveries in Biological Psychiatry was organized at Taylor Manor Hospital in Baltimore by Frank Ayd, who thought carefully about what to do about the MAOI slot. He approached Roche for its view of what happened and on the basis of that and his own impressions, having known both Saunders and Crane, he opted to have Kline make the presentation as the discoverer of the antidepressant effect of the MAOIs. He received no protest from Saunders, Crane, or anyone else.[74]

Ever since, Kline has generally been credited with being the discoverer, but this may be in part because so many of those involved feel they owe such a debt to him. It may also in part owe something to the flamboyance of a man, who in the mid-1960s was featured on the front cover of *Fortune* magazine as one of the ten best-known men in America. Doubts linger, fueled by the opinion of some that Kline wanted the Nobel Prize and would have done almost anything to get it, including pick up someone else's ideas and run with them. There are others, such as Heinz Lehmann, who saw him as a concerned researcher, regularly seeing his state patients very early in the morning before going on to tackle research issues or work at his private clinic. Others saw him as quite divorced from clinical issues, a view perhaps caught best in a quip he was wont to make when asked who did the work when he was away so often at meetings—the same people who did it when he was there!

But who did what? Saunders's claim was made on the basis that he had proposed the idea that monoamine oxidase inhibitors might be important. This idea was not unique to him—Al Zeller had raised this possibility as early as 1952. Indeed just this possibility had led Zeller and colleagues, in 1956, to treat a group of seven withdrawn schizophrenic patients with iproniazid, finding a marked increase in energy and sociability. The idea that MAO inhibition was important appeared to be settled in following years with the development of the non-hydrazine MAOIs, phenelzine, and tranylcypromine. This strongly supported the belief that it was monoamine oxidase inhibition which produced the antidepressant effect and not something else to do with hydrazine. But as will become clear in the next section, the story is actually much more complex.

Kline's claim for priority in the discovery lay in the area of speculation that there might be a category of drugs that he called psychic energizers which, despite the similarity of that term to the term psychostimulant, would differ substantially from the psychostimulants. Whether such a compound was the same thing as what Kuhn meant by an antidepressant is another matter. As late as 1964, Kline was still distinguishing between psychic energizers and antidepressants, in a manner peculiar to him.[75]

KUHN OR KLINE?

Kline's thinking was as American as Kuhn's was European. From the start American psychiatry has dealt with broad concepts, whereas European notions have been much more restricted. For Americans, the impulse disorders and the neuroses have been part and parcel of psychiatry, with notions of a continuum between normality and abnormality. This was an orientation that was receptive to the tenets of psychoanalysis, which in contrast never took hold within mainstream European psychiatry. European psychiatry was asylum psychiatry, and the prototypical mental illnesses were dementia praecox and manic-depressive insanity.

As the excitement about meprobamate (and the subsequent widespread use of the minor tranquilizer Librium) in the United States demonstrated, American psychiatrists wanted drugs that would be useful in office psychiatry rather than drugs of benefit to asylum inmates. Kline's ideas about a psychic energizer fitted their perceptions of what was needed much more than did Kuhn's thymoleptic. The Prozac phenomenon, far from being remarkable, is in many respects directly within the Klinean tradition both in the mass use of the drug and also in the portrayal of Prozac as something closer to a psychic energizer than a thymoleptic.

A number of factors led to the eclipse of Kline's thinking. Within a year of iproniazid's widespread use, reports began to emerge of a number of problems, in particular jaundice. Roche decided to withdraw the drug.[76] Of note, Smith, Kline & French subsequently introduced an MAOI, tranylcypromine, which also ran into trouble. But unlike Roche, when the FDA asked for it to be withdrawn Smith Kline fought the issue and had it reinstated, even though it was subsequently to be associated with a number of problems and in particular the "cheese effect" (Chapter 4). Thus it would seem that Roche was never fully behind Kline with regard to this drug. It hadn't been patented as an antidepressant, unlike tranylcypromine, for instance, and so the amount of money the company could expect to make from it was strictly limited. On the withdrawal of iproniazid, they introduced isocarboxazid, patented as an antidepressant, and biochemically a good monoamine oxidase inhibitor, but it never had much impact. Kline was never convinced that it was an antidepressant.

A second factor was the demise of psychoanalysis and accordingly the theoretical framework from which Kline had drawn his ideas. Notwithstanding Kline's efforts to think his way forward in psychopharmacology from an analytic perspective, analytic thinking could not easily accommodate itself to psychopharmacology. The development of the catecholamine hypothesis of depression in 1965 sounded the death knell for analysis, as we shall see in Chapter 5. It did so because though it was in many ways far more simplistic and unidimensional than analytic notions about depression, it provided an alternative framework that was seen by many both within the profession and among the lay public as having a better chance than analytic theories of accounting for one of the new central facts about depression—its response to drug treatments.

The emergence of Prozac brings back echoes of Kline. Quite apart from its mass use, Prozac has been advocated for and indeed has been shown to be useful for a wide range of conditions other than endogenous depressions. Its efficacy in phobic, panic, and obsessional states as well as for depression, and its usefulness in these states even in circumstances where subjects cannot be shown to have an underlying depression, in many ways is more consistent with Klinean thinking than with Kuhn's ideas. The basis for this profile of Prozac, ironically, was laid by another Geigy compound, clomipramine (see Chapter 6).

Thus the question of whether Kuhn or Kline made the more important contribution to the discovery of the antidepressants would appear to have been settled in different ways at different times. In the late 1950s and early 1960s, Kline seemed to be the discoverer of the antidepressants, both by virtue of the earlier public presentation of the data and because he was the first to persuade the world of the discovery. From the mid-1960s through almost to the present, Kuhn appeared to be the true discoverer. The officially endorsed WHO antipsychotic is chlorpromazine, and its antidepressant is imipramine. For most pharmacotherapeutically oriented psychiatrists working through the period from 1960 to 1990, psychopharmacology boiled down to chlorpromazine, imipramine, diazepam, and lithium, with the rest simply being variations on a theme. And for most of that time the majority opinion has been that imipramine is a treatment that is somehow specific to the syndrome of vital or major depressive disorder, as was argued by Kuhn.

OR SOMEONE ELSE?

With the emergence of the SSRIs (selective serotonin reuptake inhibitors, such as Prozac), the pendulum may be swinging once again in a manner that will restore to Kline the primacy for antidepressant discovery that until recently has rested with Kuhn. But if this occurs, consider the following. On June 4, 1953, a paper was read before the section on nervous and mental diseases at the 102nd annual session of the American Medical Association in New York detailing the effects of isoniazid in forty-one patients suffering from anxiety and depression. Two-thirds of those to whom it was given improved, and among those improving were a number who had previously only responded to ECT. Improvement took up to three weeks to appear. The authors were Harry Salzer, an assistant professor of neurology at the Cincinnati College of Medicine, and Max Lurie, an instructor in psychiatry at the same university. Both were attending psychiatrists at the Cincinnati General Hospital.[77]

Encouraged by these results, Lurie and Salzer went on to study more subjects. Subsequently, on April 13, 1954, at the annual meeting of the Ohio Psychiatric Association meeting in Columbus, they reported on the effects of isoniazid in a further forty-five depressed patients. These were depressed rather than schizophrenic patients, with a substantial proportion having shown a prior response to ECT and others having a clear bipolar disorder. They were seen in private practice. None had tuberculosis. Again two-thirds recovered—roughly the rate Kuhn was reporting. Again response took two to three weeks to appear, as Kuhn was later to find. Patients tended to relapse if the drug was tapered too quickly. Patients appeared to respond regardless of the type of depression they presented with.[78]

On the face of it, these studies were larger and more adequately reported than those of either Kline or Kuhn. Even though the studies were American, the findings could not have been discounted in Europe, given the clinical features of the patients studied. But the primary implications were for office psychiatry. Lurie and Salzer's papers avoid the convoluted terminologies of both Kuhn and Kline and are more readable today than those of Kline, in particular. The kind of "antidepressant" they are referring to is the antidepressant that most clinicians think they are prescribing today. Both authors were clearly in the business of discovering an antidepressant, and both were somewhat surprised in later years at the lack of recognition they received.

Why are Harry Salzer and Max Lurie not better known? There are a number of reasons. The more comprehensive of their two papers appeared in the *Ohio Medical Journal*. This has echoes of Gregor Mendel, who one hundred years earlier had published his seminal studies on genetics in an obscure Czech journal, where they were to languish for fifty years. Lurie and Salzer did not push their work. They weren't known to Jonathan Cole when he established the Psychopharmacology Service Center with congressional monies two years later. They never joined the American College of Neuropsychopharmacology, although Max Lurie later took part in trials for tranylcypromine (Smith, Kline & French's Parnate) and nortriptyline (Lilly's Aventyl). In 1962 they, like many other private practitioners, were forced out of private patient clinical trial work following a change in liability laws regarding outpatient clinical trials on private patients. Although on the faculty of the University of Cincinnati, they held clinical appointments and were not on a full-time salary. Being in private practice and working as individual clinicians, not under the auspices of a research department either at the university or at a large private hospital, they lacked the backing and publicity that such institutions provide. The other source of backing might have been the pharmaceutical industry. But as we have seen, neither Geigy or Roche, at the time, was terribly interested in an antidepressant. Isoniazid could not have been patented for depression.

Given the Mendel precedent, few will be surprised by Lurie and Salzer's story. Their story attests to the persistence of both Kline and Kuhn and to the difficulty in establishing scientific facts if they do not coincide with interests that are sufficiently powerful. Perhaps only the romantics or the innocents will think Lurie and Salzer were the real discoverers of the antidepressants and should be acknowledged as such, even though, given that Delay's group had come out with similar findings in 1952, there is little reason to think that Lurie and Salzer got things wrong. Their use of isoniazid does, however, have further implications for the other stories. Isoniazid, as was noted earlier, is not a monoamine oxidase inhibitor. This may say something for the importance of a convenient piece of biomythology to go with drug development. It certainly says something for Saunders's claims about his role in the discovery of iproniazid. It is also highly inconvenient for modern theories about the way antidepressants work.

The surprise for many, however, will not be that Lurie and Salzer have been neglected but rather the fact that the claims of all these investigators, Kline, Kuhn, Lurie, and Salzer were based on clinical impressions rather than on double-blind randomized placebo-controlled trials (RCT). Prozac, for example, was not discovered by some expert investigator who decided that it was an antidepressant. Once the efficacy of a drug is discovered by RCT, its subsequent progress surely no longer depends on the status or charisma of individual investigators or the idiosyncrasies of their theoretical framework. The problem for this point of view is that in 1955 there had already been a placebo-controlled randomized double-blind study conducted at one of the most prestigious psychiatric facilities in the world and reported in the pages of the *Lancet,* which had shown that yet another compound was "antidepressant"—but no one paid any heed! The investigator was Michael Shepherd. The compound was reserpine. The patients were outpatients suffering from anxiety and depression.[79]

Clearly Shepherd was not the discoverer of the antidepressants in that, like isoniazid, reserpine is a compound without issue. Perhaps his methodology was simply too far ahead of its time. Perhaps the study was neglected because he didn't trumpet the discovery of an antidepressant. Shepherd's story, like Lurie's, however, has implications for the other accounts here in that Kline, at least, knew about his work. Kline had provided him with all the findings on reserpine before the trial began, and after completing his reserpine study Shepherd attended Kline's presentation to Congress. Similarly if the antidepressant credentials of isoniazid are inconvenient for modern biological theories, those of reserpine are all but incomprehensible. The amine theories and a generation of drug development have been predicated on an understanding that, far from being an antidepressant, reserpine makes people suicidally depressed.

In addition to the stories of Kline, Kuhn, Lurie, and Shepherd, there is yet another possible discovery narrative, according to which the important breakthrough was the discovery of depression by the pharmaceutical industry in the late 1950s. Consider the following. In 1958, Merck had approached a number of U.S. investigators, including Nate Kline, Doug Goldman, Fritz Freyhan, and Frank Ayd to look at amitriptyline, a molecule with a three-ringed central structure that bore resemblances to a phenothiazine and, as it turned out, was almost identical to imipramine (see Figure 2.3). They wished to have it investigated for possible antischizophrenic properties. In the course of the study, probably influenced by his awareness of what Kuhn had found with imipramine (in which the initial observations had been of mood elevation in schizophrenic patients), Ayd, who had been one of the twelve in the audience at Kuhn's talk, felt that amitriptyline might be a similar compound. His coinvestigators were somewhat skeptical, but Merck agreed to a trial of the compound in depression.[80]

Ayd gave the drug to 130 patients and reported in 1960 that it appeared to be of benefit for very much the same kind of patients Kuhn had argued were helped by imipramine. It was effective in much the same dose range as imipramine, had a very similar profile of side effects and like imipramine it often took several weeks for the therapeutic response

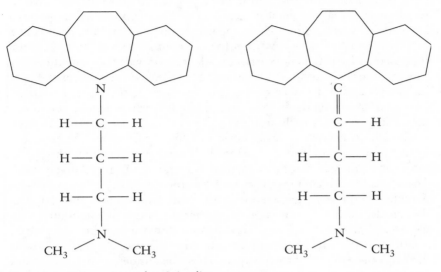

Figure 2.3 Imipramine and amitriptyline

to appear.[81] This was not surprising in that the two molecules differed only in the presence of one nitrogen atom. Amitriptyline was launched in 1961.

There were a number of other features of the amitriptyline story, however, which explain why its advent was so important. Generally, as is now clear, a second compound can be expected to make a market larger, rather than eat into the market share enjoyed by the first compound. Furthermore, in the case of amitriptyline, three companies, Merck, Roche, and Lundbeck, stumbled on independent processes to produce it at nearly the same time.[82] All three developed it and brought it onto the market. The combined efforts of all three companies meant that amitriptyline developed a worldwide distribution very quickly. Both Merck and Roche were larger than Geigy and both showed a greater marketing sophistication. When it subsequently became clear that there was money to be made out of antidepressants, these two companies clashed over who really owned the compound. Merck won because it had filed for a patent on amitriptyline specifically as an antidepressant.

But Merck had done more than file a patent. It had approached Frank Ayd, who had written a book, *Recognizing the Depressed Patient,* which was published by Grune and Stratton in 1961.[83] They bought 50,000 copies of this work and distributed it worldwide. Where Kuhn had argued that imipramine revealed the shape of a particular disorder, Merck was keen to reveal the shape of depression to as many physicians as possible as quickly as possible. The company was particularly interested in the fact that Ayd's book was written for the nonpsychiatrist. In his opinion (one shaped by personal experience), depression was not something that was found only in asylums but rather could be diagnosed on general medical wards and in primary care offices. In essence, therefore, Merck not only sold amitriptyline, it sold an idea. Amitriptyline became the first of the antidepressants to sell in substantial amounts.

The 1960s, however, was a world in which Librium and Valium were kings, which meant that compared with current perceptions the size of the nonhospitalized depression market was thought to be relatively small. There was in such circumstances no room for a diversity of concepts—for psychic energizers, thymoleptics, and antidepressants—and so even though clear differences between the different groups of compounds were recognized clinically, these were collapsed down into a broad-spectrum notion of antidepressants, sufficient to sustain sales, until the eclipse of the benzodiazepines led to an expansion of the antidepressant market and the broader "discovery" of antidepressants that only came with the SSRIs.

It was the discovery of amitriptyline that finally led to the acceptance of imipramine. With the emergence of the tricyclic group of antidepressants and the later eclipse of the MAOIs, a premium was put on ideas that were congruent with the proposal that these were treatments that were specific for depressive illness in contrast to ideas of nonspecific psychic energizing. In the 1960s a number of figures, such as Martin Roth, Max Hamilton, Herman van Praag, later to become some of the best-known names in international psychiatry, came up with formulations that appeared to coincide with the action profile of the tricyclics. Hamilton produced a rating scale that became the gold standard for the assessment of antidepressant effects, which is generally conceded to fit almost hand-in-glove with the profile of imipramine, so much so that there are concerns that its use as a standard rating instrument may be inhibiting the development of compounds that are unlike imipramine.

Roth, from Newcastle, a former pupil of Willy Mayer-Gross, formulated with his colleagues the issues surrounding vital and nonvital depressions, in a manner that everyone thought could be understood. They came up with operational definitions to distinguish between endogenous and reactive (or neurotic) depression. Endogenous depression was characterized by vegetative disturbances such as early morning wakening, loss of appetite, and diurnal variation of mood as well as retardation of thinking and feeling. The implication was that this was a biological disorder, probably stemming from constitutional or genetic factors, given that it supposedly arose out of the blue. Accordingly it was only appropriately treated pharmacologically or with ECT. Neurotic, or reactive, depression in contrast stemmed from adversity and could probably be managed psychotherapeutically in the main.[84] This formulation fitted very nicely with the amine theories of depression when they emerged in 1965 and with the fact that the tricyclics were amine reuptake inhibitors. Vital or endogenous depression appeared to have been confirmed as a disease and as appropriate a target for specific drug treatments as the latest regulations from the FDA could have wished for.

3 Other Things Being Equal

As was shown in the last chapter, chlorpromazine and imipramine were discovered in the 1950s by sensitive observations of their effects on patients, a method of discovery that is very different from the means by which the antidepressant credentials of Prozac (fluoxetine) were established. It and all other candidate antidepressant and antipsychotic drugs now have to go through the procedures of what is known as a double-blind randomized placebo-controlled trial (RCT), the approach Michael Shepherd took with reserpine, before they can be declared to have been discovered. It is a moot point whether the modern era in medicine has been created more by the development of specific therapeutic technologies or by the development of procedures such as the RCT—technologies to assess the effects of therapies of various sorts, whether old or new.

At a meeting in Cambridge in 1989 to look at aspects of the history of the psychopharmacological era, when the question of why so few new psychotropic drugs have been discovered since the mid-1960s was raised, Roland Kuhn made an impassioned plea to an audience that included Michael Shepherd and Martin Roth to the effect that far from advancing the discovery of new drugs, RCTs get in the way. "I had 40 patients in 1957 . . . , without any control study, without placebo. It was only by clinical observation that I found something. The whole legion of researchers who work now throughout the world cannot come up with anything new. That is the reality . . . with the methods that are employed today, it is impossible to find something new. You can only find an immense amount of facts, but facts which do not serve any practical result."[1]

Such feelings clearly point to a fundamental tension in medicine and indeed perhaps in science generally. This is the clash between theory and data, between qualitative and quantitative approaches, almost one is tempted to add between materialist and spiritual world views. The importance of this to our story is that depression, perhaps more than any other medical disorder, straddles these divides. Against this background it is important to review the evolution of the clinical trial in some detail, as this is the rack on which all therapeutic developments, even those in psychotherapy, are currently stretched. The dilemmas that trials pose are not novel; recent debates on these issues aired in the course of the Osheroff case in publications ranging from *Time* magazine to the columns of the *New England Journal of Medicine,* as we shall see, could have come straight from debates first conducted in the nineteenth century.

The various components of the randomized trial—randomization, placebo-control, and blind assessment—evolved separately and came together in the 1950s. Putting them together was an act as innovative as discovering the new magic bullets and, as with the discovery of the antidepressants, there are disputes about priority. The adoption of the randomized control trial by the FDA as a standard in the field of evaluating therapeutic developments decisively favored the medical empirics and isolated Kuhn and those who think like him. In the 1990s, we have perhaps begun to appreciate more fully the consequences of this settlement, which is having an increasing effect on medical practice and on the larger culture in which we live.

The Archaeology of the Clinical Trial

As it originated with the Greeks the foundation of scientific knowledge, *scientia,* was the logical syllogism—a method of constructing propositions so that the conclusions followed inescapably from the premises. Until perhaps the 1950s, this remained the goal of all scientific striving—a demonstration of what was necessarily so. In such a system, mathematics was the queen of the sciences because it was seen as a means of generating pure and certain knowledge. The triumph of Newtonian mechanics, when even the planets seemed to obey the logic of Newton's equations, seemed to confirm this.[2]

In medicine the basic premises were the four humors, as laid out by Galen and reviewed in Chapter 1. All sorts of deductions could be made from these about the balance of humors within the body or about a balance between the internal humors and influences from the environment.

Medical practice involved efforts to reconcile the logic of the system with the observed state of the patient, and so people who looked flushed were bled. But given a choice between the certain knowledge provided by the system and the uncertainty of striking out alone and going by the signs visible in the patient, physicians chose to adhere to the system, even though bleeding, for instance, regularly led to circulatory collapse and death. To choose to do otherwise was to be an empiric—which was a term of opprobrium.

The disdain of medical practitioners for mere *opinio* and for empiricism underpins the famous split between physicians and surgeons, which emerged in France in the fourteenth century. In contrast to the physicians, the surgeons were open to the developments in anatomy and therapeutics introduced by Vesalius and others at the start of the Renaissance. As the education of surgeons became more empirical and practical, it contrasted more and more with that of the physicians, who studied the ancient texts of Galen and others and above all prized the ability to reason logically.

A growing discontent with the rates of success of classical or Galenical medicine, however, began to emerge during the Middle Ages, which in the longer term, favored the empirics. Petrarch noted that he had heard a physician claim that "if a hundred or a thousand of men of the same age, same temperament and same habits, together with the same surroundings, were attacked at the same time by the same disease, that if the one half followed the prescriptions of the doctors . . . and the other half took no medicine but relied on Nature's instincts, I have no doubt as to which half would escape."[3]

Picking up on these stirrings, Paracelsus introduced the idea that physicians should look more closely at their patients and at nature before prescribing. "Theory and practice should together form one . . . for every theory is also a speculative practice and is no more and no less true than active practice. What would you do if your speculation did not jibe with findings based on practice? Look at the carpenter: first he builds his house in his head. But whence does he take this structure? From his active practice and if he did not have this, he could not erect his structure in his mind. Both theory and practice therefore should rest on experience."[4]

Paracelsus rejected the doctrines of Galen, Avicenna, and Aretaeus, based as they were on preconceived ideas of what had to be wrong, in favor of a new doctrine of signs to be built up from contemporary observations of nature and man's responses to her remedies. "My proofs derive from experience and my own reasoning and not from reference to authorities."[5] He introduced the idea that medical practice might be a matter of opinio rather than scientia—"study my books and compare my opinions with the opinions of others; then you may be guided by your own judgment."[6] It was this rejection of authority that underpins the violence of his expulsion from Basel; he had effected a breach within the contemporary world view if anything even more shocking than that of Galileo.

The factor that was ultimately to swing the balance in favor of the empirics and bring the pragmatic aspects of medicine to center stage came with the development of probability theory by Pascal and others and the subsequent elaboration of technologies to assess probabilities. By means of these, it would ultimately be possible to weigh the merits of various opinions and contrast the success of physicians and surgeons—or more to our point the successes of psychotherapists and drug therapists.

So much do we now take it for granted that different options should be compared and that decisions should be based on the outcome of such comparisons that there is perhaps some difficulty in reimagining a world view that might see things otherwise. The shock of what was being proposed, however, may be conveyed to some extent by the fact that the origins of modern probability lie in Pascal's wager on God in 1658. Where before, the existence of God and morality had been a matter of certain and logical deduction, Pascal made them a matter of probability and faced up to the consequences for behavior of so doing. He concluded that it made sense to behave morally, because even if there was only a small chance that God existed more would be lost by not believing in him than would be lost by believing.[7]

Notwithstanding the fact that the force of this argument is lost to some extent today, it should be clear that there was something distinctively modern about seeing the question of God and behavior in this light—something that is opposed to fundamentalism. Indeed, the notion of fundamentalism, with its connotations of pathological certainty, only began to take shape after Pascal. The mathematical and philosophical implications of the wager were developed after 1660 by Huygens, Leibniz, Bernouilli, Laplace, and others. The most far-reaching implication was that scientific knowledge could never be certain and that at some point the scientific question would become one of how to make scientific judgments rather than simply how to do scientific experiments.

The technology to actually weigh practical probabilities was developed to solve problems concerning the issuing of annuities and insurances. The idea emerged in the seventeenth and eighteenth centuries that states could raise money by offering to insure their citizens. If individuals invested money in bonds on the promise of a certain return when the bond matured, the state had money with which to finance projects. The problem then was one of setting a rate of return that would not only encourage investment but was also likely to be one that could be honored by the state when the bond matured. This latter depended to some extent on how many people survived to claim their return. The need to solve such issues led to the development of statistical records—the process of notifying authorities of deaths and diseases, which began around this time and the consequent development of public health records. Such data became grist to the mill of the insurance industry. The collection of records also gave rise to the idea that the predictability of outcomes did not necessarily stem from the logic of a set of initial premises; it might rather be something that held true—other things being equal. If this was the case, then comparing records might reveal the impact of interventions of one sort or another.

LA MÉTHODE NUMÉRIQUE

In the 1720s in France, one of the first recognizable modern interventions, the possibility of inoculating against smallpox, emerged. The practice at the time was to take matter from the pustules of an infected person and inject it into a vein of someone unaffected. This was termed variolation. It usually produced a mild case of the disease, but in some it brought about the full-blown illness and death. James Jurin made a comparison of the outcomes of inoculation with the effects of letting the illness take its course and found that where one in every five cases died from naturally contracted smallpox, the death rates were on the order of one in fifty following inoculation.[8]

Despite these findings, however, there was an outcry against variolation that put a stop to it. Objections were raised about this practice on the grounds that it was unethical to do harm even were good to result. One of the main objectors was La Mettrie, who argued that inoculated subjects were themselves potentially contagious, so that while the acute death rate might be lower, it was less certain what would happen in the longer term. Variations of this argument are still heard to this day from those who argue against the use of antidepressants and antipsychotics. Ironically La Mettrie was later to be vilified for introducing the idea of a godless medical engineering of society—but he no less than his contemporaries had difficulty in recognizing substantial progress in medical engineering when faced with it. Perhaps the overall idea that health was a matter of balance brought with it a certain fatalism—the cure of one problem would lead to the eruption of another one so that the sum of imbalances or disharmonies would always be the same. Eventually an epidemic of smallpox in England in 1746 led to a public demand for variolation, although it remained banned in France until the faculties of theology and medicine in Paris in 1769 accepted its use.[9]

Less dramatically, comparative methods had begun to be employed by physicians on board British naval vessels in the course of the ocean voyages of exploration then being undertaken. In 1747 John Lind, on the *Salisbury,* noted that of a group of subjects with scurvy, those who were given citrus fruits recovered, whereas those who didn't receive fruit didn't recover. This led to his recommendations for the treatment of scurvy. Robert Robertson on the *Juno* in 1776 ran out of fever bark (Cinchona) and compared the period before and period after this with subjects acting as their own controls. Collecting the figures of episodes and length of fevers, he was able to demonstrate that the bark made a significant difference.[10]

By 1800 the benefits of collecting data on outcomes had made a sufficient impression to lead Philippe Pinel, one of the founding fathers of psychiatry, to claim that for an "experiment to carry weight and to serve as a solid basis for treatment, it has to be made on a large number of patients."[11] Collecting figures seemed to some to be a way out of the dilemma of having to chose between Galenic doctrines on the one side and a "blind empiricism, which prevents medicine from acquiring the character of a true science," on the other. For Pinel, therapeutics had to rest on something more substantial than the intuition of a physician, however seductive that might be.

This approach culminated in the work of Pierre Louis in France who, in 1836, outlined a *méthode numérique*: "In any epidemic . . . let us suppose five hundred of the sick, taken indiscriminately, to be subjected to one kind of treatment, and five hundred others, taken in the same manner, to be treated in a different mode; if the mortality is greater among the first than among the second, must we not conclude that the treatment was less appropriate, or less efficacious in the first class than in the second."[12] When he applied this method to bleeding, in the course of an epidemic, he produced something of a crisis in therapeutics: "The results of my experiments on the effects of bleeding in inflammatory conditions are so little in accord with common opinion that it is only with hesitation that I have decided to publish them. The first time I analyzed the relevant facts, I believed I was mistaken, and I repeated my work but the results of this new analysis remains the same."[13]

Far from being greeted with pleasure, Louis's findings were seen as destructive of therapeutics. Opposition was widespread and deep-seated, as this quote from 1849 illustrates: "The practice of medicine according to this view is entirely empirical, it is shorn of all rational induction, and takes a position among the lower grades of experimental observations and fragmentary facts"[14]—a remark that bears striking similarities to Kuhn's at the start of the chapter. The debate was particularly vigorous in the French Academies of Medicine and Science. A common theme of those opposed to Louis's method was that generalization of the type he sought was a flight from rather than an embrace of objectivity. Physicians, it was argued, were faced with real patients in all their idiosyncrasies. Numerical methods required the physician to "strip [the patient] of his individuality in order to eliminate any accidental qualities from the question."[15] Many felt that when it came to therapy it was more important to weigh the data.

Risueno d'Amador introduced a variation on this theme by claiming that medicine was both an art and a science. Although collecting the statistics of public health was clearly beneficial, when it came to therapy, he argued, a medical judgment needed to be made as to how best to apply the fruits of scientific inquiry to the particular patient. The doctor had to act like a juror, who had to make up his mind on a particular case "more by the real value of the witnesses than by their number." He railed against practicing in accordance with a numerical method, which would be to substitute "a uniform and blind mechanical routine for the action of the spirit and the individual genius of the artist." It was uncertain how much physicians could have to do with such a method, he argued, as "the problem for the numerist is not to heal a particular patient, but to heal the most possible . . . this problem is essentially anti-medical."[16] Added to this was the civil liberties question of whether medical practitioners should be collaborating with civil authorities in a manner that would lead to increased government regulation of human behavior.

The debate was never resolved. It was brought to a close in France in the 1860s, when Claude Bernard threw the weight of his prestige against the méthode numérique.[17] Bernard's input to the debate came at a time when physiology was making dramatic progress and the sciences of bacteriology and organic chemistry were being born. He argued against a statistical approach on the basis that physiology, the scientific foundation of medicine, needed to be and could be established as certainly as physics or chemistry. To retreat to statistics, in his opinion, was a counsel of despair. Rather than establish the frequencies with which particular illnesses caused death, he argued that the exact means by which death was brought about needed to be determined. Such was his standing in France at the time that the méthode numérique was relegated from orthodox medical awareness and only reappeared in the 1950s in the shape of imports from Britain and America. Even then the success of Kuhn in introducing antidepressants and Delay in introducing antipsychotics without numerical methods reinforced a belief that such methods were in some way inferior.

While Louis and his method were eclipsed in France, numerical methods played a significant part in the development of antisepsis and anesthesia elsewhere. In 1847 Ignaz Semmelweis noted that there were marked differences between the death rates on two different wards in the obstetric hospital in Vienna. Mortality was much higher on the ward run by physicians and medical students compared with the one run by student midwives. He also noted that the differences between the two wards had only emerged, in 1841, after courses in pathology were included in medical training. He suspected that physicians and students were coming to women in labor with particles of corpses from the dissection room still on their hands. He got them to wash more thoroughly with a *chlorina liquida,* coincidentally a disinfectant, and compared the figures for deaths before 1841 and after 1847 with death rates on the medical ward for the period between 1841 and 1847. He was able to show that antisepsis made a difference, as was Joseph Lister in Glasgow Royal Infirmary in 1860.[18] While these examples of the kind of enlightened developments that could be brought about by simply collecting figures—documenting outcomes—may seem uncontroversial now, Lister's findings remained controversial for up to thirty years afterward. Semmelweis's conclusions were essentially rejected, and he subsequently went mad.[19]

Another application of the numerical method may bring home just how shocking it seemed to many practitioners. In 1846, William Morton demonstrated for the first time that ether could be used to anesthetize a patient undergoing surgery.[20] Very quickly thereafter, chloroform and nitrous oxide were also found to be effective anesthetic agents, and the practice of anesthesia spread. A number of hospitals and surgeons, however, refused to adopt the new techniques for a variety of reasons. One argument was that a depressing agent like anesthesia must necessarily be also depressing other functions and that this would militate against recovery from surgery. A second argument was that dulling the pain of surgery and of childbirth, for instance, could not be morally right. Public concern about Queen Victoria's decision to have an anesthetic during childbirth indicates how deep-seated were popular prejudices.

But the critical argument for many was an ethical one. If it was conceded that surgery could proceed under anesthesia and even that fewer people were likely to die because the surgeon had more time and could do a better operation, there still remained the fact that some people would die from the anesthetic itself. Could a doctor, whose duty it was to save life, intentionally jeopardize that life? Could evil be done even though good might result? These issues were finally resolved to the satisfaction of the majority by something of an ethical calculus. Supporters of anesthesia asked whether a doctor could advise a patient to undertake a procedure, knowing that there was an alternative method that would on balance expose the patient to less risk. Could one wash one's hands of risk and still practice medicine? Was it ever possible to avoid having to choose between the lesser of two evils? Which approach is more ethical—to take the path that one feels from first principles is morally better or to take the path that produces the better outcome? These questions haunt debates of the comparative effectiveness of psychotherapy and pharmacotherapy to this day.

One retreat from this dilemma in the 1860s was to argue that the "natural" course of action must be better—that the more artificial and further removed from nature the action proposed, the more likely it was to be wrong. Anesthesia was very artificial compared with other procedures such as bleeding or purging, which in essence were efforts to mimic nature's ways of healing. Anesthesia and the surgical developments that stemmed from it were a triumph of outright artifice and technique over nature. The argument that the more natural approach is likely to be the morally better one seemed much less obviously correct in the nineteenth century, when infant mortality was extremely high, but today when an individual's risk of premature death from infectious or other diseases has been dramatically reduced, the appeal of this argument has grown and the ethical dilemma at the heart of medicine as we know it has reemerged as a matter of public concern.

The introduction of anesthesia and antisepsis present some of the issues that stemmed from Louis's méthode numérique in a very salient and dramatic manner. Louis, himself, had a more difficult problem. During a typhoid epidemic, he advocated comparing bleeding and purging as treatments. His results showed that fewer died with purging than with bleeding but that a great number of people still died, nonetheless. For many observers there was no great numerical reason to choose one treatment or the other. What Louis lacked in 1836 was a method of difference to prove that one treatment was superior to another. It was this that the clinical trial, which developed over a century later, was to provide.

When the clinical trial did become part of medical practice, however, it was to open up a series of acute dilemmas that had been first articulated by Poisson, who had come to Louis's defense in 1836. Poisson argued that probability assessments could be used in therapeutics on the basis of common sense. "If a medication has been successfully employed in a very large number of single cases and the number of cases where it has not succeeded is small compared to the total number of cases of this experiment, it is very probable that the medication will succeed in a new trial."[21] His view was that the problem with practicing in this way was not the loss of individuality but the requirement for morality. Faced with the evidence, the challenge to clinicians would be to transcend their personal opinions, possibly at the cost of forsaking beliefs that gave coherence to their personal world.

A thought experiment will bring the issues up to date. What if double-blind studies of homeopathic remedies in allergic conditions were to indicate some efficacy—what should the medical practitioner do if faced with such findings? A demonstration that homeopathic remedies work would mean that either most of modern medical theory is wrong or else there is something wrong with the RCT. Alternatively, what is someone who believes in psychotherapy for depression, for instance, to do if the results of studies come out in favor of drug treatments? These issues are at the heart of Chapter 7.

A METHOD OF DIFFERENCE

The developments that finally emerged to test the efficacy of the magic bullets of modern pharmacology took root in the same breakthroughs that shaped modern notions of disease. While the emergence of bacteriology at one level gave a very deterministic image of disease—a target germ to be eliminated by a magic bullet—it very quickly became clear that even infectious diseases were much more complex than that (as Pettenkoffer's example illustrated in Chapter 1). A disease does not start simply with the entering of an organism into a body. Even in the case of HIV, some people seem immune. For other infections, there is a complex interplay between the general health of the person, her state of mind, and her susceptibility to a bacillus. There is also the question of whether the person has been previously exposed to that or a related strain of bacillus. After the organism actually infects the person, there are then a range of factors that determine the outcome, such as the issue of where in the body it takes root, physiological responses to the insult, the speed with which remedial measures are effected, and the confidence of the individual in those measures.[22]

Diseases, once established, can run a variety of courses and can have a number of possible outcomes. Attempting to map what is likely to happen is more like trying to predict the weather than it is like predicting the outcome of a chemical reaction. There is an essentially accidental or historical quality to a disease that can only be captured by statistical modeling. Bernard's hopes for physiology could never apply to disease. Even in the case of a post mortem, there would always be a need to make a judgment as to whether the findings on balance revealed what the person had died of or whether they spoke to what people can live with. Against this background, statistical thinking began to return to therapeutics.

With Pasteur's discoveries, the possibility of developing vaccines opened up. It became routine to test these out first on animals, using treated and untreated, or control, animals. When it came to the diphtheria antitoxin, Emil Behring applied the numerical method. He found that six of thirty infected patients treated with diphtheria antitoxin died compared with the usual mortality of 50 percent. Fibiger, in Denmark, went further and gave diphtheria antitoxin to alternate patients, before satisfying himself that it really did make a difference.[23]

This type of assessment crept into use between the wars. Digitalis had been introduced for the treatment of lobar pneumonia in 1916. In 1929 Wyckoff and colleagues attempted to assess its efficacy by comparing patients treated with it or a serum preparation or nothing. Most important, assignment to treatment was on an alternate basis determined by time of arrival to the hospital and nothing else. The death rate for those getting digitalis was 25 percent higher than among those not getting it. Another fashion was stopped by a study of the benefits of sanocrysin in tuberculosis conducted by Burns Amberson and colleagues at Maybury sanitarium in Michigan. In their view it was difficult to "distinguish the natural fluctuations of the disease from those induced by a substance or procedure under trial." They assembled two matched groups and determined which group would get sanocrysin by the flip of a coin, with the patients remaining unaware as to which medication they were on. The sanocrysin-treated group did worse.

It would seem that as the prospect drew nearer of having drugs that actually worked, it was simultaneously becoming increasingly clear that clinical trials were needed to sort out what did and what didn't work for specific conditions. No trial was needed to establish that the sulfonamides had worked for Gerard Domagk's daughter when she walked out of the hospital three weeks after admission for septicemia. But tying down what these drugs could be routinely expected to work for was another matter. Sulfonamides didn't work for tuberculosis, for instance.

The critical breakthrough that led to the modern trial was the recognition of the importance of randomization. This provided the foundation for the method of difference sought by Louis and others. Up until the 1930s, statistical studies had been conducted on large populations. When smaller samples of subjects were used, the concern was that the sampling technique would yield groups that were unrepresentative of the target populations. In 1923 Ronald Fisher developed a very different approach and a method of analysis called the analysis of variance. This enabled studies to be run with much smaller numbers in the different experimental groups,[24] provided there had been a prior randomization of subjects or items to be compared in order to exclude bias of any sort.

Fisher's views initially met with hostility. Randomization, it was argued, would simply increase the variability in the data, the noise in the system. In addition, it did not seem ethical to treat people in the indiscriminate way that was required by proper randomization. Randomization as an experimental technique came to be accepted, perhaps in part because, at least at first, it was agricultural studies that were being talked about and for this purpose it was extraordinarily convenient—and the question of treating humans like animals didn't arise. Even so Jerzy Neyman and Egon Pearson were quick to put forward the argument that Fisher had discovered a means of extracting "significant" findings from small group comparisons, raising the specter that the validity of the conclusions that could be drawn using this method for the population at large was uncertain.[25]

Fisher's method was first adopted in a Medical Research Council (MRC) trial of streptomycin for tuberculosis begun in 1946 and reported in 1948.[26] This study was the brainchild of Austin Bradford Hill, who is now commonly cited as the father of the RCT. He assigned patients to treatment groups by random number distribution. Assessments were done blindly: the outcome measure was the X-ray of the patient's lung, which was read by a radiologist, who was blind to the treatment. There were considerable ethical concerns about "treating humans like animals" in this way, but these were managed on the basis that in the immediate aftermath of the war conditions of scarcity prevailed, and it would not have been possible to provide more people with treatment than were provided anyway. In addition, those not getting streptomycin received the treatment that was then accepted as the best possible treatment. This trial broke the taboo of withholding from some patients a treatment believed to be active. Even so, two years later, when a trial of streptomycin was undertaken in the United States, it was felt unethical to withhold a drug strongly believed to be active from subjects. Death rates after streptomycin were compared with those in the preceding period, instead.

The one missing ingredient from the streptomycin studies was the concept of placebo. Many of those involved in introducing any of the new drugs, whether for respiratory or psychiatric conditions, were skeptical of the need for a placebo control. It seemed clear that if nothing available had been of any use for chronically withdrawn and deteriorated schizophrenics or melancholic, stuporous, and possibly deluded depressives, for instance, then nothing itself was hardly likely to be of any use.

Placebos already existed, but they were only in use as a control in studies using healthy volunteers with agents like caffeine.[27] It was one thing to use a placebo as part of a scientific experiment in healthy volunteers, but it was quite a different thing to propose giving it in place of treatment to someone who was ill and at risk of death if not given treatment. It took one of the giants of medicine to make such an idea respectable. Harry Beecher was an anesthetist involved in military surgery during World War II. Morphine at the time was thought to be critical to the success of the type of operations involved, both as an analgesic but also because it minimized the risk of cardiovascular shock. At one point, during a period of particularly heavy casualties, supplies of morphine ran out, threatening the life of the patient on the table. In desperation, one of the nursing staff injected saline instead and to Beecher's surprise the patient settled and full-blown shock never developed. Impressed at this, Beecher had the opportunity to confirm the observations subsequently when supplies again ran low. He came back from the war convinced of the power of placebos and collected around him at Harvard a number of colleagues keen to map out the boundaries of the placebo phenomenon.[28] On his retirement, however, the various members of the department disbanded and pursued other interests. Without his presence, it was difficult to make an academic living "selling" placebos.

Another of the pioneers was Harry Gold from Cornell, whose research interest was anginal pain. He began using double-blind methods—neither he nor the patient knew who had received what—as early as 1935.[29] The only difference between some of Gold's studies and modern RCTs was a lack of initial randomization. He had even come up with the idea of placebo controls independently, it would seem, of anyone else. His work on angina had made it clear to him that not only were many fashionable treatments no better than a placebo but, more important, that placebos appeared to exert therapeutic benefits in their own right. Retrospectively, it can be seen that angina was a particularly good condition to reveal this fact. Depression is similarly responsive. But where many would argue that this response to placebos justifies not using antidepressants too quickly in depression, few if any make this argument for angina.

Although it is now clear that depression is placebo-responsive, this was not clear in the 1950s. And regardless of the response in depression, many were certain that schizophrenia would not be placebo responsive. On this basis there was a concern about the trend toward placebo-controlled studies. Heinz Lehmann decided to test some of these issues out on three of the most mute and deteriorated schizophrenics on one of the Verdun hospital's back wards. Nursing staff and patients were told that the patients were going to be given a new experimental hormone by injection. The injection site was painted with a disinfectant that left a prominent red stain. The injections were given twice a week for two weeks. By the third week, two of the three had begun to talk and were talking rationally. The injection was a placebo.[30] In some important sense, then, placebo is not just a control for specific treatments, but may be a treatment in its own right. This is an issue that will remain with us to the end of this book.

The Psychiatric Discovery of Trials

One of the first psychiatrists to take on board the new ideas was Linford Rees. Working in Cardiff, he looked at electronarcosis, which had been introduced as a new method for treating schizophrenia.[31] It was like ECT without the fit; it involved continual stimulation of the brain for periods of up to seven minutes. A number of experts were enthusiastic advocates for it. As early as 1949, Rees ran a study comparing it with ECT and insulin coma, in which the groups were selected to be similar for age, sex, and clinical features. This research was presented at a Royal Medico-Psychological Association meeting (a forerunner of the Royal College of Psychiatrists) and created a stir. Although the study was not fully randomized, nothing comparable had been done before in British psychiatry. Rees was talking a new language.

Subsequently in 1950 and 1952, Rees ran placebo-controlled double-blind studies looking at desoxycortisone acetate and later cortisone in the treatment of schizophrenia.[32] There was a theoretical rationale for using cortisone, in that there had been reports of atrophy of the adrenal glands of individuals with schizophrenia, but in practice the enthusiasm for cortisone rested on its recent synthesis and the dramatic effects it had been shown to have for arthritis. Using these methods, neither compound was found to be of any use, despite enthusiastic endorsements from the United States. The fact that the outcomes were negative perhaps served to obscure the dramatic breakthrough that had occurred—a new technology was being used to assess a psychiatric treatment.

When it comes to deciding who performed the first fully randomized study, there are four different studies, two on chlorpromazine, one on reserpine, and one on lithium, to choose from. Starting in 1953 and reporting in 1954, Joel Elkes, then professor of experimental psychiatry at Birmingham, and his wife Charmian studied twenty-seven patients with schizophrenia, giving them either a placebo followed by chlorpromazine or chlorpromazine followed by placebo, and found that the active drug made a difference.[33] In 1954, reporting in 1955, Linford Rees did a study of one hundred anxious patients randomized to either placebo or chlorpromazine, and found that the drug was anxiolytic, albeit at a cost of demotivating side effects.[34] Starting in 1953 and reporting in 1955, David Davies and Michael Shepherd at the Maudsley Hospital in London reported on the effects of reserpine in anxious depressives.[35] Starting in 1952 and reporting in 1954, Mogens Schou and Eric Stromgren ran a placebo controlled study of lithium in mania, finding it to be of benefit.[36] So who discovered the clinical trial in psychiatry?

Elkes was subsequently persuaded to come to Johns Hopkins in Baltimore, where he became an influential figure in establishing psychopharmacology in the United States. He was the first president of the American College of Neuropsychopharmacology, an influential figure in the CINP, and a regular contributor to international symposia, at which his eloquence and charm played an important role in encouraging others to enter the field. His colleagues in America during the later 1960s developed the impression that he had effectively done the first double-blind study in psychiatry, while he had been in Britain. His presence at the Conference on the Evaluation of Psychotropic Drugs in 1956 was important to the future development of RCTs in psychiatry, because he was a witness to the fact that such studies could be done.

The impression that the Elkes had conducted the first RCT in psychiatry possibly owes something to Judith Swazey's influential 1974 book on chlorpromazine, in which she suggested that the 1954 trial had been sufficiently innovative for the *British Medical Journal* to comment on the trial design in an editorial.[37] When one reads the editorial, however, it is clear that this is a statement about chlorpromazine, occasioned by a growing number of both British and international reports—the Elkes's study was one of three published in the course of two weeks in the *BMJ*—rather than an editorial about trial design.[38] Neither Joel nor Charmian Elkes featured prominently as clinical trialists after 1954, and the cross-over design they used was superseded by 1960. Joel Elkes's efforts were far more focused on the interplay between brain functioning and behavior, in contrast to Rees, Shepherd, and Schou, who focused their energies more clearly on developing and applying RCT methods.

Linford Rees took his study to the first international meeting on chlorpromazine at the Val-de-Grâce hospital in 1955, where he presented the only double-blind studies there, among the 150 papers.[39] Although this study was later than the Elkes' study, Rees had been doing this kind of investigation for some years, and his presence in Paris took the new methodology to the world stage.[40] He subsequently attended the first CINP meeting in 1958 and presented the first double-blind study of iproniazid, which demonstrated that it worked but that it was less effective than Kline had claimed.[41] Following that he ran the first trials of phenelzine and of haloperidol in Britain, along with studies of imipramine and amitriptyline—finding the latter to be ineffective in the population studied. He did double-blind studies of minor tranquilizers, beta-blockers, and a range of other antidepressants and neuroleptics.

In many ways the most intriguing study was done by Shepherd and Davies at the Maudsley using reserpine.[42] They looked at sixty-seven outpatients referred with anxious depressions and using a randomized double-blind placebo-controlled method, found that reserpine was superior to a placebo. Despite the eminence of the authors and their institution, this trial has vanished from the literature. A number of reasons can be offered. One is the obvious one that reserpine and the literature that went with it were eclipsed by chlorpromazine.

A second, more ironic reason stems from reserpine's later reputation as an agent that caused depression. Based on a series of anecdotal reports, it subsequently became an article of faith that reserpine both depleted catecholamines and caused depression, while antidepressants both reversed the depletion of amines and lifted mood.[43] Indeed, one of the anecdotes on which reserpine's later reputation was founded is located on the facing page to Shepherd and Davies's report in the *Lancet* in 1955.[44] A further reason offered by Shepherd is that the reserpine studies were so unlike anything that anyone expected to see at the time that no one noticed them. Studies in therapeutics generally but in psychiatry in particular simply didn't come with tables and statistics.[45]

Whatever the reason, Shepherd's and Rees's studies were the first major RCTs in psychiatry to employ a parallel group design. This, rather than the cross-over design used Elkes and Schou, is the method that is used today. Shepherd subsequently became more interested in the technology of evaluation than in the drugs themselves. Along with Linford Rees, during the 1960s he was one of the leading exponents of clinical trials in Britain and indeed in the world. He coordinated a multicenter Medical Research Council trial that reported in 1965, which as we shall see in Chapter 4 all but finished the MAOIs. He also became involved in a battle over the prophylactic effects of lithium with Lehmann, Kline, Coppen, and Schou in the late 1960s that hinged on clinical trial methodology. His subsequent commitment to the principles of randomization and placebo-control was perhaps more comprehensive than that of any of the others, but the lessons he was ultimately to draw from these methods were at odds with the mainstream views of what clinical trials demonstrate.

Finally there was Schou's study, which like the study on reserpine, though positive, was similarly eclipsed by the lack of interest in lithium. This study, as the Elkes' had been, was a double-blind crossover study, which it is now clear is probably not an optimal design to pick up therapeutic effects.[46] This trial did, however, embody principles of randomization and placebo-control and was the first to do so and obtain a positive result that has stood the test to time. It would have been published even earlier, in 1953, and possibly in a more prominent journal such as the *British Journal of Psychiatry,* except that neither the method of evaluation nor lithium itself was of much interest at the time.[47] Thereafter, the fact that it was designed and conducted outside the English-speaking mainstream of postwar psychiatry and conflicted with English claims to the creators of randomized methods of assessment probably contributed to the failure to celebrate it as a significant breakthrough. The question of Schou's priority also, as we shall see in the next chapter, became entangled in bitter disputes about the prophylactic effects of lithium, where to Shepherd at least Schou seemed bent on skirting the principles of randomization and placebo-control.

But whoever one decides introduced the RCT, another lesson about introductions is that, just as with Kuhn's discovery of imipramine, there is never one moment in time to which one can point and say a treatment or procedure has been discovered. It took a controversy about insulin coma, which began in 1953, and involved all of the most important individuals working in the field, to really brand into professional consciousness just how open to bias the evaluation of psychiatric treatments is.

In November 1953, Harold Bourne, a young trainee psychiatrist had a paper published in the *Lancet* entitled "The Insulin Myth."[48] Insulin coma had been introduced by Manfred Sakel in 1933. At the time insulin was being used as an agent to stimulate appetite in the belief that building people up physically might benefit their mental state. It was also somewhat sedative when given in the regimes required for this purpose. The procedure was hazardous, because some patients could be expected to be hypersensitive to insulin and to go into a coma at low doses. This could be reversed by using glucose, but the treatment was still a hazard.

Sakel, however, noticed that some patients who had been precipitated into a coma appeared somewhat better when they were brought around. He systematized this approach in insulin coma therapy. The procedure required intensive nursing support and the dedication of a ward or, as it often was, a villa on the grounds of the hospital to its use. Considerable expenditure went into setting up such units. Given the costs and hazards involved, care was taken in the selection of patients. In general, younger patients in good physical health and with an illness of recent onset were chosen. The results seemed to warrant the expenditure. There were recoveries. The outcomes were explained in terms of insulin's being a hormone naturally produced by the body.

Bourne questioned all this. He argued that insulin coma had not been subjected to a critical evaluation; that it simply wasn't known whether it did what was claimed for it—"the evidence for the value of insulin treatment is unconvincing": "Can anyone, who is not possessed by furor therapeuticus, as Freud called it, and not hypnotized by palaver, syringes, coma and the terror of therapeutic impotence, really believe that the risk and great expenditure are worth it?" His critique was answered by nearly all of the eminent figures in British psychiatry—Willy Mayer-Gross, William Sargant, David Davies, and even Linford Rees.[49] All protested their certainty that insulin coma treatment worked. A randomized trial was set up to determine the reality of the matter. This reported in 1957 that, when patients were randomized to insulin coma or treatment with a barbiturate, Insulin was not convincingly better than the control treatment.[50]

It didn't, however, take an RCT to discredit insulin coma. Before the 1957 study reported, most psychiatrists had decided that, however good the procedure might be, it was not as good as chlorpromazine—or at least that the risks involved in giving the treatment were no longer warranted in the light of the newly available treatments.[51] Many experts associated with insulin coma continued to believe that it worked through to the 1970s.[52] The majority, however, such as Linford Rees, changed their mind and adopted Bourne's position. They recognized that in the course of the evolution of the treatment, it had been reserved for those most likely to have a good outcome anyway—the young, healthy subjects with an illness of recent onset—added to which were the element of special attention that went with being separated off from the rest of the hospital patients in a ward with a much better staff-to-patient ratio than others, and the mystique of a treatment that unlike any other was expected to work.[53] This lesson was taken to heart and the beachhead that properly randomized clinical trials had established in Britain was secured—at least in psychiatry.

In France, the legacy of Claude Bernard meant that, despite the example of Louis, statistical approaches were slower to develop than elsewhere. The tradition remained the one to which Kuhn adhered—that the proper discovery of a medicine was by the sensitive observation of empathic clinicians. They had the example of Delay and Deniker to show for it. It was only in the 1960s that numerical methods regained a foothold in French psychiatry. Meanwhile in the German-speaking countries, as the quotes from Janzarik and Kuhn indicate, there was a deep-seated hostility to the hijacking of psychiatry by the empiricists.[54]

The Psychopharmacology Service Center

In the United States, this was the period when Nate Kline and others were lobbying Congress for funds to evaluate the new treatments. When funding came through, reactions were mixed. There were those like Robert Felix, the head of NIMH, who felt that psychiatric treatments, even the new drug treatments, were still not at a stage at which they could be evaluated systematically. There was skepticism from both the biological sciences side of psychiatry and the psychoanalytic side. For the emerging neuroscientists, the clinical area was simply not one to which the "scientific method" could be applied. There was too much that was uncontrollable. And wasn't medicine supposedly an art anyway? The analysts regarded drug treatment as inferior, a necessary evil, the evaluation of which was as peripheral to psychiatry as were the drugs themselves. The idea that their own treatments might also need to be subjected to systematic evaluation was not even contemplated.[55]

There was a group, however, who felt that proper evaluation was both feasible and necessary. The monies from Congress came to the NIH and were held in grant form in the first instance by Seymour Kety. He convened a steering committee that included Joseph Brady, an early animal behavior pharmacologist, Ralph Gerard, a physiologist from Michigan, Louis Lasagna, one of the first clinical pharmacologists, and Jonathan Cole, who was the scientific administrator for the National Research Council. None, except Cole, was a practicing psychiatrist. All, however, believed that evaluation was possible and necessary—if only to contain the phenomenon of Kline.[56] Cole had had the advantage of seeing double-blind studies in action, when training as a medical student with Harry Gold, while Lasagna had worked with Henry Beecher and was actively involved in designing trials in other branches of medicine. The steering group invited a number of observers to join them, including Michael Shepherd, who at the time had a traveling scholarship and was visiting the Department of Public Health at Johns Hopkins.

The final decision was to channel the money from Congress into the establishment of a Psychopharmacology Service Center, with Cole in charge. This would begin to function following an inaugural conference, which was organized in September of 1956 by Gerard and Cole. The brief of the conference was to explore all aspects of evaluation, from patient recruitment to clinical trial methodology to the use of preclinical animal behaviors as screening methods or the use of biological measures to determine the mode of action of the new drugs.[57] It was a conference that took a "long hard and skeptical look into the future," as Joel Elkes put it, and discerned "the rhythms of development of a new discipline . . . [in which] doubt and enquiry are resolved into transitory certainties and the vessel of certainty is cracked by expanding doubt."[58]

What was unique about the American input was the collective approach to the issues. This was to issue in a number of coordinated programs linking basic science testing and clinical evaluation and in the establishment of multicenter studies in Veterans Administration hospitals, state hospitals, and private clinics. As Seymour Kety put it, clinicians were all doing large-scale but uncoordinated research, whether they knew it or not, which posterity would finally evaluate.[59] The risk was that only positive results would be reported, creating the illusion of effectiveness. If they let this happen, it would be akin to a scientist studying ten cases and only reporting the positive ones. The Americans have ever since led the way in coordinating large-scale studies. Cole and Klerman put together a nine-hospital study of chlorpromazine, which reported in 1964 and finally settled conclusively the question of whether the new antipsychotics worked or not.[60]

When it came to questions of methodology for the new trials, consider the following: in 1852 Pierre Touery proposed that activated charcoal could act as an antidote to strychnine. The medical community demurred. In response, on the floor of the French Academy of Medicine in front of a large audience, Touery drank ten times the lethal dose of strychnine, without any ill-effects. Given an endpoint as clear as this, clinical trials of the type used in psychiatry are simply not necessary. They become more important, the more ambiguous the endpoints. In tuberculosis or cardiac disorders, the outcome variables may be simple—death, for example. In psychiatry more than in any other branch of medicine, many of the endpoints are either unclear or the subject of intense ideological division.

There was a general belief that it was not possible to measure subjective change of the kind that seemed to be involved in psychiatric illness. The studies of Beecher, Gold, and Lasagna on analgesia, however, were probably critical in helping physicians and researchers realize that what had previously seemed impossible might after all be feasible. Pain, after all, was just as subjective as depression. Stemming from this, one of the key issues at the conference was the use of rating scales for the mapping of clinical change. A number of rating scales had been produced by Dick Wittenborn and others. One complaint from some of those who opposed evaluation centered on the issue of how one could capture the complexity of clinical reality. How could a process that was so subtle, nuanced, and involved interactions between the subject and significant others be captured by an instrument? Where were the X-rays on which standardized blind assessments could be made?

Nate Kline was equally opposed to rating scales but viewed the problem in an entirely different light. The only clear-cut outcome as he saw it was that either the patient was discharged or that she moved from a back ward to a more open ward. Kline was opposed to a reliance on rating scales. He argued that outcome measures other than discharge from the hospital risked producing a variation of the rabbit out of the hat effect. Rating scales were a means of first putting the rabbit into the hat, which when it was subsequently pulled out would fool the investigator as much as anyone else into thinking that something wonderful had happened.[61]

With the closure of the old asylums Kline's simple endpoint of discharge from hospital is no longer a practicable research goal. But it is not clear that efforts to find some such goal should not be pursued. Most randomized trials of the type discussed so far and most of the rest that will come up in the remainder of this book have involved the intensive study of one hundred or two hundred patients, with a large number of ratings and blood tests. Such studies are nearly impossible to combine with normal clinical practice. They depend critically on standardized classification systems and a strict selection of patients. In contrast, Kline's ideal was to do studies involving a few clear outcomes, such as whether the patient lives or dies. This approach only took off in any branch of medicine in 1981 with the first International Study of Infarct Survival (ISIS), which looked at the effect of a beta-blocker on survival after a heart attack. It involved 245 hospitals in fourteen countries and enrolled 16,000 patients. Beta-blockers were shown to be useful.

This study caught the imagination of many. It was followed up by ISIS-2, -3, and -4. In ISIS-3, a seventeen-country 41,000-patient study, a novel thrombolytic agent (clot-dissolving), TPA, was compared with an older drug, streptokinase, and found to be no more effective, and indeed in some respects to be less effective. There were a number of interesting as-

pects to this. One was that TPA was much more expensive than streptokinase but nevertheless many centers, particularly in the United States were
using it rather than streptokinase. TPA was a new kind of drug, derived
from genetic engineering, and the implication was that because it was
more natural, it would be more effective. The emergence of this kind of
drug had even prompted some regulatory authorities in Europe to consider licensing such agents without the customary requirement for
placebo-controlled evidence of efficacy—a strategy that the results of
ISIS-3 call into question. ISIS-3 also challenges easy assumptions about
the benefits of natural treatments. No one at present, however, has attempted to run such a study in psychiatry.

The Standardization of Psychiatry

Kline lost the argument, possibly because the state of ideological division
in psychiatry was such that virtually everything had to be made as explicit
as possible. In the case of depression, one or two instruments subsequently caught on. The foremost of these was the Hamilton Rating Scale
for Depression, first published in 1960.[62] This was constructed of seventeen items, some of which required answers from patients as to how suicidal or guilty they were, for instance, and some of which were rated by the
interviewer on the basis of observable behavior—how agitated or slowed
up in themselves patients appeared. For many the scale seemed problematic because it appeared to abstract from the richness of clinical reality.
Hamilton's answer to this was that the point was undoubtedly valid in
some cases, but that all too often, far from abstracting from what was
done clinically, employing the scale would mean that a basic minimum of
assessment was done that might not otherwise have been done in a busy
clinical schedule. Using the scale, then, would ensure a basic level of quality to the clinical interaction.[63]

A second point that caused problems was the idea of adding up the
scores on different items to produce a total and acting as though that
total had some real meaning. Many clinicians had difficulties with the
idea that items of very different meaning could simply be summed.
Should early morning wakening be counted in the same balance as guilt
or suicidality? The sleep and appetite disturbances that occur in depression have always been seen by many, even the majority of psychiatrists,
both psychologically and biologically oriented, as peripheral to the core
problem of depressed mood. Were they to count for as much as depressed
mood?—and indeed for more if the Hamilton scale were used? For
Hamilton, the simple answer was that the scale worked.[64]

In a sense, there is an element of rabbits and hats to Hamilton's scale.
Its items, which cover sleep and appetite, could not have been better de-

signed to demonstrate the effects of tricyclic antidepressants like imipramine, which are somewhat sedative and appetite stimulant even in people who are not depressed. There is widespread acceptance that the scale is not as suited to demonstrating the effects of the MAOIs and SSRIs. Hamilton's scale languished for a few years after its publication. Clinicians finally began to use it largely because the pharmaceutical industry had adopted it. One of the key figures here was Alan Broadhurst from Geigy, who incorporated the scale into the clinical trial protocols for imipramine done in the Britain. As the tricyclic antidepressants triumphed in the 1960s, they in turn brought the scale with them.[65]

If there is an element of rabbits and hats to the Hamilton scale, however, this is arguably no more than there has been with its principal competitor, Beck's Depression Inventory (BDI).[66] The BDI focuses only on cognitive aspects of depression. It is difficult to imagine anyone who has been through a cognitive therapy program not being biased toward answering the particular items on the scale in a different manner after therapy by virtue of the procedures of cognitive therapy, quite independently of any effect those procedures may have on depression. In the case of the BDI, however, just as with the Hamilton scale, there has been a group interested in accepting this scale—cognitive therapists. Another aspect of successful scales, then, is their need to coincide with certain interests if they are to survive.

The Hamilton scale samples a slightly wider domain of relevant pathology than Beck's, but the fact is that both are intensely individualistic instruments. There are large areas of interpersonal functioning that are not captured by either scale. Neither rates the extent to which subjects make eye contact, for instance, which is actually a good predictor of response to treatment. The exclusion of items like eye contact could be accepted if these instruments clearly worked, but there are increasing concerns about just how well these scales function. There are significant chemical differences between some of the antidepressants now available, yet current clinical trial procedures do not seem able to pick up differences between treatments either in terms of effectiveness or in terms of clinical distinctiveness. One must wonder whether the problem hasn't partly to do with the methods of measuring effectiveness.

Max Hamilton, for the most part, was fairly summary in his dismissal of criticisms of his scale, but he never claimed a particular virtue to the use of his or any other scales. He was aware that the change in culture involved in the new approach was substantial—"it may be that we are witnessing a change as revolutionary as was the introduction of standardization and mass production in manufacture. Both have their positive and negative sides."[67] While the use of rating scales is sometimes seen as contributing to an increase in objectivity in the evaluative process for new drugs, another aspect to their use is that they contribute to the industrialization of both psychiatry and the evaluation of its practices. Randomized control trials of the kind that are used to register pharmaceuticals demand standardized rating scales, operational criteria, and the standardization of psychiatric diagnoses. After 1962 a standardization of diagnostic practice was all but inevitable. The change, when it came, was most marked in America and most clearly symbolized by the publication of the third *Diagnostic and Statistical Manual* in 1980.

This change has had a number of consequences. It has contributed to the establishment, to some degree, of an American hegemony in world psychiatry. The fact that a pharmaceutical company will almost always be aiming for registration in the American market has led to the adoption of American diagnostic criteria (DSM-III, -IIIR, and -IV) in drug studies worldwide. From this beachhead, the use of DSM-III has spread out to studies in other areas of psychiatry and even to research on psychotherapy. In addition the FDA requirements for studies with sufficient power to be conclusive have contributed to a globalization of American terminology. In the 1960s, it was still possible for individual investigators to conduct studies on populations attending the hospital in which they worked. Now studies have to be multicentered and often multinational in order to enroll sufficient numbers. This has meant that protocols need to be agreed on by investigators from different countries, who are therefore forced to face up to regional or national differences in prescribing and forced to adopt a common terminology.

The FDA has come to occupy something of a magisterial role on the world stage. As with the magisterium of the Catholic Church, it acts not to say what must be believed but to adjudicate on certain claims that are made, allowing some of those to stand and refusing to legitimize others. In one sense this is a very minimal role, but the strategic positioning of the FDA in the current world psychiatric economy means that this minimum can be extraordinarily influential. Yet the FDA is in place to regulate an industry, not to arbitrate on science. Is the standardization that is appearing in psychiatry the result of scientific progress or a consequence of an industrial process?

The emergence of the FDA into this narrative stems from the fact that the thalidomide tragedy happened at a critical moment in the evolution of events. The question of whether it would be better to do randomized trials with operational criteria and batteries of rating scales or randomized trials in the large simple format had not been settled when the FDA was forced to act to establish what form of test would constitute an acceptable demonstration of efficacy. In essence there were two separate issues—what form of study would best tackle the scientific issues involved in therapeutics and what form of study would allow the FDA to license a claim regarding a particular compound. The actions of the FDA, coming when they did, led to these two questions being conflated—at least in psychiatry. Thalidomide, arguably, acted not only as a teratogen on developing fetuses but had potent effects on the embryonic science of evaluation.[68]

Considering the implications of what had happened, the FDA clarified its claim that agents must be shown to have substantial effectiveness before registration. Five different types of studies, it was argued, might show evidence of efficacy: uncontrolled studies; studies employing a historical control, as with the American streptomycin studies; studies which indicated a dose response, that is, the higher the dose the greater the response; studies comparing a new agent with an active comparator; and finally placebo-controlled studies.[69] In addition, whatever the type of study, effectiveness was to be tied to medically legitimate ends.

By the early 1960s there was a general consensus that placebo-controlled studies were the most sound. There were voices that still complained about the dehumanizing aspects of RCTs, but for every one of those, there were others who argued that it would be unethical to let ineffective treatments onto the market—that the public must be protected against its own insistence on being poisoned. The problems posed for some by the use of placebos in the case of the antidepressants were minimized to some extent by the trial design adopted for registration purposes, which only called for a six-week trial exposure. Ordinarily, six weeks is sufficient to show changes in Hamilton Rating Scale scores and significant differences from placebos, and therefore the FDA could accept that evidence from such a trial could be used to sustain a claim of antidepressant efficacy. In order to achieve this, however, the industry had to adopt, and the FDA to sanction, Fisher's model complete with operational criteria and standardized rating scales rather than Kline's large simple trials. Arguably the relatively brief period of treatment involved in this design, or the widespread confusion between regulatory and scientific studies, or both factors combined to create an impression of therapeutic specificity—depression was rather similar to a bacterial infection, something that could be knocked out by a short sharp course of treatment—an impression that, as we shall see in subsequent chapters, is quite misleading.

While placebo-controlled studies were called for, in practice company-sponsored studies of antidepressants moved toward the use of "active" rather than placebo controls. Submissions to the FDA for the most part consisted of trials in which the new compound was compared to amitriptyline or imipramine, and the lack of a significant difference between the two was taken as evidence that the new drug worked just as well as the older one. This retreat from a hard line was justified on the basis that it was unethical to withhold a known active drug (and indeed this is the position of the Helsinki Declaration on Human Rights) from those who might benefit from it. If it was just a matter of different response rates to active and placebo treatments, that would be one matter, but what was involved in antidepressant trials was exposing people to a known risk of suicide.[70]

Such considerations held until 1980, when Paul Leber, who had joined the FDA in 1978, pointed out to all concerned that given the numbers involved in a traditional antidepressant trial, the lack of a difference between old and new treatments was not convincing evidence that the new treatment worked.[71] The clearest outcomes from such studies would concern differences in the side-effect profile of the two compounds; therefore all the trials were doing, in one sense, was providing good marketing copy for the companies. Despite general disbelief, Leber pushed through a formula that all submissions must contain evidence of at least two pivotal studies. Though not spelled out in detail, it is understood by all parties that pivotal ordinarily means placebo-controlled. The drug development plans of a number of companies were thrown into turmoil. Compounds that had come onto the market in Europe and become best sellers, for instance, never made it to the United States for this reason.[72]

One possible conclusion that can be drawn from this sequence of events is that none of the antidepressants, prior to the first of the SSRIs, zimelidine, has strictly speaking been shown to be an antidepressant—at least to the FDA's satisfaction. The fact that the older compounds have not been withdrawn from the market, however, tacitly indicates that everyone "knows" that they are antidepressants. There is a sense in which, then, in the real (nonregulated) world randomized placebo-controlled trials are not necessary to prove that a compound is an antidepressant.

The new formula, furthermore, is not without its ambiguities. For instance, antidepressants, such as Prozac, have been licensed on the basis of demonstrated superiority to placebo in patients with moderate rather than severe depression. But this superiority to placebo says nothing about how good the compound is. Superiority to placebo legitimizes a marketing claim that the compound has antidepressant efficacy, but is it as good as older compounds?[73] A number of studies that have been done by independent groups, notably the Danish Universities Antidepressant Study Group, have compared some of the newer compounds with clomipramine in more severely depressed populations and in most cases so far clomipramine has been found superior. As we shall see in Chapter 7, the question of how an antidepressant does against the more severe forms of depression is of no small importance to any definition of what an antidepressant is and indeed what depressive disease might be.

There are further ambiguities. The demonstration that the SSRIs, for instance, are of benefit in moderate depression is consistent with an assumption that they will also be useful for milder depressions, but strictly speaking this has not been shown. As outlined in Chapter 1, moreover, there are good theoretical grounds to argue that some milder depressions are different from the more severe depressions—they are depressive neuroses rather than depressive diseases and as such might not be expected to respond to antidepressants. How well then should we expect the results of such studies to generalize to the real world of primary care depression? There is increasing evidence that patients with milder depressions do not always benefit from antidepressants.[74]

At present there are a number of campaigns to improve the rates at which primary care physicians detect depression. Such efforts have been based on the assumption that the demonstration of effects in an RCT will translate into a real difference in the real world. But if they don't, where does that leave us? All treatment involves a trade-off between risks and benefits. Where the benefits are substantial, as in anesthesia, considerable risks will be undertaken. In moderate to severe depression some risks seem justified, but are the same risks justified in mild depression when the benefits may not be forthcoming? The act of diagnosis, by "medicalizing" experience without being able to remedy it, may in itself be an iatrogenic injury.

Proceeding down the route prescribed in 1962 has had a number of other consequences. For instance, what the antipsychotics most clearly do is reduce tension and agitation, regardless of diagnosis. But they are never put into a clinical trial for this indication—trials are always for a disease indication such as schizophrenia and they are conducted on agents that will end up being available on prescription only. This has happened in great part because the 1962 amendments encouraged the development of compounds for indications that experts thought appropriate. Tranquilizers, tonics, and treatments for conditions such as halitosis were to be discouraged. If the rules under which the "game" has been played for the last thirty years were different and the drugs were produced for sale over the counter, where tension reduction rather than the cure of a disease would be more likely to sell the product, it seems quite probable that an entirely different package of study designs and rating instruments would have come into being. RCTs in conjunction with the restriction of prescribing to physicians, therefore, have powerfully reinforced categorical and medical models as opposed to dimensional models of psychiatric disease.[75]

A number of the problems alluded to here stem from the fact that most studies conducted seem as much if not more oriented to the registration of a compound than they are aimed at answering a question of scientific importance. In this sense Kuhn's comments at the meeting in Cambridge seem apposite and unsurprising—the result of the labors of vast numbers of researchers using ever more sophisticated scales has been the registration of compounds but not the advancement of science. Because companies are not in the business of proving that there are differences among their compounds, the result is a marketplace stuffed full of agents that are biologically quite heterogeneous, but whose clinical differences are minimized, or indeed left blatantly unexplored, for the sake of grabbing a share of the big indication—which in psychiatry since 1980 has been depression.

Pharmaceutical companies, however, should not be seen as being uniquely evil in this regard. The trials that the industry has sponsored have established that certain compounds with particular biochemical profiles are at the very least antidepressant principles, if not specific magic bullets for depression. The various psychotherapies, in contrast, have conspicuously avoided testing the different therapeutic components contained in the treatment packages they offer to see which contribute what to the resolution of depressive or other nervous disorders. In short, a great deal of what a study reveals depends on who sponsors it, and the community at large has not for the most part been in the business of sponsoring independent studies in the therapeutics of nervous disorders.[76]

The Challenge of AIDS

Just when the arguments about placebo-controls seemed to have been "won" in the mid-1980s, the AIDS epidemic posed another threat to the legitimacy of the methods.[77] Many of those infected with HIV, faced with requirements from the medical profession that they enter placebo-controlled studies of zidovudine and other drugs and accept possible randomization to placebo, refused to participate. Retrospectively, it is clear that one of the factors involved was that HIV sufferers formed a community in a way that most other patient groups do not—a community that was well versed in the latest research and the nuances of medical research. Another factor was that inequities in the U.S. health care system denied treatment to many sufferers unless they entered a double-blind trial. Many entered, but large numbers of those then tested their medication to find out what they were on and pooled medicine with others, so that the trial design was subverted.

The arguments put forward were often poignant. So what, one would hear, if the drug in itself is useless, it brings with it hope; what right do researchers, more interested in fitting the pieces of a research puzzle together than in ministering to sick human beings, have to deny hope? Researchers who, it was felt, paid heed to the principles of the Helsinki Declaration in much the same way that seasoned air travelers paid heed to the safety announcements before plane flights. Research designs that had initially been conceived in an effort to protect subjects from abuse by zealous therapists all of a sudden began to be seen as a means of apparently denying a benefit to ill subjects.

It began to seem to many that the idea of clinical trials was in trouble—that they had evolved at a time when science was shrouded in mystery, a time when health care was seen as a benefaction rather than a right and when professional judgment was unquestioned. In the face of the AIDS crisis, extraordinary responses were called for and were provided. In many places researchers, medical ethicists, and AIDS activists sat around the same tables and hammered out agreements about what needed to be done and what could be done.[78] There now appears to be some degree of consensus that research is legitimate when there is uncertainty as to the outcome and perhaps most legitimate in conditions like AIDS or mental illness, when the enthusiasm of experts or companies can too easily lead the community down the wrong path.

What the question of AIDS seems to indicate is that science is a community activity. It is not something that can be conducted solely by research institutes or by private corporations. If an activity designated as scientific is not serving the interests of the community, there comes a point when it will be rejected. I would go further and argue that the vaunted objectivity of science lies not in any method so much as in scrutiny by a community—ordinarily a community of scientists, in the first instance, but in the final analysis by the community at large.[79]

This raises the question of how much community involvement there is in company-sponsored drug trials. On the one hand, these are aimed at getting a license rather than advancing science. On the other hand, the registration process permits the community to insist on randomization and placebo-control as part of a demonstration of efficacy. There are a number of ambiguities here. One concerns the question of whether the community at large or the pharmaceutical industry has the greater interest in RCTs. Another has to do with the fact that while RCTs may get a compound onto the market, a considerable proportion of its subsequent use may be off-license. Indeed, greater medical and scientific advances may stem from this off-license use than from anything sanctioned by a clinical trial. Whether these tensions produce a problem or not probably depends a great deal on how far the pharmaceutical industry is perceived by the medical profession, in the first instance, but also by the community at large to be working in the community's interest.

The Information Revolution

I preempt our story somewhat in jumping from the 1960s to the 1990s, but this leap is worth taking because even though it will reveal some details of the plot, my hope is that it will make even more dramatic the story, which will unfold over the next four chapters, of how we got from where we were in 1962 to where we are now.

If the 1950s was the decade when new treatments transformed the face of medicine, the 1990s looks like the decade when information about those treatments will radically resculpt its character. A number of different inputs to the current interest in medical information can be identified. One is the need for containment of health care costs. A second input comes from a sense that the information revolution may facilitate possibilities of which we are at present only dimly aware. A third has to do with increasing public awareness of variation in medical practice and responses to that. The new technologies are spawning new jargons about outcomes and evidence-based medicine. Many of these inputs have political aspects.

Since World War II health care costs have increased dramatically. Following the introduction of Medicaid and Medicare in the United States in 1966, the American budget for health care went through the roof. In 1969 Richard Nixon drew attention to a developing problem. Some have argued that he was biased against the growth of medical and other science for political reasons, but in retrospect it is hard to argue with the fact that difficulties were emerging.

The problem lies in trying to pinpoint the origin of the increasing costs. They could stem from increasing hospital costs, and a substantial body of evidence points to inefficiencies in the hospital system. It could be the "evil drug companies" milking the system. The industry tends to be the easiest target for reformers, although in a regulated system like the United Kingdom's health market drug costs, as a proportion of total costs, have remained steady over decades. A further possibility is simply the increase in the number and intensity of medical services available. It is a truism in health care management that appointing a new consultant increases the amount of work to be done rather than brings about a more efficient handling of the current volume. This increase is not necessarily by virtue of inappropriate medicalization; it may equally reflect a legitimate expansion of the market into areas of previously unmet need.

One method of capping growth would be to insist that those who are behind new developments demonstrate that they will actually lead to benefits. This idea, however, has spawned two quite different schools of thought. One of these has been termed the outcomes movement. A focus on outcomes, through the information offered by databases, has been offered as one way to get round the ethical difficulties of RCTs. The kind of efficacy shown by RCTs is one thing, the argument goes, but what happens in real life may be quite another. Building up large databases of actual outcomes might be another way to do research.[80]

This kind of statement leads to heated debate. Those against outcome studies argue that losing the element of randomization would fatally expose research to the effects of bias. Outcome studies are fine, but they don't show what has brought about the outcomes observed. Those in favor of outcomes studies argue that randomization is critical to a particular kind of small sample research, but it is less clear that it is needed when applied to population samples. Indeed, randomization runs the risk of trivializing science by throwing up statistically significant values that in real life are insignificant. In addition, just how many people are fooled by placebos? Outcome studies have ecological validity, their proponents argue, but more to the point, with developments in information technology, they have become possible.[81]

A study by Susan Jicks and colleagues in Boston illustrates the possibilities and the pitfalls of this approach. This group has recently interrogated the databases of primary care physicians in the United Kingdom and followed up 170,000 patients given antidepressant prescriptions to see who had committed suicide. A major selling point of the newer antidepressants (the SSRIs and others) has been that they are safer in overdose than the tricyclic antidepressants, imipramine and amitriptyline. Surprisingly, then, the Boston study indicates that there were a greater number of suicides in those who had been prescribed nontricyclic compounds than tricyclics. The problem with outcome research is that it tells nothing about what has produced this outcome—was it because primary care physicians tended to prescribe the newer drugs for patients they perceived to be at greater risk of suicide or were there more suicides because the newer drugs are less effective or did the newer drugs get given to milder depressives who might not otherwise have been treated and were some of them made suicidal by treatment?[82] While it is not clear from this study what has actually happened, one advantage of research like this is that it may expose seemingly scientific claims as marketing copy.

An almost diametrically opposite use of information is found in the current craze for evidence-based medicine (EBM). One of the first to draw attention to this area was Archie Cochrane, who argued as early as 1971 that many medical services and procedures had not been tested for their effectiveness and that others, despite being tested and shown to be unsatisfactory, still persisted.[83] The implication was that if everything were submitted to RCTs, effective clinical procedures and services would in due course emerge and a great deal of expense could be saved by withdrawing funding from treatments not shown to provide health gain.

Databases of clinical trials have since been established. This has led to a recognition of the need for what are now called systematic reviews. Since World War II, there has been a burgeoning of medical literature. Many of the most cited articles have been review articles, but it is now clear that many of these reviews were biased. Some were biased of necessity because, until quite recently, the reviewer of an area could not possibly be in possession of all the relevant trials conducted. As a result, many reviews were all too often lengthy articles put together with impressive reference lists but espousing only one point of view. They were essentially rhetorical rather than scientific exercises. Systematic reviews, in contrast, are constrained by an obligation on the reviewer to account for all the data. They are scientific, in other words.

Useful though such developments might be, the concept of evidence-based medicine was criticized as early as 1979 by Thomas McKeown, who argued that of even greater importance than the question of efficacy was the issue of direction.[84] One could imagine a health service, McKeown argued, in which everything was of proven efficacy but yet the service as a whole in its orientation to high-tech management to the neglect of preventative measures was actually less effective than another service that might utilize many unproven therapies. He went on to propose that taking an evidence-based approach was likely to lead to a neglect of both external influences, such as sanitation, and the personal behavior of subjects, such as smoking, diet, and so on, which he contested are the predominant determinants of health. In the case of psychiatry, child abuse would clearly be one such factor.[85]

An intertwined development has stemmed from a growing awareness of variations in treatment. Clinical audits have made it increasingly clear that, in addition to expected variations among countries in the recognition of diseases and the availability of treatment, there are marked variations within countries regarding the adoption rates for treatments. There are enormous variations in ECT usage and in the dosages of drugs used to manage violent behavior, for instance. Concerns with both quality and effectiveness have led to an interest in clinical guidelines. Guidelines are systematically developed statements indicating what the panel of experts who drew them up believes to be the better treatment options in particular health care situations. They have become a focus of intense debate. Clinicians worry about their clinical judgment being circumscribed and the possible compulsory imposition of guidelines by managed care dictates or following courtroom settlements. Others answer their concerns by arguing that it is necessary to start closing the gap between knowledge and practice. The script for both camps could come straight from the debates between Louis, Risueno d'Amador, and Poisson in the 1830s.

Guidelines for depression have been elaborated by the Public Health Service Agency and the American Psychiatric Association.[86] Applying the PHSA guidelines in primary care settings has, however, not produced a significant enhancement of outcomes, perhaps for reasons that were alluded to above—there is a lack of evidence regarding the management of milder depressions and perhaps even fundamental misunderstandings about the nature of such depressions.[87] There is a further problem that we will return to in Chapter 7, which is that at present the evidence used in the construction of guidelines for the most part is derived from RCTs, in the belief that RCTs offer evidence of specific efficacy. But as we shall see in Chapter 7, RCTs in depression in fact suggest that placebos can be expected to work for the majority of primary care depressions.

Placebos haven't been shown to work compared with anything else—they can't be. They simply work—as Harry Gold and Harry Beecher found out. The large placebo response to almost all psychiatric treatments means that responses are heavily shaped by factors such as the quality of the interaction between physician and patient, the circumstances in which help is sought, and other nonspecific factors. Beecher was struck by the fact that soldiers horribly wounded on the beach at Anzio complained of pain much less than civilians less seriously injured. He put this down to the different meanings a wound may have—for the civilian it may be a disaster for her career, for the soldier it may be a blessed escape from hell. These elements of the therapeutic encounter, though they can be evaluated, cannot be subject to randomized control procedures or made a matter of guidelines. It is difficult to see how there could ever be guidelines for locating the right moment to give a drug in the unfolding drama of a disorder set against the background of a particular constitution and temperament and life story. Although clearly therapists do need to be held to account for what they do, as the Osheroff case dramatically illustrates, there is also a matter of skill and wisdom involved in therapy.

The mobilization and evaluation of nonspecific factors is central to the management of depression. It is also central to the politics of health care. When assessing evidence-based medicine, it is well to remember that this is a concept in which governments and the pharmaceutical industry have an interest. In the new politics of health, the issues of who owns the data, who controls the means of producing the data, and who decides what data count are paramount. Whereas in Chapter 1 the notion of a disease as a pathological process was contrasted with that of an illness as experienced, the current situation suggests that in practice disease is coming to be defined, neither as a pathological process nor in terms of unwelcome experiences, but rather as what third-party payers will reimburse.[88]

Reflecting, in 1966, on the development of RCTs—long before the arrival of evidence-based medicine—Bradford Hill suggested that if RCTs ever became the only method of assessing treatments, that not only would the pendulum have swung too far, it would have in fact come off its hook.[89] Has it come off its hook? As orthodox medicine moves toward providing evidence of efficacy for what is being done, consumers, far from being impressed, appear to be deserting orthodoxy and seeking out a range of alternative approaches. What is being rejected? One possibility is that evidence-based medicine, RCTs, and guidelines all reflect a belief in the notions of specific disease states and specific semimechanical therapeutic interventions, and what is being rejected is the idea that this is all there should be to medicine.

I will argue strongly at the end of the book that an understanding of the placebo has the potential to offer a way out of this stand-off. The crucial point to realize is that the placebo phenomenon is not something left over after the effects of specific treatments have been accounted for, nor is it some throwback to older humoral models of disease—although clearly the phenomenon has a lot to do with the why of a disease rather than the how. The placebo and the general role of nonspecific factors in therapy are an even newer discovery than that of the magic bullets,[90] which at present have not been exploited by either orthodox or alternative treatment approaches and were never suspected before the advent of specifically effective treatments revealed their existence.

It is difficult, however, to see a corporation gearing itself up for the sale of placebos. And as some of the succeeding chapters will suggest, the current scientific marketplace is such that if a product or a concept can't be sold, it doesn't exist. Until things change, we are likely to be left with the present situation in which the majority of patients are uninformed about their treatments, or inappropriately informed—and in the clinic they meet clinicians who may be as inappropriately informed. These clinicians may be overinformed, but they have not learned to incorporate the relative risks and benefits that stem from modern treatments into the act of therapy. They have been misled by demonstrations of specific efficacy into thinking that the art of therapy comes down to an ability to match a particular diagnostic category to a supposedly specific agent and to prescribe accordingly. Such clinicians, far from transcending their prejudices in response to data, are drowning in data. It is as though, instead of leading on to the glories of Rembrandt, the discovery of oil paints had led to the eclipse of art and its replacement by technical drawing. The next four chapters on one level, perhaps, can be read as an account of our fall from grace.

Because of the early studies that I had been carrying out with reserpine and then chlorpromazine, I became involved in the other great "revolution" of the time, namely psychopharmacology . . . Many workers were beginning to focus their attention on the possibilities that were opened up by the prospects of effective drugs for mental illness. And this almost meaningless word, psychopharmacology, gradually expanded and sucked in a mixed bag of people. —Michael Shepherd, 1993[1]

4 The Trials of Therapeutic Empiricism

We have seen in previous chapters something of the deep divide that can exist between theoretical and empirical approaches to medicine. Psychopharmacology, however, has subsumed both approaches under its heading. The term *psychopharmacology* was proposed by members of the theoretical school to describe what they were up to. Jean Thuillier, one of the founders of the Collegium Internationale Neuropsychopharmacologium, suggested the term in 1957 on the basis that it was now possible to both provoke psychoses with drugs, such as LSD, and cure them with chlorpromazine, and therefore for first time it was possible to speak about an experimental or scientific psychiatry.[2]

Unbeknownst to him, Seymour Kety, Ralph Gerard, and Jonathan Cole, who were involved in organizing the Conference on the Evaluation of Psychotropic Drugs held in September of 1956, had also come up with the term.[3] For the purposes of this conference, psychopharmacology was something of a portmanteau word that provided a focus for a disparate group of clinicians, statisticians, animal psychologists, physiologists, and others who gathered to evaluate the therapeutic possibilities of the new drugs. Cole's use of the term, which essentially referred more to psychopharmacotherapeutics, seems closer to the modern usage than Thuillier's. But between these poles, the experimental and the therapeutic, pharmacopsychology and psychopharmacology, the difference between Kuhn and Kline, hangs a tale of tension and acrimony.

Pharmacopsychology

There had been two earlier combinations of psyche and pharmacology, of which most people in the 1950s seem to have been unaware. In 1892 Kraepelin had called research in which he had begun giving drugs to human subjects pharmacopsychology, and then later in 1921 David Macht had used the term psychopharmacology for essentially similar work carried out in Johns Hopkins with animals.[4]

The foundations of this kind of pharmacopsychology were laid by Hermann von Helmholtz, who in 1850, at a time when electrical transmission along nerves was not generally accepted, established not only that it happened but the precise speed at which it happened—between 25 and 40 meters per second. Helmholtz's research assistant Wilhelm Wundt married these findings to the then new concept of a nervous reflex and began a program of mapping various functions in the brain on the basis of speed of reaction times. Because the speed of nervous transmission was known, the length of time one association took to call up another was taken as indicating something about the distances that might exist between brain regions in which different concepts were stored. Although appointed to a chair in philosophy in Leipzig in 1875, Wundt was clearly not a philosopher; indeed, on the basis of this experimental approach, he is usually credited with being the first psychologist.[5]

Kraepelin came to Leipzig in 1883 to train with Wundt in the new research methods. He introduced a new idea which essentially was to see whether drug treatments modified reaction times: "Under Wundt's influence, I began to work more on experimental psychological problems . . . I planned . . . tests with drugs [morphine, cocaine, alcohol, bromine, and trional], coffee and tea and to measure the mental reactions of psychiatric patients . . . We not only wanted to identify the behavior of different mental processes in mental disease but also the external and internal influences."[6] In 1892, summing up what had been achieved, he coined the term pharmacopsychology: "Herein lies, it seems to me, a not insignificant benefit of this pharmacopsychology, in that it may lead us at times to recognize the true nature of certain psychological processes from the special effects of an already accurately known drug . . . Following this school of thought, the study of psychic medicines will provide information for psychology."[7]

Whereas Kraepelin's ideas about the classification of mental diseases swept all before them, his pharmacopsychology had much less impact. It was traditionally believed that only poisons brought about changes in mental functions. As such, the general view was that it was difficult to be certain what if anything could be learned about normal mental functioning by poisoning the brain—an attitude that reflects very clearly Descartes's division of the mind into a mechanical brain and an all but immaterial soul and of course the fact that the idea of chemical transmission in the central nervous system was still over half a century away. Criticizing Kraepelin's research in 1912, Jaspers argued that "hardly any of the results stand up to keen criticism as the relationships are for the most part so complicated."[8] The pure Kraepelinian program of trying to dissect out psychological functions using drugs remained largely undeveloped outside Germany, until it was picked up again in the work of Hannah Steinberg in London in the mid-1950s and later by Gordon Claridge and others.[9]

From the mid-1950s, a related body of research that also goes under the heading of psychopharmacology, although it might more appropriately be called pharmacopsychopathology, flourished hectically for just over a decade, stimulated by the discovery of LSD. This branch of the subject has roots that every writer in the field refers to—the use of plants and alcohol to alter consciousness for religious, medical, or social purposes. The widespread use of drugs in this manner has suggested to many that there must be something of a basic human appetite to seek altered states of consciousness.

The development of a science from these uses took shape slowly during the nineteenth century. In 1799 Humphrey Davy experimented with the recently produced nitrous oxide and noted its consciousness-altering properties. Subsequently Jacques-Joseph Moreau (de Tours), a student of Esquirol's and a frequenter of the Club des Haschichins in Paris in 1845, put forward the view that the loosening of associations, clouding of consciousness, and the semi-psychotic states that hashish could produce might have implications for our understanding of mental illness.[10]

Despite the widespread use of a variety of consciousness-altering agents during the nineteenth century, Moreau's idea was too radical. It was a century before it was picked up again. This is perhaps somewhat surprising, in that nitrous oxide, for example, was used as a laughing gas at circuses or parties. Coca was available and from the late 1880s, cocaine. Opium was relatively freely available and used widely to sedate children and adults. Cannabis was also available, and in 1869 Horatio Wood won an American Philosophical Society prize for describing its effects. Mescaline had also come to the attention of both physicians and the public after its isolation from the peyotl cactus in 1896. In the same year, Weir Mitchell wrote a much praised paper describing its effects on aspects of his own mental functioning.[11] The use of psychotropic compounds was sufficiently widespread and interest in their effects was great enough for William James to include a discussion of the subject in his *Varieties of Re-*

ligious Experience in 1902, in which he concluded that religion had indeed something to learn from the altered states of consciousness brought about by drugs.[12]

Against this background the relative failure to use these observations to further an understanding of mental illness is striking. Moves to restrict the use of opiates, cocaine, cannabis, mescaline, and other drugs toward the end of the nineteenth century and in the early years of the twentieth century may have played a part in this. The rise of both behaviorism and psychoanalysis, neither of which could satisfactorily accommodate data on drug-induced altered states of consciousness, may also have had something to do with it.[13]

The discovery of LSD in 1943 and later PCP and the relationship of their effects to classical compounds such as mescaline, however, finally brought about in the 1950s the results that Moreau had hoped for in 1845.[14] The concept of a model psychosis—a drug-induced change in mental state—came into being. Efforts went into establishing the clinical features found in the psychoses induced by different drugs and into determining how closely these states resembled the clinical psychoses of schizophrenia and paranoia. As with any other classical (theoretical) science, the hope was that psychiatry and psychopathology would proceed by setting up models, manipulating those models, and determining the degree of correspondence between the model and reality. The idea of the model psychosis took hold in Germany in particular, and of course it was to Germany that Kuhn, as a German-speaking Swiss, would have looked.

The discovery of chlorpromazine was initially seen by many in Germany as just another drug on which to base a model or with which to test the boundaries of already established models. Its ability to modify the effects of drugs like mescaline and LSD, at least temporarily, appeared to strengthen the hands of those interested in the model psychoses. In 1956 Wolfgang de Boor, one of the rising stars of postwar German psychiatry and another of the founding fathers of the CINP, produced a book entitled *Pharmacopsychology and Psychopathology,* detailing all that could be inferred about the workings of the psyche from the research then in progress. This research was largely, though not exclusively, German. De Boor's book is about a new era—one that never occurred.[15] Psychopharmacology ended up traveling a different road.

It took time for this to become apparent. Both LSD and Phencyclidine (PCP), which was produced by Parke-Davis in 1959, were quickly pressed into pharmacopsychological use. There is a rich literature from the late 1950s through the mid-1960s detailing these drugs' effects on aspects of cognitive functioning in healthy volunteers—effects, for instance, on perception, memory, the ability to learn and to concentrate—exactly the characteristics that Kraepelin had been looking at. The principles of doing such experiments scientifically were laid out by Gordon Claridge in 1969,[16] but by then the field was all but dead. What had happened?

A number of factors can be pointed to. One which will be explored further in the next chapter is that while classical Kraepelinian psychopathology was categorical rather than dimensional, work on the model psychoses was dimensional rather than categorical. It emphasized continuities between the normal and the abnormal. In some ways therefore it could be seen as being in line with the critiques of psychiatry from antipsychiatrists and others during the 1960s. Their criticisms were also predicated on continuities between the normal and the supposedly abnormal. Such work was never likely to find favor at a time when psychiatry was moving to a more hard-line medical model. The problems were compounded by the fact that human psychology at the same time was turning away from the nuts and bolts of actual brain functioning to become for the subsequent two decades a discipline of flowcharts (a boxology, as it has sometimes been called).[17]

A second factor involved consent and insurance. In the 1950s, few investigators thought about the issues of informed consent or insurance coverage for volunteers or patients. But the climate began to change in the 1960s, as possible abuses of trust in many branches of medicine came to light. In psychopharmacology, the protocols that volunteers or patients were put through have been one issue but another has been a linkage between research funds and the CIA, which was interested in what the new pharmacopsychopathology might reveal about brain-washing technologies. A climate began to develop in which doing anything other than helping patients to get better was seen as questionable.[18]

One consequence of the lack of research this has occasioned is that to this day it is not possible to distinguish the drug a volunteer may have taken—whether a major tranquilizer, minor tranquilizer, or antidepressant—on the basis of their performance on tests of cognitive function. The studies haven't been done. Volunteers will know clearly what they have had because of a drug's distinctly different subjective effects, but it will not be possible for a blind observer on the basis of volunteers' performance on tests of cognitive function to know what they might have had. Given that modern drugs, far from being poisons, act to modulate the levels of chemicals already known to be in the brain, there would seem to be something of a remarkable gap here in our knowledge. This gap in turn affects our ability to extrapolate from what we know the drugs do to the nature of the conditions for which they do something.

One party for whom underwriting this kind of research would be little problem is the pharmaceutical industry. Clinical trial work, involving many more subjects, both patients and healthy volunteers, taking larger doses of drugs, is all insured by the companies conducting studies for regulatory purposes. Where new compounds are involved, however, companies are now required to be able to submit to the FDA every piece of data produced by giving their drug to humans, and this makes them reluctant to permit studies by independent investigators for purposes unrelated to the market goals of the company. As regards underwriting studies with older drugs, already on the market, the industry seems to have little interest in such an exercise.

In this sense, the industry has de facto little interest in the question of what the drugs—which now have a specificity one can only imagine Kraepelin drooling over—might reveal about the workings of the psyche. At a time when whole new disciplines, such as pharmacoeconomics and pharmacoepidemiology, are being created, largely by industry funding, this seems extraordinary. These latter disciplines, of course, are more closely related to market purposes than is pharmacopsychology.

A further important factor in the sidelining of pharmacopsychology, however, was the advent of imipramine, which neither caused nor alleviated a psychotic state. This made it of little use to the model builders. Indeed it played a considerable part in the demise of some of the better-known models, as we shall see in the next chapter. In contrast, it fit very well with the overwhelming interest of the public at large, as well as state governments and clinicians, in seeing what could be done to get people well. It fit a science of therapeutics, rather than pharmacopsychopathology, and this led to the eclipse of the model psychoses long before LSD came under a cloud in the late 1960s. As Janzarik put it, there was a triumph of therapeutic empiricism over theoretical examination and interpretation of the nature of psychiatric disorders.[19] The empirical camp, however, was riven with its own disputes about the proper standards of scientific proof and interpretation. The randomized trial had entered psychotherapeutics, and it was about to pit a new breed of skeptics against the therapeutic enthusiasts.

The Demise of the MAOIs

As mentioned in Chapter 2, almost immediately after Kline's launching of iproniazid, reports emerged of jaundice in subjects taking it. It would seem that upward of 400,000 patients were prescribed the new drug in the year after its launch as an antidepressant. Kline has estimated that, given these numbers, one might have expected roughly 100 cases of jaundice—and there were 127 reports.[20] The figure of approximately 100 refers to the population at large, but older subjects and especially women are both more likely to get jaundice and more likely to get depressed. Did 127 then represent a problem? Kline thought not, but Roche, for perhaps a number of reasons, withdrew the drug.[21]

As noted in Chapter 2, a number of other MAOIs followed iproniazid onto the market, of which the best known were phenelzine and tranylcypromine. In 1961 there was a report in the *Lancet* of a patient who had a fatal subarachnoid hemorrhage while taking tranylcypromine. Subarachnoid hemorrhages are uncommon but occur frequently enough to make it all but impossible on the basis of a single case to implicate a drug that someone might have been taking. Reporting such a case in a journal like the *Lancet,* however, indicates a certain degree of suspicion on the part of the clinicians involved. This was one of seven such cases reported between 1961 and 1963. Some were fatal, but in those instances the subjects were taking more than one drug, making it impossible to finger tranylcypromine.

It also seemed that primary care physicians were noticing an increased occurrence of headaches in patients taking monoamine oxidase inhibitors. Examination of such patients suggested that some had blood pressure elevations, which could potentially tie in with the subarachnoid hemorrhages. Barry Blackwell, a psychiatric trainee at the Maudsley, drew attention to the possibility in a letter to the *Lancet,* which caught the eye of a hospital pharmacist in Nottingham named Rowe. He wrote to Blackwell detailing the occurrence of headache and hypertension in his wife, who had been taking a monoamine oxidase inhibitor, after she ate cheese. Could there be something in cheese which caused a problem for people on these drugs? Blackwell and his colleagues were amused at the suggestion and dismissed it, not knowing that some of the American clinical trialists for tranylcypromine had noted headaches as a side effect.[22] Max Lurie even suspected an interaction with the food that people were eating.[23]

Gerald Samuel, working with one of the manufacturers of an MAOI at the time, was less skeptical than the Maudsley doctors because the company had received two other suggestions of a similar nature. His remarks encouraged Blackwell and a colleague to take tranylcypromine for a week and then have cheese—nothing happened. The notion might have died at that point but that weekend Blackwell was called to see a lady who had taken phenelzine and had developed headache and hypertension—after a cheese sandwich. A patient in the hospital taking tranylcypromine agreed to take cheese with lunch and several hours later developed the increasingly familiar symptoms, as did two other patients in the hospital the afternoon after cheese made its weekly appearance on the dinner menu.

In late 1963 Blackwell reported the proposed connection to the *Lancet.* Few were persuaded. As Blackwell was later to note, an idea must fit in with current common sense for it to be accepted, and this didn't. Patients apparently recognized the connection because of the immediacy of their personal experience. This is of great interest, because current wisdom would suggest that in general most of us are fairly poor at making judgments of causality in situations such as this one.

The medical profession only began to take the observations seriously when a scientific account could be given for what was happening. Ultimately it was shown that cheese contained tyramine, which appeared in the bloodstream after it was eaten. This could indeed increase blood pressure. The implication was that the inhibition of monoamine oxidase in the wall of the gut allowed more tyramine than usual to enter the bloodstream and hence the problem. The "cheese effect" was born. A variety of other foods were then shown to be problematic as well, including both wine and beer, beans, and other vegetables. Subsequently there were reports of possible fatal interactions between tricyclic and MAOI antidepressants and interactions between MAOIs and over-the-counter pills, anesthetics, and analgesics.

All of these problems occurred against a backdrop of the thalidomide disaster. They were clearly not good news for makers of the MAOIs. Fatal though the "cheese effect" could be, the identification of what was involved led to a judgment that the ratio of risks and benefits was such that it was possible to keep the drugs on the market, provided the right kind of warnings were issued to patients. Indeed, as the more recent example of clozapine has shown, the whiff of danger and the extra precautions needed to contain risks may even be a factor making for better responses. Worse news, however, was to follow.

As early as 1959, a double-blind, placebo-controlled randomized study had been undertaken in Columbus, Ohio, comparing iproniazid, placebo, and psychotherapy. All depressive disorders, rather than just vital depressions, were included—a sample no less reasonable than Kline's. No differences between placebo, therapy, and iproniazid were found.[24] Despite this there was no great concern, as a number of other studies by some of the most notable clinical trialists in the field, such as Linford Rees, had clearly indicated that the MAOIs were efficacious.[25] In 1965, however, reports of a Medical Research Council (MRC) trial comparing antidepressant treatments appeared that dealt an almost fatal blow to the MAOIs.[26]

Following the streptomycin studies in 1948 and as it became apparent that a stream of new agents was entering all branches of medicine, the Medical Research Council in Britain set up a clinical trials division to coordinate trials in all areas of medicine, including psychiatry. Bradford Hill was its chairman and Michael Shepherd the secretary. Starting in 1960, their committee began planning a multicenter study to compare imipramine, phenelzine, ECT, and placebo. Some of the most senior psychiatrists in the country were approached to support the study, which they agreed to do at least in part because clinical trials were clearly the latest fashion. There were some demurrals to the effect that regardless of the results, they still expected their bedside experience to dictate their clinical practice. A series of planning meetings subsequently followed, a number of which were attended by Gerald Klerman, who had been recruited from Harvard to Washington as Jonathan Cole's assistant in the Psychopharmacology Research Center (PRC), as the Psychopharmacology Service Center had been renamed.[27] Through Klerman, this MRC study therefore influenced the design of the subsequent PRC chlorpromazine study and the 1980s NIMH study comparing drug therapies and psychotherapy in the treatment of depression.

Patients were recruited from 1962 through 1964 and the study was reported in 1965. The best-known names in British psychiatry, Michael Shepherd, Linford Rees, Martin Roth, and Max Hamilton, were all involved. Two hundred and fifty patients were entered in the study on a randomized basis. ECT and imipramine were both shown to be significantly superior to a placebo, while phenelzine was not and in women, indeed, phenelzine did not even have a response rate as positive as that of the placebo.[28]

As they had predicted, a number of eminent clinicians paid no heed to the result. William Sargant, in a 1965 article that looks back to the debates about the méthode numérique in France in the 1830s and forward to the Osheroff case, wrote that "there is no psychiatric illness in which bedside knowledge and clinical experience pays better dividends: and we are never going to learn how to treat depressions properly from double-blind sampling in an M.R.C. statistician's office"[29] Despite comments such as these, the bottom fell out of the MAOI market. As a consequence, imipramine was left as the gold-standard antidepressant.

Companies with MAOIs were left to market their compounds as best they could. As will become clear in Chapter 6, this marketing has been nothing short of inspired. Subtle differences in meaning between the notions of a reactive or atypical and neurotic depression were opened up that contributed in time to the redrawing of the boundaries of psychiatric nosology. But it was to take almost twenty years for these efforts to bear fruit. In the meantime, clinicians were faced with the need to reconcile the cognitive dissonance created by having a treatment many of them were certain worked apparently disproved by the best attested methods of the day. Many disparaged the way the study had been conducted and an antagonism toward Shepherd developed that was to be fueled by disputes over the efficacy of lithium which broke out a few years later.

Other studies appeared in the early 1970s indicating that the dose of phenelzine that had been used in the MRC and other trials, 45 mg, was too low and that if 60 or 90 mg were used a clear response was found.[30] There were also some interesting data from as early as 1962, where

Michael Pare and Linford Rees, reviewing the responses of depressed subjects over a number of episodes, found that those who responded to MAOIs on one occasion but not to imipramine were more likely to respond to MAOIs in a subsequent episode, as were their relatives.[31] This suggested that while the majority of patients may respond to both groups of drugs, some patients may only respond to one or the other.[32]

But none of this was enough to save the MAOIs. The cavalry had arrived too late. The tricyclics came increasingly to be seen as the standard antidepressants. The conceptual complex of endogenous depression, as advocated by Roth, the Hamilton Rating Scale, and imipramine responsiveness became mainstream. For reasons that cannot be easily explained, it would seem that the reputation of the MAOIs was fatally compromised. Although the older generation of clinicians used to prescribing them continued to do so despite the "evidence" of inefficacy, for younger psychiatrists the mere mention of the term MAOI was enough to trigger thoughts of hypertensive crises and deter them from prescribing in a way that hasn't happened, for instance, following proposals of a link between Prozac and suicide.

The pharmacologists, meanwhile, were left to wonder how best to overcome the biochemical drawbacks of the MAOIs. In 1968 J. P. Johnston discovered that there were two forms of the MAO enzyme, MAO-A and MAO-B.[33] This opened up the possibility of producing drugs that would target only the A- or the B-form and potentially do so in a reversible manner that would avoid the cheese and other interactions. Two drugs were quickly produced, clorgyline, which was an irreversible inhibitor but relatively selective to MAO-A, and pargyline, which was also an irreversible inhibitor but selective to the B-form. Clorgyline seemed more antidepressant, suggesting that MAO-A inhibition was more important. This impression formed the basis for the development of RIMAs—reversible inhibitors of monoamine oxidase-A—so called because companies involved with them want to differentiate themselves as clearly as possible from the negative reputation of the MAOIs.

A number of compounds have since been introduced that inhibit MAO-B, in particular deprenyl. Work on deprenyl has suggested that it may be useful in the treatment of Parkinson's disease and indeed may even be neuroprotective. But the fascinating aspect of deprenyl use is that it is prescribed readily in Parkinson's disease by neurologists, primary care physicians, and even psychiatrists who would have distinct qualms about using an MAOI for depression.

In the early 1970s both Ciba and Geigy had a new antidepressant, maprotiline (Ludiomil), which was a very selective inhibitor of catecholamine reuptake. In the course of testing its reuptake properties and screening for other compounds, Ciba's Peter Baumann and Peter

Waldmeier discovered a compound that inhibited both norepinephrine and serotonin reuptake and in addition was an inhibitor of MAO-A. The compound was a benzofuranylpiperidine.[34] Taking this as a starting point, Raymond Bernasconi produced a series of three hundred analogues of the new compound, all of which inhibited 5HT reuptake and MAO to some extent. Out of this came a group of 5HT reuptake inhibitors, of which CGP 6085 was one of the most potent 5HT reuptake inhibitors ever synthesized—substantially more so than fluoxetine. Another group of compounds in the series inhibited 5HT reuptake weakly but MAO-A potently. Ciba's CNS group recommended one of these—brofaromine—for further development. There was surprise both within the company and outside it. Given the general perception of MAOIs, it didn't seem to make sense to develop another. Waldmeier and colleagues persisted and eventually the drug was put into human volunteer studies in 1977, but with no sense of urgency.[35]

It was 1982 before this part of the development program was completed—an inordinately lengthy period of time. The results, however, were good—the drug was a good inhibitor of MAO-A in humans. It didn't cause a cheese effect. It went into the early clinical trial phase of development (Phase II), but again with seemingly little sense of purpose. It was only in the mid-1980s, when Ciba learned that its neighbor, Roche, was also working on a similar compound, that the pace of development picked up.

Roche's path into the area came through a group of antipsychotics called the benzamides. The central nervous system effects of these drugs were first discovered in the 1960s by Alexandra Delini-Stula working in Budapest.[36] In 1970, on screening tests, Roche found that a benzamide compound belonging to a series that seemed predominantly antiviral in its activity, also had antilipid activity. A number of different series were constructed with the aim of finding a compound that would have a more significant antilipid profile. In the course of this, a compound which was neither antilipid nor antiviral emerged. Because of its benzamide structure it was screened for CNS activity on the basis that it might be an antipsychotic. It was devoid of conventional indicators of antipsychotic activity, but the biochemistry suggested it might be an MAO-A inhibitor. The compound was moclobemide.[37]

The key people in CNS development in Roche at the time were a distinguished group. There was Alfred Pletscher, who had collaborated with Bernard Brodie at Bethesda in demonstrating the paradoxical effects of iproniazid in animals when coadministered with reserpine (results that had so caught the imagination of Kline), Moshe da Prada, a veteran in the field of catecholamine biochemistry, and finally there was Willy Haefely, one of the most gifted behavioral biologists in Europe. All three were internationally known. All three gave their backing to the further development of moclobemide. After all, MAO-inhibition had once been something of a Roche preserve.

In contrast to Ciba, however, Roche, although initially well behind in the race, around 1987 made the decision to go with moclobemide. The company pushed it through a comprehensive clinical trial program, which resulted in its launch in Europe as early as 1989 in Sweden. In the meantime Ciba stepped up its development program, but the date for the expiry of the patent life on the compound was closing in. The only hope seemed to be to find a different basis on which to repatent it. Social phobia seemed like a good bet, and early clinical trial work in this area began to yield very promising results—some of the clinical investigators involved in the trial work felt that the compound was among the most potent in anxiety states they had ever used. Company intentions were frustrated, however, when the possible use of a RIMA for social phobia was mentioned in an independent scientific publication, preventing efforts to save brofaromine by patenting it for social phobia. Ciba's development of the compound came to a full stop.

Roche has since faced considerable difficulties marketing moclobemide owing to a number of factors. One is the lingering MAOI mythology. A second factor is that it is the only company with a compound of this type on the market, and this is a drawback. The example of the SSRIs shows that it often takes the combined inputs of three or four large companies to make a market. Far from competing with each other, if there are several compounds of a particular class the promotional efforts of each company benefit all, it would seem. All the indications, however, are that moclobemide is a good mainstream antidepressant but one that will probably have to be sold for something other than depression if it is to have a significant impact. Its sales potential as a treatment for social phobia, as we shall see in Chapter 6, is one of the most important stories in psychopharmacology—the story of how the driving forces of therapeutic empiricism will usually find a way to surmount blocks imposed by theoretical skepticism.

Lithium and the Natural History of Depression

Lithium, an alkali metal, was first isolated by August Arfwedson in 1817, from stone—hence its name. With the benefit of extraordinary retroscopic vision, references to suggestions in the works of Galen to the beneficial effects in mania of spring waters, which might be expected to have a higher lithium content, can be cited as a forerunner of modern lithium use. Its isolation in pure form, as was the case with many metals in the nineteenth century, led to its inclusion in a number of patent medicine combinations. It was advocated by Alfred Garrod for the treatment of gout in 1859, on the relatively sound basis that it could be shown to dissolve the urates that form in cartilage in gout. However its inclusion in patent preparations led to an increasing awareness of problems associated with its use. It had been noted to increase urine flow, produce a tremor in the hand, upset the stomach, but, more worryingly, lead to cardiac problems. Its use declined and finally problems caused by its inclusion as a substitute for salt as part of a salt-restriction diet for patients with cardiac failure led to its being banned by the FDA in 1949.[38]

In the same year, John Cade began research on lithium in an isolated hospital in Bundoora, near Melbourne.[39] Cade worked from the hypothesis that manic patients overproduce something, depressed patients underproduce it, and therapy involves trying to get a balance. Believing that the substance in question might be excreted in urine, Cade began to inject patients' urine into guinea pigs. The urine of manic patients seemed notably more toxic. It turned out that urea was the lethal substance but that the greater toxicity of manic urine was due to some other factor, which Cade never identified. In an effort to see whether uric acid might be the substance in question, he dissolved it in lithium and found that co-injecting that with urea greatly reduced the toxicity of the urea. He found that lithium on its own produced strikingly tranquilizing effects in the guinea pigs. He tried it out on himself and subsequently on a number of manic, depressed and schizophrenic patients. The manic patients responded.[40] The world paid no attention.

Cade's reports had not gone entirely unnoticed, however. In 1952 Eric Strömgren, professor of psychiatry in Aarhus, Denmark, a figure held in universally high esteem, suggested to a research associate at the hospital, Mogens Schou, that it would be worth exploring the usefulness of lithium. Schou did so, using a placebo, double-blind techniques, and specially devised rating scales. He randomized forty manic patients to either lithium or placebo. This study is a serious candidate for the award of the first RCT in psychiatry.[41] Schou reported in 1954 that lithium could be of benefit for some manic patients. A number of other studies followed by Ronnie Maggs in England,[42] Samuel Gershon at Ypsilanti,[43] who had come there from Melbourne, and Ronald Fieve in Columbia, which all essentially pointed in a similar direction.[44]

Problems were beginning to emerge with the methods of investigating lithium, however. It seemed clear by the late 1950s that chlorpromazine was also a very effective treatment for mania. Was it ethical to do further studies with lithium that would involve randomizing people to lithium or placebo, denying them the benefits of a treatment that clearly worked— even though there was no RCT evidence that chlorpromazine worked for mania? A particular problem with studying mania was that manic episodes are often relatively short lived anyway—much shorter than untreated major depressive episodes. It was one thing to study patients who were hospitalized for virtually chronic mania, but was it possible to prove that lithium further shortened the episodes of individuals who were having remitting episodes anyway? Finally, it seemed clear that people who responded to lithium often relapsed quickly once it was halted—did this mean that there was a withdrawal effect? If it did, then any evidence that there were fewer problems on lithium than off it might be a consequence of withdrawal rather than therapeutic effects.

These issues came to center stage when Mogens Schou went on to suggest that lithium might have a particular role in the prevention of episodes of both mania and depression. This idea had emerged from a number of studies, both those of Fieve and another Dane, Poul Baastrup in Copenhagen, and Toby Hartigan at Canterbury in England. There was some uncertainty as to whether lithium was an antidepressant, in addition to being antimanic, but there was an increasing hunch among clinicians using it that manic patients who responded to it and stayed on it were subsequently having fewer episodes of either mania or depression. Schou went on to suggest on the basis of this that all available treatments for mood disorders were mood stabilizers rather than antidepressants. ECT, he argued, was beneficial for both mania and depression. Lithium seemed to be also. And despite Kuhn's discovery of imipramine being triggered partly by the experience of manic-type reactions to imipramine, Schou noted that a number of studies had suggested that imipramine might be useful for mania.[45]

It was the claim for a possible prophylactic effect of lithium that really led to problems. Michael Shepherd took exception to this claim—was perhaps almost forced into taking exception to it. Shepherd and Schou were speaking on the same platform at a meeting in Göttingen in 1966. Shepherd spoke first on the principles of clinical trials, Schou spoke later on the prophylactic effects of lithium. Schou's claims were based on non-randomized clinical observations. Behind them lay the fact that a number of his relatives were manic-depressive and were at this stage maintained on lithium, and Schou was all but certain that lithium had forestalled a number of family disasters. These latter issues weren't apparent to Shepherd, who was asked by the chairman to comment on Schou's presentation. He acknowledged the importance of the proposed therapeutic breakthrough but suggested that the point had not yet been proved. The audience was split as audiences have been ever since.[46] Schou explained the family circumstances to Shepherd after the meeting and the basis for his conviction—all too well in one sense; Shepherd was left with the impression that Schou was a "believer."[47]

Baastrup and Schou went on to publish a joint study looking at the number of manic-depressive episodes patients had had in the period before going on lithium compared with the period after starting it. They reported a marked fall in the number of episodes.[48] But in order to enter the study, patients had to have had a number of episodes in the previous two years. Baastrup and Schou thus were not, as Shepherd, Blackwell, Lader, and Saran, all from the Maudsley, were to argue, studying manic-depression in its natural state. They were picking patients who had had a particularly bad two years and seeing what happened then—in the ordinary course of events one might expect that a proportion of these would have a better period after their bad spell. And if Schou's study wasn't just misinterpreting the natural course of the illness, where was the evidence that lithium was acting as any more than a psychological crutch, with its availability rather than any actual physiological action contributing to reduced rates of relapse?[49]

The issue became a cause célèbre. For Shepherd the principle of randomization was at stake. Proper randomization would have driven a stake through the heart of insulin coma therapy, despite the "hunches" of the most eminent establishment figures, if only it had been undertaken when the treatment was first introduced. Proper randomization had also appeared to kill off the MAOIs despite the hunches of many eminent figures. Lithium prophylaxis, to Shepherd, seemed like the latest treatment to come in on the back of enthusiastic advocacy rather than cast-iron proof of efficacy. For those outside the Maudsley, the fuss was one more instance of what appeared to be a growing Maudsley hostility to biological treatments, famously caught by Aubrey Lewis's quote that if he had to choose between the new drugs and the new community developments the drugs would have to go. There was also a feeling that the Maudsley was more interested in pure science than in getting patients well.[50]

In scientific terms, the debate became extraordinarily bitter, with Shepherd, for instance, citing the following quote from Asher. "If you admit to yourself that the treatment you are giving is frankly inactive, you will inspire little confidence in your patient . . . and the results of your treatment will be negligible. But if you believe fervently in your treatment, even though controlled tests show that it is useless, then your results are much better, your patients are much better and your income is much better too. I believe this accounts for the remarkable success of some of the less gifted but more credulous members of our profession, and also for the violent dislike of statistics and controlled tests which fashionable and successful doctors commonly display."[51]

He followed this up with: "The problem created by Professor Schou was of his own making: he appears to have believed so firmly in his own judgement as to have concluded that independent assessment was unnecessary . . . the history of physical treatment in psychiatry has unfortunately demonstrated too often the folly of relying on uncontrolled studies alone, however eminent and enthusiastic the clinical observers."[52]

In response to an editorial from Kline in the *American Journal of Psychiatry,* which was run as part of a campaign to get lithium licensed, Barry Blackwell wrote that the editorial "will be highly prized by collectors of original enthusiasms tempered by subsequent experience. Only time will tell whether your author's eulogy will earn him the fate his analogy deserves—to join Cinderella's godmother in the pages of mythology. To transform 'just plain old lithium' into the elixir of life, on the evidence available is an achievement second only to converting a pumpkin into a stagecoach." Kline replied that "Dr Blackwell's delightful letter reads as though it were written by one of Cinderella's spiteful sisters . . . There is even a feeling of faint personal familiarity with Dr Blackwell's warning which reminds me of some of the caveats concerning the introduction of both antipsychotic and antidepressant agents. Of course it may also be that Dr. Blackwell is also not convinced that any of these drugs have been demonstrated to be of any use."[53]

Paradoxically, these disputes perhaps did more than anything else could to bring attention to the question of lithium. It needed attention drawn to it, if it was to survive, owing to its uneven history and to its lack of company support. No company could get a patent on it, and accordingly there was little money to be made on it. This was a particular problem in America, where unlike in Europe, lithium was still officially unavailable. Following its use by Gershon and Gerard at Ypsilanti State Hospital in Michigan, Rowell Laboratories, a small company from Minnesota, became interested in producing the compound made up for investigators and soon, while the FDA was not prepared to license it formally, it found that it was giving investigator licenses (INDs) to an ever larger number of clinicians keen to use it.[54] Pressure began to build for its registration.

While this happened, the war between the Maudsley and what almost seemed at one point to be the rest of the psychopharmacology confraternity continued. One side plotted on a beach in the Caribbean.[55] In the early 1960s, Nate Kline had successfully treated a Mrs. Denghausen for depression. The grateful, and wealthy, Mrs. Denghausen was keen to make a contribution to scientific research and was persuaded to fund a study group that would meet once a year on a Caribbean island to have a brain-storming session. The group was small and selected by Kline. It included Arvid Carlsson, Heinz Lehmann, Jules Angst, Linford Rees, Merton Sandler, and others in addition to Alec Coppen and Mogens Schou. Coppen's interest in lithium lay in the fact that he and David

Shaw had shown that there were quite dramatic salt and water changes that took place in the bodies of patients swinging into either depression or mania—changes that to this day have not been explained. Schou was obviously also interested.

There were two issues to be tackled. One was the question of the natural history of manic-depression and the other was the issue of whether a trial of the sort that Shepherd was advocating could be put together. The issue of the natural history was picked up by Angst, who studied the course of manic-depressive illness presenting to the services in Zurich and was able to demonstrate that the natural course of the disorder was such that the illness was made of cycles with the duration of the cycle—the time between the onset of one episode and the onset of the next—decreasing with age and the number of previous episodes. This finding supported the assumptions Baastrup and Schou had made about the course of the illness.[56] Angst went on to participate in a study using the same mirror-image design that Baastrup and Schou had used and got the same results that they had.[57] Baastrup and Schou ran a discontinuation study, randomizing patients who were better to either lithium or a placebo, and found that those remaining on lithium did much better than those switched to a placebo.

Shepherd and his colleagues remained unimpressed. Until patients were entered randomly into a placebo-controlled, double-blind study, there could be no proof that lithium did what was being claimed for it despite Angst's studies on the natural history of bipolar disorders. The findings from the discontinuation study might simply be caused by lithium withdrawal. For supporters of lithium there were problems with randomization. It was one thing to use a placebo in an acute study, but to continue giving it for several years, when there was strong suspicions that the alternative treatment was effective, seemed ill-advised. Heinz Lehmann, for instance, broke the blind during a Canadian study with just this design, following a number of serious suicide attempts, and found that all those involved were on a placebo. While the debate was taking place publicly in print, however, it didn't happen in person. A number of symposia were set up aimed at airing both points of view, but none actually happened. When it came to a crunch one or other of the key combatants would cancel.

The Denghausen group decided to design a four-hospital prospective study in the London area, including the Maudsley group and thus involving Shepherd, Edward Hare, and others. Ronnie Maggs and Alec Coppen hammered out a protocol acceptable to the Maudsley group in a series of meetings. The design randomized patients to lithium or a placebo for two years. The issue of suicide risk was managed by letting the clinicians involved in the care of the patient give whatever other medication or ECT they thought fit to give during the course of the study. Outcomes would be based on the relative frequency of admission to the hospital from the lithium and the placebo groups, as well as on the amount of extra medication each group had needed. The outcomes were conclusive. Those getting lithium, whether they were unipolar or bipolar, had many fewer episodes and much less other medication over the two years than those on the placebo. The Maudsley group analyzed the results independently and found nothing to argue with. The results were reported in 1971.[58]

In the meantime Paul Blatchley, Nate Kline, and others had begun to work on public opinion in America through television and radio to generate interest in the compound. Kline pushed the American College of Neuropsychopharmacology to consider applying for a license for it, in default of a company application—an unheard-of move. Finally Smith Kline & French, in part, it is said, because of their acknowledged lead role in psychopharmacology in the United States following the introduction of chlorpromazine, submitted an application for registration to the FDA, as did Pfiser, and licenses were granted.

The results of the various studies that came out of the Denghausen group were conclusive, however, in ways other than had seemed immediately apparent in the first instance. The one uncontested finding was that the outcome of depressive disorders was much worse than had previously been thought. A great number of patients relapsed within a year or two. Coppen went on to establish a lithium clinic and tracked the outcomes over twenty years. He found that in those who took lithium consistently, the suicide rate dropped from seven per thousand to less that one per thousand. This was an outcome measure of the kind that Nate Kline believed in—one that didn't involve putting the rabbit into the hat in the first place. It became increasingly clear that if a patient had experienced previous episodes then further episodes were the norm.[59]

But if this was the case, might not other antidepressants equally prevent against relapse? One of the lithium studies, run by Bob Prien, had included an imipramine comparison group in addition to placebo and had found that both imipramine and lithium reduced rates of relapse. This was relevant because Shepherd and Blackwell's argument proposed that there was no evidence that lithium was uniquely prophylactic. While Coppen's study was running, Shepherd was running another in which patients who had responded to amitriptyline or imipramine were randomized after response to continuation treatment with either the active drug or the placebo. Those who stayed on the active drug had a much lower rate of relapse.[60] A later, even larger study compared amitriptyline, lithium, and placebo and found that both active drugs significantly reduced the rate of relapse; lithium demonstrated no apparent superiority over amitriptyline.[61] Shepherd used such findings to justify his original claims that in fact there was nothing specifically prophylactic about lithium.

A number of other studies were also run but it must be said at something of a cost in terms of higher suicide rates among those taking a placebo. Is there something specifically prophylactic about lithium? To this day opinion remains divided, even bitterly divided.[62] Its nature as a natural element suggests that it might act in a manner quite different from that of other agents. But against this consider the case of G32883, another Geigy tricylic compound, synthesized by Schindler who had made imipramine. This had been made from another tricyclic G26301, which had been made in 1953 and had been found in clinical testing to be anticonvulsant. G32883 was later developed as carbamazepine, which in the 1980s became the most widely used anticonvulsant. In the course of its use, the conviction grew among many that it had mood-stabilizing or prophylactic properties that were rather like lithium's. This has led to its use as a mood stabilizer in patients who prove intolerant to or unresponsive to lithium—but its structure is that of a classic tricyclic.[63]

Whatever the verdict, the result of all this was to put the question of prevention on the map. It began to seem as though the initial successes with antidepressants during the 1950s and 1960s had been misleading to some extent, in that for a time clinicians had gotten the impression that bringing about an initial recovery from a depressive episode was all there was to treating depression—whereas this was not the case. Against this background it is worth remembering that Kuhn in 1958 had clearly stated that treatment with imipramine was often only symptomatic and needed in some cases to be continued for months or even years.

These findings have led to efforts in recent years to reconceptualize depressive episodes. At present distinctions are drawn between responses leading to remission and responses leading to recovery as well as between relapses and recurrences. The initial response to imipramine for instance can be seen as inducing a remission. If the person becomes depressed again on discontinuing treatment, what has happened?—it could be that the initial episode had never cleared up and discontinuing treatment reveals it or, alternatively, perhaps it had resolved and this is a new episode.[64]

Distinctions such as these have led to recommendations that antidepressant treatment should ordinarily continue for at least four months after apparent recovery. Even then, in the case of severe episodes of depression, it is clear that recurrences of depression are common—up to 80 percent can be expected to relapse in two years. A particularly influential study, run from Pittsburgh by David Kupfer and colleagues, has shown that active treatment maintained for up to five years can significantly reduce this rate of relapse. It also demonstrated that a psychotherapeutic procedure, in this case interpersonal psychotherapy, can reduce the rates of relapse.[65]

This raises a number of issues. Not many people are happy to either take or prescribe antidepressant drugs for such periods of time. The same concerns do not arise to the same extent with the use of aspirin prophylactically to reduce the risk of heart attacks and stroke or the use of antihypertensives or cholesterol-lowering agents. Chronic treatment with these agents seems broadly acceptable in a way that the ongoing use of antidepressants is not. The behavior of prescribers and patients regarding recurrent depressions often seems to come closer to views about the ongoing use of minor tranquilizers for chronic anxiety disorders—something that should not be encouraged (see Chapter 7).

The Standardization of Drug Development

Chapter 1 introduced the thalidomide case and the 1962 amendments to the Food, Drug and Cosmetics Act. Although there have been no further amendments to the act, there have been other regulatory developments. Between 1975 and 1977 the safety profile of Aldactone, a Searle product, was investigated by the Senate subcommittee on health. A report had indicated the possibility of an increased frequency of liver and testicular tumors with the drug. Although FDA regulations required that alarming findings be submitted to the agency promptly, this had not been done. There also appeared to have been some doctoring of the original studies submitted to the agency. The requirements were that all rats given the drug in long-term studies should have been examined microscopically at post mortem. This had not happened, however, and in particular did not appear to have happened in the case of some rats that had lesions large enough to be visible to the naked eye.

Concerns were also raised about the safety testing of two of Searle's other products, Flagyl and the sweetener Aspartame. In cases where several pathology reports had been completed on a particular tissue, for instance, Searle, it appeared, had submitted the most favorable one. When attempts were made to establish who in the company had investigated what, it was found that many of the findings were undated and unsigned and therefore it was not possible to establish exactly what had happened. There was concern on the part of the FDA that "tissue masses come and go and animals die more than once."[66]

While those of a conspiratorial bent will be concerned about this, it should be pointed out that little of this was out of line with general laboratory practice at the time. The FDA's concern was as much with a cumulative pattern as it was with any particular error. The pattern was grounds for prosecution, the FDA thought. The Justice Department didn't agree. Its view was that no one could be prosecuted for the pattern of practice that had been unearthed. The interests of justice would not be served by a handful of junior staff being convicted of crimes regarding which there could be no proof of intent on the part of senior management. The Justice Department view won out. The case led to the FDA's drawing up a detailed code of Good Laboratory Practice (GLP).

At the same time the FDA ran a survey of how clinical investigators were complying with the 1962 amendments and found that 74 percent of 155 clinicians surveyed had failed to comply with one or more of the requirements of the law. Problems encountered included not getting or failing to register informed consent from trial subjects, failing to keep accurate records of the amount of drugs received and distributed, failure of adherence to study protocols, and failure to keep adequate records on patients, so that subsequent scrutinizers could satisfy themselves that the patients had actually existed and that they had the illness they were supposed to have and had it to a degree of severity that would warrant their inclusion in the study.[67]

The Senate subcommittee pursued these issues as well and found evidence that clinical investigators might even not be qualified, that data collected in one trial was being inserted as part of the report of another, that there were fictitious subjects, that drug doses on occasion exceeded the protocol stipulations, that some subjects were being given two or more investigational drugs at the same time, and a range of other violations of what might be deemed to be good clinical practice.

The practices involved were later the subject of two celebrated legal cases. The first, against Upjohn, was filed by Ilo Grundberg, who claimed that her responsibility for killing her mother was diminished by virtue of her having been on Upjohn's hypnotic Halcion.[68] This ran in the media as a "Halcion Causes Murder" story, notwithstanding the fact up to 11 million prescriptions for the drug were being filled each year at the time. In the course of Grundberg's trial and as a consequence of surrounding media investigations, a number of what appeared to be worrying facts emerged.

One of the key study protocols, protocol 321, which Upjohn had used in its application to have Halcion licensed, contained what the company termed transcription errors, the effect of which was to minimize the apparent frequency of serious side effects experienced by takers of the drug. Another protocol, 6415, was a study carried out by a Dr. Samuel Furst, who had committed suicide in 1985 while under investigation for fraudulent activity (unrelated to the Halcion studies). Furst, it would seem, had filed case reports on Halcion in his own handwriting, signed consent forms himself, and had included patients in the study who could not be traced. As if all this wasn't bad enough, the full details of the Furst story emerged in the midst of the production of a BBC television program on Halcion. Upjohn settled out of court with Grundberg; the charge was dropped.

Although all of this emerged long after the Senate subcommittee hearings on Aldactone, these studies along with those for Halcion had all been conducted during the mid-1970s, and represent some indication of what many investigators and other interested parties were concerned about. A number of points are in order. One is that the sloppy practices involved were common to both laboratory scientists and clinicians, both those working within the industry and those not. Although some clinical investigators may have engaged in studies for the purposes of personal gain, many others were university employees, a substantial proportion of whom would have had to donate any payments from the study into research funds. Protocol 321 was one of over ninety studies that Upjohn had run with Halcion, the majority of which appeared to be reasonable projects.

Such practices are probably as common among both clinical and basic sciences investigators, engaged in studies that have nothing to do with the pharmaceutical industry. Doctoring the data is not something restricted to doctors. The invention of data is commonplace in all branches of science, as is the omission of inconvenient data, selective reporting of facts, and profound biases in interpretation, which after the event may appear inexplicable other than by intent to deceive. To expect that industrial drug development would be free from such difficulties is perhaps somewhat naive.

For example, one of the most prominent critics of Halcion from academia was Anthony Kales, who even before the difficulties in the late 1980s had reported that it caused rebound anxiety and rebound insomnia. Between 1974 and 1985 Kales and his colleagues produced thirteen publications, each apparently outlining hazards associated with the use of Halcion. Closer reading of the studies, it has been argued, however, reveals that all thirteen papers depend critically on data drawn from just one investigation, an initial one done on seven patients for Upjohn. The later reports include details of new methods and findings that must be of uncertain value, in that the ratings concerned were not included in the original protocol. The findings were also compared with controls drawn from entirely different studies.[69] If this is all true, then by changing the number of subjects apparently being reported on, changing data analyses, drawing different conclusions from essentially the same data set and other methods, Kales and his colleagues have arguably created an impression in a manner every bit as dubious as any investigators involved with either Searle or Upjohn.

The question, then, is not so much was there demonstrable error or even fraud in the Upjohn studies, as whether the level of error was of a different order than that found in other studies at the time, whether related to regulatory submissions or not. In 1994, in the second legal action, then the longest libel action in British legal history, the British High Court ruled against another critic of Upjohn, Ian Oswald and the British Broadcasting Corporation. The court decided that by the standards of the time the Upjohn studies were unremarkable.

Unremarkable though they might have been, the FDA decided that such practices were not good enough and just as the Searle events led to the establishment of GLP, so also the Senate subcommittee investigations led to the creation of Good Clinical Practice (GCP). Trialists undertaking to do a study are now warned, wherever in the world they are based, that FDA representatives may appear at any time during the course of the study and inspect the study documents as well as the clinical notes and otherwise seek to establish that the patients actually exist and have the disorders it is claimed they have.

In addition to the introduction of GCP and GLP, the early 1980s, as we saw in Chapter 3, witnessed a renewed insistence on the necessity of placebo-controlled as opposed to active controlled studies. The effect of all these changes had a dramatic impact on the process of drug development. In 1958, Lundbeck were able to release chlorprothixene for clinical use three months after it had first been tested in the clinic. In the early 1970s, antidepressants took two to three years to move from the laboratory into clinical use. In contrast, Prozac, for instance, was first investigated in the laboratory in 1974 but only released for use clinically in 1987. All of this was happening at a time when most companies had called in business consultants to advise on how drug production might be standardized. It was a time when medical and scientific staff at the managerial level were being replaced by business managers, who on some occasions interpreted the "worthlessness" that had been visited on studies by regulatory changes in terms of an incompetence in their clinical trials division. Morale suffered.

Morale continued to suffer through the 1980s as the processes by which drugs had been developed were reengineered. In the 1960s and 1970s screening was done using animal models and the insights of behavioral pharmacology. The originator of this was David Macht at Johns Hopkins University, who in 1921 proposed that "the effect of drugs on psychological functions and animal behavior has been the subject of remarkably little investigation on the part of either physiologists or phar-

macologists. With the exception of two substances—alcohol and caffeine—the contributions to this subject have until very recently been few
and meager, so the field of what we may term 'psychopharmacology' is
virgin soil, full of possibilities."[70]

With his associates, Macht developed two conditioned behavior
tasks—coordinated running across a rope and the monitoring of the behavior of rats in a circular maze, which was a task first developed by John
Watson, the creator of behaviorism. A variation of the rope test, in which
animals were trained to climb a rope to a platform for food, became famous in the 1950s when it proved particularly sensitive to the effects of
chlorpromazine. The rope test became the first important screening test
for the development of psychotropic agents.[71]

The rise of B. F. Skinner's operant behaviorism also provided a framework for drug and behavior studies. Skinner himself looked at the effects
of caffeine and amphetamine on conditioned responding. But the experiment, which laid the basis for future investigations, was developed by Bill
Estes. This was the conditioned emotional response (CER), which was a
test that looked at the capacity of a stimulus to suppress a rat's behavior
when paired with a footshock. Joseph Brady, working in Chicago, found
that electroshock attenuated the CER but not as one might have expected, by simply inducing amnesia. ECT seemed to reverse an emotional
disturbance.[72]

As was true for many psychiatrists of the time, the Korean War made a
research stint in the Public Health Service seem more attractive to a number of experimental psychologists than an outward-bound stint in the Far
East. This led to the development of behavioral pharmacology and the establishment of a core group of interested investigators at the Walter Reed
Hospital in Maryland. When reserpine was shown in the mid-1950s to at-
tenuate the CER, it was thought that this test might be of use in screening
for new compounds. Faced with the problem of giving away the monies
that followed Kline's lobbying of Congress in 1955, Jonathan Cole gave
substantial grants to a Behavioral Pharmacology Program at the University of Maryland and to a similar program at the University of London,
involving Hannah Steinberg and Roger Russell.[73] These and other groups
rapidly developed useful animal models for screening for drug effects at a
time when not enough was known about the biochemistry of the brain to
permit biochemical assays to be used for drug detection. The industry,
therefore, turned to the behavioral pharmacologists.

This collaboration led to the elaboration of a great number of tests of
behavior and to the selection of drugs on the basis of their performance
across a range of tests. In the course of this work, phenomena such as
learned helplessness were discovered. Learned helplessness was based on
the CER. It involved exposing a pair of animals to a shock, in conditions

where one of them could escape but the other couldn't. The fact of being unable to escape seemed to produce profound changes in behavior. Learned helplessness generated widespread interest for a number of reasons. It marked the entry of cognitive psychology into the realm of animal behavior, which had previously been dominated by a strict behaviorism. It also appeared to offer a good model of human depression. In addition, the behavior of helpless animals was shown to be sensitive to the effects of antidepressants. Such models were advocated on the grounds that understanding the disease process that antidepressants might modify seemed a good idea: on this basis the process of theoretical development within psychopathology was temporarily harnessed to the engine of drug development.

This was a policy that the pharmaceutical industry adopted during the 1970s and 1980s but that it now seems to have abandoned.[74] Developments in the field of molecular biology have made it possible to target specific receptors or proteins. This is sometimes popularly portrayed as a move toward rational drug development, toward a world of designer drugs. It is certainly a way to standardize drug development; whether it is rational would seem to depend on how much we understand of the disease process. If anything the efficacy of psychotropic compounds stemming from this process seems to be falling off. The significance of this will be chased further in the next chapter.

THE COLE EFFECT

In addition to problems of development, two other problems hang over every drug after it is launched. One is the "irrationality" of the marketplace (the Cole Effect) and the other is the stain of thalidomide, with its implication of cover-ups. The trouble with drug-induced problems is that when reports of side effects emerge it may be as difficult for a company to decide that its drug is actually causing the problem as it was for Geigy to decide that imipramine was an antidepressant. New concepts may be needed to accommodate the observations. The reports may initially come from clinicians as obscure as Roland Kuhn once was, whose capacity for clinical observation may be accordingly downplayed.[75]

The drug may have been co-prescribed with something else and may only produce problems in those circumstances, or the patient may be taking an over-the-counter preparation that he or she has not brought to the attention of the physician. Alternatively, factors specific to a particular locality, such as diet or climate, may play a part. In the case of the contraceptive pill, there were reports of blood clots in association with usage, but the reports were from women over the age of thirty-five in a part of the world where women were often obese and tended to smoke cigarettes, both of which increase risk. Another possibility is that clinicians may be prescribing the drug in doses greatly in excess of the original recommendations. In such cases the intentions of prescribers may be good, although one might criticize their therapeutic heroism as naive, but should the manufacturer or the regulatory authorities be held to blame for any problems that result?

There may be a failure to register side effects owing to a belief that they could not happen. At the time of the thalidomide affair, many believed that the fetus could not be affected by drugs. In the case of diethylstilbestrol (DES), the effects emerged twenty years afterward, which was completely unexpected. A further problem is that the side effects of a drug may occur naturally or as part of the disorder being treated. Many drugs cause nausea or dizziness or headaches, but so too do many illnesses. Even the phocomelia brought about by thalidomide occurs naturally, as do dyskinesias of the type that are now called tardive dyskinesia and attributed specifically to antipsychotics. Suicide is not uncommon in people who are depressed, even in those not taking fluoxetine. In addition, given that some people develop the problem and others do not, there is the question of constitutional vulnerability.

In the case of phocomelia and thalidomide or tardive dyskinesia and the antipsychotics, the difference is a matter of differences in rate. Even accepting that a particular drug has caused a particular problem in a particular patient, when does one decide that a drug is causing the problem frequently enough that warnings should be issued regarding its use? What happens if a drug, such as fluoxetine, makes depressed people less likely to commit suicide but is suspected of causing some people who might otherwise not have committed suicide to be more likely to do so?[76] And if that question can be answered, there is a further issue to do with the mechanism that leads to suicide. Fluoxetine, for instance, might lead inevitably to suicide in some cases, so that persons, once they have had the drug, if they are liable to this side effect, will kill themselves. Alternatively, it may make people more aroused, more "wired," and this may lead to their feeling suicidal, in which case the issue becomes one of whether this is a side effect that can be managed with appropriate warning and monitoring of the patient.

If the decision to remove a drug is made, should the company be left to remove the compound from the market? It is the company, after all, which has its reputation to preserve, which is an important determinant of future business success. It is unclear whether companies can be relied on to take a compound off the market as early as might be thought necessary by outside observers, but consider the following case. In 1962 Wander, a German company, produced a novel antipsychotic, clozapine. This was put through a clinical trials program. A number of the trialists were convinced that clozapine was an exceptional compound, in that it appeared to produce an antipsychotic effect without triggering the extra-pyramidal side effects that were associated with so many other antipsychotics. Early trials, however, suggested that in terms of therapeutic efficacy it was not much different from other available compounds. In 1975 reports emerged from Finland of a series of deaths in subjects who had been taking clozapine. Wander had by this stage been taken over by Sandoz, who sent an investigator. The deaths were linked to an agranulocytosis—the loss of white blood cells, which would leave an individual prone to a lethal infection. Sandoz wished to withdraw the drug and was only persuaded, by the insistence of a number of the most eminent psychopharmacologists in the field, to permit its continued use for a number of patients for whom there was clear evidence that they would not do as well on other neuroleptics.[77]

Several years later, in 1988, Herbert Meltzer, John Kane, and others persuaded Sandoz to seek a license for the compound in the United States for patients with treatment-resistant schizophrenia. This led to Study 30, comparing clozapine with chlorpromazine in groups of treatment-resistant patients. The complex nature of the findings from this trial have been debated elsewhere, but in brief it became clear that clozapine produced significant benefits. This led to its licensing in the United States provided that adequate precautions were taken to safeguard takers against the possibility of death. Subjects taking clozapine therefore must now have regular blood tests to monitor their white cell counts—a fact that may contribute to the good responses it produces. Even though its use is supposed to be restricted to treatment resistant populations, which it was initially supposed would be a relatively small group (and hence its price was set high), clozapine is now widely used. It is one of the greatest profit makers among psychotropic drugs and, more important, is generally thought of as having redirected the course of antipsychotic drug development.[78]

In contrast to clozapine, consider the case of mianserin. Mianserin was an antidepressant introduced by Organon in Europe in the 1970s. It was significantly different from existing compounds in that it inhibited neither norepinephrine nor 5HT reuptake. Evidence that it was an antidepressant in fact led to a significant revision in theories about the mode of action of antidepressant drugs. It appeared significantly safer in overdose than older drugs, which led to its widespread use, so that by the mid-1980s, it had become one of the two most commonly prescribed antidepressants in Britain. Reports appeared, however, linking it to lowered white cell counts in older subjects—raising the specter of a clozapine-like problem. Warnings to this effect were issued to clinicians in many European countries (mianserin had not been licensed in the United States). There was pressure to have the drug removed from the market in the United Kingdom, even though there was no conclusive evidence that mianserin brought about more white cell problems than did other antidepressants available at the time.

In contrast to Wander, Organon chose to contest the issue on the grounds that problems stemming from one side effect needed to be weighed in a balance containing all effects of the drug and weighed against a similar complex of effects for other compounds. The question of death from overdose with antidepressants had at the time become topical. In 1987 John Henry from the National Poisons Institute in London had introduced the idea that a fatal toxicity index (FTI) could be constructed by assessing the number of fatalities from overdosing with particular compounds per million prescriptions. When this figure for each antidepressant was added to the number of unexplained deaths associated with each million prescriptions for that antidepressant, mianserin came out rather well. Arguments like this played some part in persuading the British Committee on Safety of Medicines to allow the compound to remain on the market.[79] But despite the "free publicity" regarding its safety in overdose and beneficial side-effect profile, mianserin has all but vanished.

How rational is the marketplace—even the restricted market of pre-scribers? The usage of psychiatric drugs and problems with their usage play in the media, and the consequences, as Jonathan Cole has pointed out, are almost entirely unpredictable. Reports of a widespread precipitation of suicidality or homicidality—real or imagined—may have little impact on the sales of one compound, while rare reports that another may have precipitated epileptic fits in occasional cases may kill that compound off.[80]

Finally, consider the fortunes of Astra. In the 1980s, of all the companies in the psychotropic field, Astra had a reputation second to none in terms of bringing genuine therapeutic developments to the marketplace. The company was critically involved in the development of 5HT reuptake inhibitors and brought the first one, zimelidine, into clinical use, five years before Prozac. Within a few months of its launch, however, zimelidine use was found to be associated with the development of Guillain-Barré syndrome, a polyneuritic condition of unknown origin that can lead to profound muscular weakness and even death. Zimelidine had to be withdrawn from the market.

Astra's next foray into the psychotropic market was with remoxipride, an antipsychotic. Before clozapine, the dominant theory of antipsychotic action was that these drugs worked through the D-2 receptor. But none of the compounds that were in use was particularly selective for this receptor—they all blocked a range of other receptors as well. Remoxipride, when it was launched in 1992, was by far the most selective compound available, which according to the dominant views of the time held out the promise of greater efficacy and greater freedom from side effects. But within months of its release, there were reports of aplastic anemia in patients and remoxipride too had to be withdrawn. Whereas in the 1960s a company could quickly substitute another drug for a compound that had to be withdrawn, the process of drug development is now such that this is no longer possible. At the very least a new clinical trial program requiring up to five years has to be initiated. The loss of two flagship compounds in the manner of zimelidine and remoxipride, therefore, brings home the fact that substantial financial risks are associated with drug development and that companies have less room for maneuver than they might once have had. All of which is highly germane to the fate of Ciba-Geigy.

THE DEMISE OF CIBA-GEIGY

During the 1960s, following the introduction of imipramine and the subsequent introduction of amitriptyline, while the MAOIs were self-destructing, further tricylics came on stream—desipramine, nortriptyline, dothiepin, doxepin, protriptyline, and butriptyline. Whereas the methods of drug development for an MAOI were relatively straightforward—look for an increase in brain norepinephrine and serotonin levels—biochemical methods to detect other compounds that might be antidepressants were nonexistent. It was only with Fridolin Sulser's Beta adrenoreceptor down-regulation test, in the late 1970s, that a reference method became available.

For the most part detection involved a range of animal screening tests, starting with efforts to find out whether the compound blocked the effects of reserpine and moving on to more complex tests of behavior. The compounds that were screened were largely selected on the basis of structural similarity to known antidepressants. In the mid-1960s, the biochemistry of the brain was still essentially unknown and for all investigators knew the question of a compound's actual physical shape might be the critical determinant of its effects. Clozapine had been discovered by Hanns Hippius on the basis of a consideration of its structure.[81]

To many, this appeared to point the way forward. The classical neuroleptics and tricyclics had consistent differences in structure, which became more apparent with 3-D modeling—as opposed to on paper, where the formulae for the compounds imipramine and chlorpromazine, for example, looked all but identical. Considerations such as these led Max Wilhelm and Paul Schmidt at Ciba, liaising with Roland Kuhn, to the development in 1967 of maprotiline, a new nontricyclic compound.[82] It was one of a series of compounds, of which the structural formulae of the day suggested some should be more neuroleptic and others antidepressant—

and this was borne out. Curiously, Geigy developed the same compound at much the same time—who got there first is now unclear. What is clear is that Ciba filed for a patent first and shortly thereafter, in 1972, Geigy merged with Ciba.

The drug was given to Roland Kuhn to study, and he confirmed that in his opinion it was an antidepressant. He used the methods he believed appropriate for the task—studying patients intensively on a single-case basis, giving the drug for at least three months ("observations of less than three months are so subject to misinterpretation as to be valueless"),[83] giving it in both inpatient and outpatient settings, and vigilantly attempting to discriminate between antidepressant effects and the possible confounds of sedative, anxiolytic, euphoriant, or neuroleptic effects. On the basis of observations on 320 patients, he concluded that maprotiline was a "good specific antidepressant."[84]

Maprotiline was launched in 1972. At a scientific meeting organized to mark the launch, the talk immediately after Kuhn's was given by Dr. J. Welner from Copenhagen, who presented findings from one of the first multinational, multicenter studies that have since become a feature of drug development.[85] Maprotiline was compared with amitripytline but not with a placebo. Two hundred and eleven patients were included and no difference was found between the two drugs, although there were differences, it seemed, between Dutch and Scandinavian depressives, as assessed with the Hamilton scale. Maprotiline was launched and became the best-selling antidepressant in a number of European countries and in Japan.

Following the launch of maprotiline, Ciba-Geigy set about the next phase of drug development. As mentioned, it had one of the first 5HT reuptake inhibitors in CGP 6085 and the first RIMA, brofaromine, which it took forward for development, albeit at a leisurely pace, in part because another line of development had opened up that caught the imagination of many within the company. Maprotiline had a derivative that proved to have antidepressant properties—oxaprotiline. Oxaprotiline came in two forms, dextraprotiline and levoprotiline. By this time biochemical rather than structural considerations had begun to take over as the main predictor of antidepressant activity. As it transpired maprotiline was a selective inhibitor of catecholamine reuptake, which seemed to fit the catecholamine hypothesis (see Chapter 5). So too was dextraprotiline, but levoprotiline, while in other respects similar, was not a catecholamine reuptake inhibitor. This opened up the possibility of testing the catecholamine hypothesis by running dextraprotiline and levoprotiline against each other in the same trial.

It proved impossible to run the comparison between the two compounds owing to a lack of toxicology data on dextraprotiline, but it was possible to put levoprotiline into early open studies in 1982. Initial reports appeared to be encouraging. The drug was given to Kuhn. Kuhn studied levoprotiline in his usual manner—giving it to large numbers of patients with vital depression and observing closely what happened over a period of time. He came to the conclusion that the new compound was an antidepressant. A number of small- and medium-sized studies followed that appeared to confirm his impression, but these were not placebo controlled. The clinical trials division in Ciba became increasingly excited. A placebo-controlled study was set up. The results, to the disbelief of many in the company, were negative.

Ciba had to decide whether to cut off development. Kuhn wrote some strongly worded letters to the management berating clinical trials methodology. Even after the drug was shelved, he insisted on referring to it as a definite antidepressant. Yet by the early 1980s it had become clear that the opinions of an expert, even one as distinguished as Kuhn, were not going to count with the FDA and neither were studies that demonstrated no difference between the trial compound and a known standard. Despite Kuhn's protests, Ciba-Geigy shelved the development of levoprotiline in November of 1990.[86]

The times were changing. In the mid-1970s, Ciba-Geigy, the company that had introduced reserpine, imipramine, desipramine, clomipramine, carbamazepine, methylphenidate, and maprotiline, the company that had effectively dominated the psychotropic market, looked set to continue that domination. It had up to twenty central nervous system products in development, including one of the first 5HT reuptake inhibitors, the first RIMA, and the first combined selective norepinephrine and serotonin reuptake inhibitor (SNRI); and it had an atypical neuroleptic savoxepine that appeared free of many of the subjectively distressing side effects of other neuroleptics and, post-clozapine, now looks as though it might have been a genuine innovation. But none of these drugs ever got to the market. Ciba-Geigy never produced another psychotropic drug. After levoprotiline and brofaromine, it effectively withdrew from the psychiatric market.[87] In March of 1996 Ciba-Geigy merged with Sandoz in a new company called Novartis.

How do we explain Ciba's demise as the market leader in psychiatry? Is its fall from preeminence simply to be explained in terms of the internal history of the company or are there larger factors that played a part in its decline? It is common to hear figures cited indicating that as few as 5–10 percent of drugs introduced offer genuine therapeutic gains and that as few as 1 percent provide important therapeutic gains. Drug development has been criticized as too often producing compounds that involve a minor modification of an existing compound, aimed at getting around patent laws. This is difficult to criticize from a theoretical point of view, when a minor manipulation turned the mild sedative promazine into chlorpromazine, or imipramine into clomipramine, or alternatively desipramine into lofepramine, which at present appears to be the safest antidepressant on the market. It is also difficult to criticize from a business point of view when minor differences, such as those between Zantac and Tagamet, can lead to a company's having larger sales than ever before recorded by any drug company. If it were all this easy to produce a drug, it would be very difficult to explain why Ciba-Geigy failed, short of a catastrophic loss of direction on the part of the company.

Has increasing regulation and bureaucracy finally killed or seriously harmed the golden goose so that few if any more eggs can be expected? The argument from regulators is that we have moved from an era when we built biplanes or even Mustangs that could be mass-produced to an era of jumbo jets and Stealth Bombers. These take more time and money, but one can wipe out a squadron of Mustangs. Fewer drugs may be produced but the proportion offering genuine advances is likely to rise from a figure of less than 10 percent. We should be cautious about complaints about red tape, regulators argue, which may say more about the mid-life crises of laboratory chiefs or clinicians than about the true state of drug development.

However, the idea that we are now producing jumbo jets, where once we just produced Mustangs, is one that most clinicians and basic scientists within psychopharmacology would probably find hard to accept. The newer compounds have, if anything, been less effective than their older counterparts. For the theory-oriented psychopharmacologist, the future looks bleak in that under the influence of a relentless empiricism, with the move away from behavioral biology, all traces of theoretical developments within psychopharmacology appear to have been abandoned. For the empirics, the outlook looks almost as bleak in that the current situation in psychopharmacology is much the same as that facing geologists hunting for oil—theory is one thing but drilling as many oil wells as possible is still the best way to find oil. Getting as many drugs onto the market as possible will remain important until such time as our understanding of what is happening in the brains of those with mental illnesses permits us to design agents that will rationally tackle the disturbances that constitute depression.[88] At present, however, the regulatory apparatus appears to be all but choking off this avenue of development.

A great deal rides, therefore, on where we have got to in our understanding of the pathophysiology of depression. Part of the mythology of Prozac is that it came about by a process of rational design. Did it?

5   A Pleasing Look of Truth

The story this chapter has to tell is as much about artistic verisimilitude as it is about scientific exactitude. Although clearly there has been some scientific achievement, readers who consider current theories about what antidepressants do may by the end feel like echoing Jerome Gaub's opinion of Leibniz' views on the relation of the mind to the body—it is a "fable whose novelty has recommended it, whose recommendation has spread it, whose spread has polished it, refined and adorned it with . . . a pleasing look of truth."[1]

The story of the discoveries that came about in the search for the mechanism of action of antidepressants and neuroleptics has also been the story of the rise of biological language in psychiatry. In the Netherlands, this rise led students to picket the early lectures on biological psychiatry of Herman van Praag and even led to death threats against both van Praag and his family.[2] Given that student anger with van Praag was unlikely to have stemmed from a feeling that he was settling for a less accurate measurement of neurotransmitters than he should, what was the fuss about?[3]

When the neuroleptics and later the antidepressants were first introduced, little or nothing was known about brain chemistry. Indeed the whole idea that there might be a brain chemistry was repugnant to many. During the 1930s and 1940s, as preliminary indications that biochemicals might be involved in neurotransmission began to emerge, orthodox opinion remained convinced that electrical transmission was the medium of brain functioning and that chemical events, if they took place in the central nervous system, were at most of marginal importance. This debate took place most keenly in Britain, where there was a long neurophysiological tradition dating back to William Cullen and Robert Whytt in the latter part of the eighteenth century.[4] In the first half of the twentieth century, studies of the nervous system were dominated by physiologists, such as Edgar Adrian at Cambridge and John Eccles at Oxford, who saw little need to interest themselves in the details of a new science called biochemistry. So much was this the case that it took almost sixty years from the first observations that pointed to the possibility of chemical neurotransmission to a final acceptance of its reality.

The difficulties in accepting the idea of chemical neurotransmission in the 1950s did not stem simply from the physiologists' lack of familiarity with the details and principles of the new science. A similar impasse had occurred at the turn of the century, when there had been a debate about the organization of the nervous system that centered on the issue of whether nerve cells were continuous or contiguous.[5] Up until the 1890s, investigators were still not certain whether the mass of what looked like tapioca was composed of cells that were confluent with each other or whether, like cells elsewhere in the body, nerve cells had junctions, or gaps, between them. The possibility remained that the nervous system was composed of a continuous net with unimpeded communication between the various nuclear elements in the net. This was the preferred view, despite the evidence that other tissues were made of discrete cells. The argument was only settled when Santiago Ramon-y-Cajal demonstrated conclusively the existence of synaptic clefts between nerve cells.

Similar disputes had followed the introduction of the notion of a reflex arc in 1823. This had for the first time divided the nervous system into radically different parts—a higher or spiritual part and a lower or mechanical part, which it seemed could still function even when disconnected from the "soul."[6] As advances in neurophysiology extended the mechanical domain further and further up the central nervous system, the implication that consciousness was of no greater functional importance than the plume of steam was to the functioning of a locomotive unsettled quite a few observers. These developments were passionately resisted by many, as it became less and less clear where the soul, which by definition could not be divisible, might reside in the body.

Common to all these cases is the theme that developments in biology with materialist or mechanical implications have appeared to threaten views of what it means to be human. The emergence of a particulate, material, or mechanical view of the mind has been resisted ever since Julien Offray de La Mettrie first proposed such a view in 1747 in his *L'Homme machine* and subsequently in *Discours sur le bonheur*,[7] the *Listening to Prozac* of its time, which proposed that human happiness might someday be a matter of medical engineering. La Mettrie received death threats and was hounded out of France in a manner that seems to foreshadow reactions to van Praag 220 years later. One of the questions that arises is how much this resistance ties in to our reluctance to take drugs and our preference for psychotherapy over drugs as a treatment for nervous disorders. It is in these terms that the issues have played in America. In Europe the debate has played more in terms of a hostility to industrial developments, of which psychotropic drugs have been a potent symbol.[8]

With the acceptance, at the turn of the twentieth century, that nerve cells were contiguous, the central question was how they communicated. Early views centered on the idea that an electrical spark or impulse might cross the synaptic cleft between two nerve cells. Several factors probably influenced this vision. One is that electricity somehow seems more vital, or consistent with spiritual identity in some sense, than does the idea of chemical reactions. Furthermore, it was recognized that nerve cells could propagate electrical impulses. From a practical point of view, also, it appeared that chemical neurotransmission was likely to be too slow to support the complex operations that would necessarily be involved in central nervous system processing. Finally, in the 1920s, the first recordings of brain waves were made, leading to the creation of a method to record them that was of clinical relevance—the electroencephalograph (EEG).[9]

The first experimental indications that chemical neurotransmission might be important, however, came as early as 1905, with the work of John Langley on the peripheral nervous system (the nerves that lead from the brain to innervate muscles, heart, lungs, gut, and so on). This system is divided into a voluntary and involuntary or autonomic nervous systems. The autonomic system, which was so called by Langley in 1895, comprises those nerves involved in the control of breathing or the movements of one's gut during digestion. Observations that some nerves leading to the heart caused it to increase its rate of beating whereas others caused a decrease led to the notion that there were complementary systems within this system. In 1905 Langley suggested calling the subdivisions of the autonomic system the sympathetic and parasympathetic systems.[10]

The year 1905 was a watershed in a number of other ways as well. T. R. Elliott, working with Langley, discovered that sympathetic nerves released a substance, which he termed sympathin. This had properties, it seemed, in terms of its effects on the heart, that resembled the effects of injecting extracts of the adrenal gland—which has led to the modern names for the substance, adrenaline in Europe and epinephrine in the United States. At first these observations didn't lead to any radical revision of views of nervous system functioning, as the sympathetic system was but one small and seemingly unimportant division of the nervous system that was concerned with physiological and apparently primitive functions rather than cognition. It remained unclear whether the release of sympathin was anything other than a curiosity. Its release as such had not been tied to a change in "behavior" in any way.

There was yet another development in 1905; Langley outlined a theory that explained how at least some nerves might function. He suggested that they liberated sympathin or other substances and that the response to this depended on whether the stimulated cell had effector or inhibitory receptive substances. This was a first outlining of receptor theory, which has crucially underpinned the idea of therapeutic specificity in psychiatry and much of the rest of medicine, with its connotations that drugs might be as specific for certain receptors as keys are to a lock.

Subsequent work by Henry Dale, in 1914, demonstrated the presence of another chemical, acetylcholine (Ach), in the nerve endings of the parasympathetic system. Dale was able to show that its effects mimicked the effects produced by stimulating the nerves themselves. He also suggested that there was an esterase enzyme present in the tissues, later called cholinesterase, which inactivated acetylcholine after its release. In 1921 Otto Loewi demonstrated experimentally that acetylcholine was of functional importance—not only was it released but its release brought about changes in the functioning of the "target" tissue—in other words, it was a neurotransmitter.[11]

But again, the implications of this for nervous transmission generally were initially not thought to be of particular importance, because all that had been demonstrated was that one neurochemical, acetylcholine, at that time found only in peripheral tissues, might have a role to play. Things heated up in the late 1920s, when Walter Cannon found that sympathin brought about increases in heart rate and blood pressure and on the basis of these observations proposed that the autonomic system played a part in the mediation of emotion, particularly the emotions of fear and anxiety.

Dale and a group of colleagues, including John Gaddum, William Feldberg, and Marthe Vogt, gradually won the argument—as far as peripheral neurotransmission was concerned. They went on to demonstrate the presence of acetylcholine and cholinesterase in the central nervous system. But despite this they fought shy of proposing that central chemical neurotransmission might be of importance, probably owing to a concern that the establishment of chemical neurotransmission in the peripheral nervous system might be compromised by any dispute that developed over the role of chemicals in central neurotransmission.[12]

Meanwhile, in 1933 a compound that appeared to promote contractions of the gut wall had been isolated from the gut by Vittorio Erpsamer in Italy. He called it enteramine.[13] What later turned out to be the same compound was also isolated from blood platelets by Irvine Page and Maurice Rapport in 1947. They found that it caused blood vessels to contract, and so they called it serotonin.[14] The chemical structure of this compound was established as 5-hydroxytryptamine (5HT), and in 1953 Betty Twarog and Irvine Page demonstrated that 5HT could be found in the brain, among other tissues.[15] Observers initially reacted to this discovery by suggesting that any 5HT in the brain was probably a residue of blood flow through the brain.[16] However, LSD had been discovered a few years previously. Structural affinities between it and 5HT were noticed following the elucidation of the structure of serotonin. This independently led D. Woolley and E. Shaw in the United States and John Gaddum in the United Kingdom in 1954 to speculate that brain 5HT might have a role to play in nervous disorders, even though it was not generally accepted that 5HT was a neurotransmitter at this point.[17]

In addition, in 1952 Marthe Vogt demonstrated the presence of norepinephrine (noradrenaline) in the central nervous system, and in 1958 Arvid Carlsson provided evidence that dopamine was present in the central nervous system and functioned as a neurotransmitter. Despite all this, there was a reluctance to wholeheartedly embrace the idea that chemical neurotransmission might not simply occur but might be the dominant mode of transmission within the central nervous system.[18]

The ambiguities in the attitudes of Dale and his colleagues were revealed at a meeting held at the Ciba Foundation as late as 1960 in London, where the role of the catecholamines in neurotransmission was the subject for discussion.[19] Some features of this meeting are worth bringing out to illustrate just how large was the conceptual shift needed and to bring home just how recently the shift was made. At this meeting, Vogt, Gaddum, and all of those who had ever advocated the importance of chemical neurotransmission—at that point an almost entirely British group—were present. They were presented with evidence by Carlsson that a range of putative neurotransmitters existed in the central nervous system and that changes in the levels of some of these correlated quite well with changes in behavior. They also had the example of the newly introduced psychotropic drugs, which presumably were working through chemical mediators at some point—and yet most of those at the meeting could not bring themselves to grasp the nettle and accept that chemical neurotransmission occurred in the brain and was probably the dominant form of neurotransmission.[20] This should give some idea of the resistance there is to a particulate or atomistic view of human beings. This fact must also be at least one important reason why, despite previous British dominance of the field, the primary breakthroughs in psychobiology were made outside Britain—in America and Sweden and by nonphysiologists.

Brodie and Axelrod

The identification of 5HT in the brain, recognition of its affinities with LSD, and suggestions that LSD might act on 5HT receptors were the first hints of a revolution to come. As of 1954 it was still too early to make definitive statements—to offer anything more than tentative speculations. The technology did not exist to decide whether the levels of 5HT or norepinephrine that were in the brain were of functional importance or not. In 1955, however, a number of events changed the view of many. One was the development by Robert Bowman and Sidney Udenfriend in Bernard Brodie's lab at the NIH of the spectrophotofluorimeter. This allowed a much more accurate detection of neurochemical levels, and the increase in sensitivity played a significant part in enabling Brodie and his colleagues to demonstrate for the first time a close relationship between alterations of behavior and alterations in the levels of a putative neurotransmitter—5HT.[21]

As mentioned in Chapter 2, reserpine had been introduced at almost the same time as chlorpromazine and was being widely used in both the United States and Europe. In an effort to establish how it worked, Brodie and his colleagues Park Shore, Sidney Udenfriend, Alfred Pletscher, and others gave reserpine to animals and found that it produced a very typical state—the animals were sedated, their eyelids drooped and sometimes went into spasm. On the basis of the idea that reserpine antagonized the actions of LSD and that LSD might act on the 5HT system, Brodie and the others decided to look at the effects of reserpine on brain 5HT. Using the spectrophotofluorimeter, they found a very convincing fall in the levels of 5HT that appeared to correlate with the change in behavior of the animals.[22]

An article outlining these findings first appeared in *Science* in 1955.[23] In many respects, this can be seen as the inaugural article of a new era. A bridge had been built from neurochemistry to behavior, and it remained for other researchers to cross this bridge and establish a beachhead on the opposite side. As will be clear in the next few pages, a great many researchers poured over this bridge, adding bit by bit to Brodie's findings. But part of the importance of the finding was that it was made in Brodie's lab. In contrast to the discovery of imipramine, where there was a considerable amount of time from the initial observations of antidepressant efficacy to the adoption of the new treatments, which stemmed in part from the obscurity of those who made the observations, Brodie's stature in the field and personality was such that nobody was in any doubt that the finding was important—even though it later turned out to be wrong.[24]

Bernard Brodie was born in Liverpool in 1907, but the family moved to Ottawa when he was four. He dropped out of school early. There are confusing stories thereafter about his being markedly shy, so much so that he would cross the street to avoid encounters with others, about his boxing for financial reward—he did become Canadian Army champion in his weight class—and about a skill at poker playing that paid for his tuition at McGill University. There he apparently discovered the appeal that can accompany successful experiments, and contrary to earlier expectations, he pursued a Ph.D. in organic chemistry at New York University. From there he moved as a research assistant to the pharmacology laboratory of George Wallace at NYU, where Otto Loewi was also working. The experimenters focused on the distribution of substances in the body after their consumption—a field in which Brodie later became the leading figure.[25]

On the floor below Wallace's lab, Jim Shannon had a laboratory. Shannon was a physician whose early research was on the physiology of the kidney. In 1941 he moved from the building Wallace and Brodie were in to head a research laboratory at Goldwater Memorial Hospital on Welfare Island in the East River. There, in 1942, he was persuaded to change his research effort from renal physiology to a search for anti-malarial drugs as part of the U.S. war effort. Japan had just bombed Pearl Harbor and the United States had entered the war. From the outset, one of the problems that loomed for the army was a shortage of quinine, stemming from the Japanese capture of large areas of the Far East where the trees grew from whose bark quinine was isolated. There were other drugs available, notably Atabrine, but Atabrine caused serious gastric disturbances that meant that few were willing to take a dose sufficient to suppress malarial attacks. The problem of its side effects had to be sorted out, or a new drug found.[26]

Shannon had recruited Brodie, Udenfriend, and others to Goldwater and assigned them the problem. Between them they developed a method to determine how much Atabrine there was in the blood, and not only the blood but other tissues such as the liver and muscles, and how much was excreted in the urine and feces. It became apparent that to get Atabrine to work, it was necessary to give a much larger initial loading dose than was usually given, followed by smaller doses than usual thereafter. The malaria problem was solved and the science of pharmacokinetics—the study of the behavior of drugs in the body—was born.

By all accounts the atmosphere in Goldwater was electric. Brodie earned himself the nickname Steve, in recognition of his preparedness to take a chance; Steve Brodie had been a twenty-three-year-old New York saloon keeper who in 1886 had jumped off the Brooklyn Bridge to win a $200 bet. Brodie, Udenfriend, and later Julius Axelrod and others posed and solved problems. The group seemed to have an ability to pose the right technical problem—one that could be solved—or else was able to come up with the instruments to solve it—culminating in Robert Bowman's spectrophotofluorimeter, without which neurochemistry would not have been possible.[27]

After the war, in 1949, Shannon moved to Bethesda to become the director of the National Heart Institute. He recruited Brodie, Udenfriend, Axelrod, and others to join him—a galaxy of talent. This may not seem surprising now, given the reputation of the National Institutes of Health and Mental Health, but at the time it was thought unrealistic to expect progressive science to come out of a government bureaucracy. Both Shannon's reputation and the opportunity to avoid being drafted to Korea by being posted to the Public Health Service played a part in overcoming the doubts of many of those who signed up.

Following the observations of Brodie's group regarding reserpine and 5HT, a string of other findings emerged. Pletscher, Brodie, and Shore experimented with the interaction of reserpine and iproniazid. Witnessing the stimulant effects of this combination probably played some part in leading Kline to conceive of the possibility that iproniazid was a psychic energizer. In 1952 Al Zeller and his colleagues demonstrated that iproniazid inhibited monoamine oxidase. The deduction that giving iproniazid would increase levels of brain monoamines—both norepinephrine and 5HT—naturally followed. For Brodie and his group, 5HT appeared to be the more important of the two transmitters.

At this stage there were a number of proposals that increasing brain amines might mediate the action of antidepressants—these early versions of the catecholamine hypothesis appeared almost immediately on the description of the antidepressant effects of iproniazid. Attempts to test the hypothesis were under way as early as 1958, when Merton Sandler and Michael Pare in England gave concurrent doses of amine precursors to subjects to see whether this would boost the antidepressant effects of iproniazid—as might be expected if its antidepressant effect was being exerted by virtue of a monoamine-increasing mechanism. Their attempts failed.[28]

A second front, however, was opening up. Arvid Carlsson, a young biochemist working in Lund, Sweden, came to Brodie's laboratory for six months in 1955. There he worked on the effects of reserpine on platelet 5HT concentrations—showing that they were reduced by reserpine treatment. While doing this, however, he wondered whether reserpine might also have effects on brain norepinephrine. Brodie thought not. Carlsson went back to Lund and ran the experiments and found that norepinephrine, like 5HT, was depleted. This finding, which prompted debates at international meetings that many remember to this day, added tremendous vigor to the field. It fostered a competitive spirit and channeled efforts and ingenuity into solving the basic facts of neurotransmission.[29]

A collaborator of Carlsson's, Nils-Åke Hillarp, who was with him at the 1960 Ciba meeting in London where Dale and colleagues failed to accept the reality of central chemical neurotransmission, in 1962 developed the techniques that allowed a mapping of 5HT and norepinephrine neurones in the brain—where they originated and where they terminated. It became clear that both sets of nerves were strategically placed in the brain to influence a wide range of cerebral functions—both vegetative and cognitive. It was probably the visual evidence that Hillarp supplied, pictures of brains with fluorescent pathways streaked through them, that challenged the unbelievers and finally established the reality of chemical neurotransmission. This finding was bracketed by findings that dopamine was deficient in Parkinson's disease and indications that L-dopa might be of help in the disorder.

Once it was accepted that norepinephrine was in the brain, dopamine had to be there also, because it is produced in the synthesis of norepinephrine. It was thought to be simply a building block, however, until Carlsson and his group attempted to reverse the effects of reserpine depletion by giving L-dopa. Believing that reserpine was acting on norepinephrine rather than on 5HT, they sought to prove the point by reversing the deficit with L-dopa, which was an even earlier building block for norepinephrine than dopamine. It worked. Reserpine-sedated animals were made alert by the L-dopa. However, when the researchers then proceeded to measure the brain's norepinephrine levels, which they confidently expected to be back toward normal, a finding that would have been very difficult for Brodie to explain, they discovered that they remained low, even though the animals' behavior had normalized. Puzzled, they developed a method to measure dopamine levels and found that they changed exactly as might be expected, if it was changes in dopamine that underlay the sedation produced by reserpine. On the basis of observations that reserpine could cause Parkinsonian symptoms, they suggested that a deficiency of dopamine might play a part in Parkinson's disease. Two years later Oleh Hornykiewicz and Werther Birkmayer reported that there was an apparent loss of dopamine-containing cells in the brains of individuals dying with Parkinson's disease.

Despite Brodie's initial championing of 5HT, opinion as to which was the critical neurotransmitter for antidepressant action had begun to swing from 5HT toward the catecholamines and in particular norepinephrine. Interest in establishing exactly what reserpine was doing was growing, because a number of reports had emerged of people who had become suicidal while being treated with reserpine for hypertension. If reserpine was lowering something, then it looked like that something was involved in mood. The emergence of the MAOIs, which raised monoamine levels, appeared to be telling a coherent story. Although, compared with the neuroleptics, the antidepressants took time to be discovered, a convincing theory about how they worked was practically established even before they had been shown to work. As mentioned in Chapter 3, however, the enthusiasm to find out what was going on swept aside one very inconvenient observation—which was that the only prospective, randomized, double-blind, placebo-controlled study of reserpine, done by Shepherd and Davies, suggested that, far from making things worse, reserpine was a passably effective treatment for anxious depressions.[30]

There was a more substantial problem, however, which lay in determining the mechanism of action of the newly emerging tricyclics, imipramine and later amitriptyline. In the first place, these drugs were somewhat sedative rather than somewhat stimulant as iproniazid and other MAOIs appeared to be. This ran against expectation. Second, the drugs did not inhibit monoamine oxidase. Indeed the first finding reported from human subjects found the action of the tricyclics to be almost entirely the opposite to that of the monoamine oxidase inhibitors. Archie Todrick, Elizabeth Marshall, and Allan Tait, working at the Crichton Royal Hospital in Dumfries, Scotland, had decided to look at levels of blood 5HT in individuals taking antidepressants. It had been reported by Pletscher that monoamine oxidase inhibitors increased blood 5HT levels, and the implication was that brain 5HT levels might also be increased. Marshall and Todrick, however, found that imipramine led to a fall in blood 5HT levels in those taking it.[31]

The discrepancies in the modes of action of monoamine oxidase inhibitors and tricyclic antidepressants were solved over the course of the next few years, largely through the work of Julius Axelrod. Axelrod was born in 1912 and came from the Lower East Side of Manhattan. He didn't get into the more prestigious Stuyvesant High School and went to Seward Park instead. After high school he went to City College, but didn't earn good enough grades to be accepted from there into medical school. He graduated in 1933. Faced with a choice between a job at the post office and a less well paid job at a research laboratory, he took the research job. He worked in first one and then another laboratory for ten years, essentially as a technician, running assays to standardize vitamins. He studied chemistry in New York University at night and got a master's degree. A laboratory accident, which cost him the sight of his left eye, kept him out of the war.[32]

One day in 1946 the head of the laboratory, George Wallace, Brodie's old boss, presented Axelrod with a problem involving people taking painkillers containing acetanilide or phenacetin: some who took these drugs appeared to get a condition called methemoglobinemia, a condition in which a methyl group binds to hemoglobin and makes it unable to carry oxygen. Wallace sent Axelrod over to Goldwater to consult Brodie. Brodie suggested that acetanilide might be transformed in the body into something more toxic. When one looked at the molecular structure, it was clear that one possibility was a transformation to aniline, which could cause methemoglobinemia. Axelrod and Brodie developed a method to measure aniline, examined patients who took acetanilide, and found traces of aniline in their urine. They then demonstrated that there was a direct relationship between the amount of aniline in the blood and methemoglobinemia. They also found that another metabolite of acetanilide was acetaminophen, which was also analgesic but didn't itself break down into aniline. Acetaminophen later went on to become the best-selling Tylenol.

Despite finding work with Brodie immensely stimulating, Axelrod had gotten the research bug and wanted to get out and pursue his own projects. He applied for a place at the National Heart Institute when Shannon went there and got it—only to find that Brodie had also been recruited and that he'd ended up back under Brodie's wing. But at Bethesda, as it turned out, Axelrod was given greater freedom by Brodie. He became interested in enzymes, including the enzymes that metabolize drugs. Working on amphetamine, he found that it was metabolized by an enzyme in the liver and that this enzyme also metabolized ephedrine and other substances. This was quite a surprise, because until then enzymes had always been thought to be specific to one substance. If you have an enzyme to break down norepinephrine, you want it to do just that and not break down 5HT or anything else that might be around. But Axelrod's enzyme was behaving differently.

What had been discovered was the liver microsomal enzyme system, a unique set of enzymes that have a multipurpose quality to them that no other enzymes in the body have and that renders them tailor made for removing unwanted substances from the body. This was a discovery worthy of a Nobel Prize. The paper was published with Brodie as the first author. Axelrod left, got a Ph.D. at neighboring George Washington University, and came back to NIH as an independent scientist. Under Seymour Kety at the newly formed National Institute for Mental Health (NIMH), he began to work on the metabolism of adrenaline and norepinephrine.

Monoamine oxidase was known to be involved in the breakdown of norepinephrine, but if one gave an infusion of norepinephrine and inhibited monoamine oxidase with an MAOI, norepinephrine was still inactivated. Working on this problem, Axelrod and his colleagues found the enzyme catechol-o-methyl-transferase (COMT), which was an earlier step in the inactivation of norepinephrine. The breakdown product of this then was handled by monoamine oxidase. At this time there were known inhibitors for both COMT and MAO, but when Dick Crout inhibited both enzymes norepinephrine was still inactivated. It seemed that norepinephrine was removed from the system in yet another way.

As chance would have it, Seymour Kety had ordered radio-labeled norepinephrine to explore a hypothesis regarding the chemical basis for schizophrenia. Axelrod asked for and was given some of the radio-labeled norepinephrine. Fresh experiments, done with Hans Weil-Malherbe, George Hertting, Irv Kopin, and others, using the radio-labeled norepinephrine, revealed that it persisted in the body despite the inhibition of both enzymes and its apparent absence from the bloodstream. Where had it gone? One possibility was that it had been taken up by nerve endings. The obvious nerve endings to look at were those of the sympathetic nervous system, which had been known for some time to use norepinephrine as a neurotransmitter.

In order to establish whether this was the case or not, Axelrod and his colleagues cut the sympathetic nervous system on one side of the body of their experimental animals, gave radio-labeled norepinephrine in the presence of both enzyme inhibitors, and X-rayed the results. They found that on the side of the body where the sympathetic system was intact, the norepinephrine was clearly present, whereas on the other side there was no visible radioactivity. This suggested that there was a reuptake mechanism in nerve endings. They then demonstrated that stimulation of the nerve caused a release of norepinephrine and that if norepinephrine was infused into the pineal gland, an area rich in sympathetic nerve endings, it ended up in those nerves.

These experiments led Axelrod to propose that there was a reuptake mechanism for norepinephrine—that it was conserved by being taken back up into nerve endings.[33] Nothing in classical pharmacological theory supposed that such a mechanism existed. As Axelrod was later to put it, if he had known more about pharmacology he wouldn't have considered the idea. However, studies with imipramine showed that it blocked this reuptake mechanism, which persuaded many who might otherwise have remained skeptical. The idea of a reuptake blockade also opened a pathway to conceiving of a theory that might link the actions of both imipramine and the MAOIs—both, it could be argued, increased catecholamine concentrations in the synaptic cleft. Subsequent experiments on 5HT showed that there was also a 5HT reuptake mechanism, and work with imipramine revealed that it blocked 5HT reuptake as well—which is what Prozac blocks—although by then 5HT seemed increasingly less important.[34]

On the basis of these studies Axelrod received the Nobel Prize in 1970. There was surprise at NIH and worldwide that Brodie did not receive a share of the prize. It was given specifically for work on catecholamines, which had not been Brodie's area. By 1970 Brodie had begun to fade from the scene owing to a combination of ill health and other factors, but the disappointment was still great. He had been awarded the Lasker Prize and most other top honors. He is to this day credited with having established the field of biochemical pharmacology—and not just because of one lucky observation but because he was one of the few older academicians working in the area who had a feel for what drugs might do and how they might be transformed in the body. A younger person would not have had the stature to create a school—to pull in Arvid Carlsson, Alfred Pletscher, Erminio Costa, and a host of others—to work with him the way he did. An older person approaching the area would have been trained as a physiologist and almost certainly as a consequence would have thought the experiments that Brodie and Axelrod conceived too naive to be worth doing. Neither Brodie nor Axelrod was a great reader of the literature. Neither knew much about physiology.

The Monoamine Hypotheses

Axelrod's demonstration that imipramine inhibited catecholamine reuptake was a significant factor in shifting attention from 5HT to the catecholamines. Although Axelrod and his colleagues were subsequently to show that there was a 5HT reuptake mechanism and that antidepressants inhibited that as well, the demonstration of their effects on catecholamines came several years earlier. Furthermore, notwithstanding the interactions between LSD and the 5HT system that had led in the early 1950s to proposals that the 5HT in our brains might be involved in keeping us sane, there was a much longer-standing association between catecholamines and emotional states (in particular, the emotions of fear and anxiety), stemming from the work of Walter Cannon, that suggested that catecholamines might have a role in mood disorders.

In addition, from a practical point of view, apart from LSD there were very few drugs available at the time with which to manipulate the 5HT system (a state of affairs that persisted until the early 1980s), whereas there were a great number of compounds available which acted on different ways on the noradrenergic and adrenergic systems. In 1963 Carlsson and his co-workers had demonstrated that chlorpromazine acted preferentially on the catecholamine system—the first observation of a series that led to the dopamine hypothesis of schizophrenia.

Thus there was increasing interest in catecholamines, along with the fact that the influence of Brodie, the champion of 5HT, was declining. He maintained his view that, notwithstanding the observations of Carlsson and others, 5HT was the critical neurotransmitter. This view was to win back some support in the 1980s as a result of studies, by another of his protégés, Fridolin Sulser, that pointed to a role for the 5HT system in regulating the noradrenergic system—a view that is proving useful as marketing copy for the post-SSRI generation of antidepressants. But before Brodie retired, he was involved in a further discovery that indirectly prepared the field for the emergence of the catecholamine hypothesis of depression.

It had been found that imipramine breaks down into the body to desmethlyimipramine, more commonly known as desipramine.[35] Sulser and Brodie found that this was able to prevent the onset of the reserpine syndrome much more effectively than imipramine itself, leading Brodie to suppose that in its own right it might be antidepressant. Brodie, Kline, and George Simpson, among others, tried it out on depressed patients—and found that it worked.[36] It took effect somewhat more speedily than imipramine, they thought, and with fewer side effects. Desipramine was subsequently licensed as an antidepressant and became at one point the best-selling antidepressant in the United States. When it became possible to measure its effects on both catecholamine and 5HT reuptake, it turned out that it blocked catecholamine but not 5HT reuptake.

These elements, when combined, led to the catecholamine hypothesis of depression, which was put forward in 1965 by Joseph Schildkraut in a paper in the *American Journal of Psychiatry,* later to be one of the most cited papers in psychiatry.[37] The idea wasn't new, in the sense that since the end of the 1950s many people had thought that antidepressants must be acting by virtue of their effects on monoamines, with opinions divided as to whether 5HT or norepinephrine was the pertinent neurotransmitter. It wasn't a unique proposal even for 1965, in that William Bunney and John Davis published a similar proposal in the *Archives of General Psychiatry* in that same year.[38] But for some reason, Schildkraut's was the right article at the right time.

For the first time, in a mainstream psychiatric journal, psychobiological ideas from various sources had been put together in a respectable manner and formulated into a specific hypothesis. The argument was that as reserpine was known to deplete catecholamines and as it was also known that reserpine could lead to depression and in some cases make people suicidal, there were strong suggestions that the depletion of catecholamines might be associated with the generation of depressive states. Because tricylic antidepressants inhibited reuptake, it could be inferred that the effects of this would be to increase catecholamines in the synaptic cleft, thereby leading to a functional increase in the level of these neurotransmitters. Because the MAOIs also led to an increase in catecholamine levels, by blocking their metabolism, the inference was that both major groups of antidepressants led to increased amine levels.

A similar argument can be made for 5HT and indeed was later made by Alec Coppen in 1967.[39] By then, however, it had been discovered that not only did imipramine break down in the body to desipramine but amitriptyline broke down to nortriptyline and both of these breakdown products appeared to be antidepressant, suggesting that imipramine and amitriptyline were pro-drugs, containing nortriptyline and desipramine, which were the actual antidepressants. Both of these were much more potent inhibitors of catecholamine reuptake than they were of 5HT reuptake, strongly suggesting that the catecholamines were the more important neurotransmitters. Although Coppen had shown that adding tryptophan, a 5HT precursor, to MAOIs appeared to enhance their antidepressant efficacy, the tide was flowing in the opposite direction.[40]

Despite the fact that such a hypothesis could have been proposed earlier and despite the many flaws inherent in it, this particular statement in 1965 came to dominate the field, set research agendas, and direct drug company efforts for the following two decades. It crystallized a split in psychiatry between biological and psychodynamic branches, each group having its own journals, its own meetings, and de facto very little to do with each other. Although the introduction of the drugs had thrown differences into relief, psychiatrists of all orientations could nevertheless use them. The catecholamine hypothesis, however, emphasized the advent of the psychiatric researcher who was conversant with the details of neurotransmitter metabolism, receptor binding, and endocrine changes—a researcher who spoke a different language from the analyst.[41]

Before considering why the catecholamine hypothesis should have had the appeal it did, it is worthwhile to point out in greater detail some of the flaws in the hypothesis. First, while there were anecdotal reports of people committing suicide who had been treated with reserpine for hypertension, it remains a fact that in the only prospective study where reserpine was given to patients and the level of their depression and suicidality was monitored, it was found that far from causing depression reserpine was antidepressant and anxiolytic, as was outlined in Chapter 3.

Second, while iproniazid inhibited monoamine oxidase and it was assumed that this was the mechanism by which it worked, this supposition in its own right was open to dispute. As we saw in Chapter 2, there are good grounds to think that a related compound, isoniazid, has antidepressant properties as well, but interestingly isoniazid does not inhibit monoamine oxidase. This opens up the possibility that there is something else in common among this group of compounds that brings about the antidepressant effect. Indeed the inhibition of monoamine oxidase has been paradoxical in that there has never been any convincing correlation between antidepressant efficacy and a compound's potency at inhibiting monoamine oxidase.

Third, while tricyclic antidepressants inhibited catecholamine reuptake, it was not clear that this was going to bring about an increase in the levels of catecholamines, in that the purpose of reuptake is to conserve neurotransmitters. Its inhibition might be logically expected, therefore, to lead to a loss or decrease of those same neurotransmitters rather than an increase. Studies of the levels of neurotransmitters in the brains of animals being chronically given these drugs suggest that a depletion rather than an increase of monoamines often occurs.

Fourth, treatment strategies based on the catecholamine hypothesis were notably ineffective. If the hypothesis were true the addition of treatments aimed at increasing catecholamine levels to standard antidepressant regimes should have enhanced their utility as antidepressants, but this has never been shown to be reliably the case.

Finally, on the basis of the catecholamine hypothesis, one would expect that antidepressants would be unhelpful in the treatment of mania. If depression was a state characterized by lowered catecholamines, then mania presumably was a state of raised amines, and if antidepressants acted to raise amines, they could be expected to induce mania and apparently did so in some cases. For many, one of the most compelling pieces of evidence in favor of Schildkraut's hypothesis was that some people given antidepressants actually did become manic. This is a striking testimony of the power of a seductive formula to triumph over inconvenient observations, as both ECT and lithium (and more controversially the tricyclics—see Chapter 4) are used to treat both depression and mania. Notwithstanding the evidence of ECT and lithium, giving a tricyclic to someone with mania very quickly became something that would be hard to defend in a court of law.

The possible role of the tricyclics in precipitating mania remained unaddressed for twenty years until picked up by Jules Angst in 1985. Angst compared the rates at which bipolar patients switched from depression to mania in the eras before the introduction of ECT, after the introduction of ECT but before the advent of imipramine, and subsequently after the widespread use of imipramine and found that the rates from all three periods were the same. In other words, there is a natural rate of switching from depression to mania in bipolar patients. As Angst commented, there is also a natural tendency to believe that changes that come about after a drug is given are brought about by that drug.[42]

Schildkraut characterized his hypothesis in his 1965 article as "at best a reductionist oversimplification" that might be "of heuristic value" that "may soon require revision due to rapid progress in the basic sciences." Despite this view and all the drawbacks, most of which were clearly apparent in 1965, the catecholamine hypothesis became rapidly established and persisted for more than two decades. Why? This was clearly no ordinary hypothesis. What was it? Thomas Kuhn, in *The Structure of Scientific Revolutions*, has argued that science operates by means of paradigms. Certain hypotheses and theories hold sway at particular times, not necessarily because they are the most comprehensive, the most sophisticated, and the best, but because they make some attempt to explain particular sets of facts that a scientific community at that time feels need explaining and they do a better job than their competitors.

The competing explanations at the time were the analytic and other dynamic hypotheses of depression, compared with which the catecholamine hypothesis was a rather unidimensional view. But it replaced them because it made some effort to account for the fact that antidepressants worked and afforded some predictions that could be tested about the mechanism of action of those drugs. By 1965 such hypotheses were needed in order to catalyze further development. Research can rarely proceed on a solely empirical basis. It is usually guided by theories of one sort or another. And indeed far from spending most of their time attempting to refute current theories, as many philosophers of science have argued they should, most scientists probably spend most of their time attempting to confirm and support pet theories.

The amine theories, as they quickly came to be called, became irrefutable. Indeed they have never been refuted, although there are many who would concede that they cannot be right. Large amounts of data were collected from depressed people in studies that looked at the precursors and the metabolites as well as the receptor densities of the various different monoaminergic systems, but no abnormalities of either catecholamine or 5HT systems have ever been replicated in a manner that has commanded widespread support. It would seem that it hasn't been possible to set up a decisive experiment that would either confirm or kill off the amine hypotheses. Even the NIMH, when it got involved in 1980 in a large-scale study of the monoamine systems of depressed patients, was reluctant to conclude that the hypothesis might be wrong, despite the fact that its own predictions were not confirmed. The NIMH chose instead to juggle the figures, appeal to the well-known heterogeneity of depression, and limply back away from findings that were "empirically robust but from a theoretical standpoint difficult to explain."[43]

This last quote illustrates as well as any in any science the clash between a paradigm and reality, between data and ideology. Why should the scientific school that conceived the inconceivable in creating psychopharmacology have found it so difficult to accept the implications of its own data? This failure echoes that of Dale and his colleagues twenty years earlier. The young Turks who had achieved academic positions in one revolution had in turn become the conservatives standing in the way of change.

A number of factors may have contributed to the impasse. After 1965 a situation developed in which just as every study appeared on the latest monoamine measure in depressed patients, another receptor or another metabolite of the amines was reported. So the negative findings of successive studies could be set aside, in the belief or expectation that abnormalities of the newly described receptor, or metabolite, or the balance between that receptor and other receptors would be found to be linked to depression.

Another reason was that the catecholamine hypotheses gave the pharmaceutical industry a clear goal to aim at. If the industry could produce drugs that inhibited catecholamine reuptake or reversed the effects of reserpine, it would most probably have an antidepressant. Once companies' production efforts turned this way, industry scientists had a vested interest in talking up catecholamines, at least until that generation of compounds had come to market. The role of such industry scientists is worth bearing in mind, because during the 1960s and 1970s an ever- greater proportion of the scientists working in psychopharmacology or biological psychiatry were associated with the industry. The catecholamine theories only began to fade when the SSRIs came onto the market in the early 1980s. Since 1980 scientific discussion has been dominated by 5HT-speak, even though the 5HT hypotheses have never had anything more going for them than the catecholamine hypotheses had.

There was, furthermore, a certain marketing-copy quality to the amine hypothesis that was also important. The idea that depression was "known" to involve low levels of biogenic amines was something that fitted neatly into the snappy format in which truths have to be conveyed on advertisements to physicians. Indeed, a feature of many sciences, but certainly psychiatry, is that at a certain point key terms succeed in pulling the field together by virtue of combining the right measure of simplicity and ambiguity. In Chapter 1 we saw how the terms *neurosis* and *psychosis* came into being; they almost certainly owed their long life to the fact that they succinctly conveyed a meaning that people thought they could grasp but that in actual fact was like a Rorschach test, something into which different groups could read whatever meaning they wished to see. Similarly the term *schizophrenia* held the right level of definition and promise. So also with the amine theories. At one level, they permitted communication between experts and the general mass of physicians, while at the same time allowing the physician to communicate with her patients. When prescribing any pill it is always useful to have a model of what the pills do to convey to patients, and for the past thirty years physicians prescribing antidepressants have found it very easy to sell a package of low brain amines, which the pills are designed to correct.

In part it has been so easy to sell this because the idea of low amines became public knowledge very quickly. From the early 1970s, magazine and newspaper articles referred to the accepted lowering of brain amines in depression. Where once lay people had gone to psychiatrists expecting to hear about sexual repression, they now came knowing that something might be wrong with their amines or with some brain chemical. The amine theories even catered to those for whom treatment with antidepressants was inimical—those who were health food oriented—in that it was possible to boost one's amines by taking tryptophan, a precursor of 5HT, which was available in high concentrations in a range of foods.

The Receptor Targets of Antidepressants

During the 1970s attention switched from levels of norepinephrine and its precursors and metabolites to a focus on receptors. The introduction of receptors into the argument about antidepressants' mode of action appeared to offer answers to some outstanding difficulties with the amine hypotheses. A number of compounds that weren't antidepressants, such as cocaine, blocked reuptake, which should have made them antidepressants—but they weren't. Furthermore, two drugs which appeared to be antidepressants, mianserin and iprindole, had been discovered that neither blocked reuptake nor inhibited monoamine-oxidase; what did they work on? As it turned out they bound to a number of brain receptors.

It had also become clear that reuptake-inhibiting antidepressants blocked reuptake mechanisms within an hour or so of the drug's being consumed, yet recovery took several weeks to appear. The idea that the blockade of reuptake might have downstream effects on a receptor system appeared to solve this, because conceptions of receptors at the time pictured them as more solid entities within the system that perhaps took some weeks of chiseling at, before they altered their number or their configuration either back toward normal or to a state that would compensate for the effects of depressive illness. It is now apparent that such views of receptors were quite mythical, in that receptors themselves are in a state of constant flux and drugs affect their number and configuration within half an hour to an hour of their being administered—that is, as quickly as they act on reuptake mechanisms. But in the 1970s this was far from clear.

The idea that the natural hormones of the body acted by binding to a particular protein, a receptive substance, had first been put forward, as we have seen, by Langley in 1905. The classical statements of receptor theory were formulated around 1932 by Alfred Clark, who worked out the principles whereby drugs could be expected to bind to a receptor and the conditions under which there could be competition at the receptor site.[44] Although receptors remained theoretical rather than substantive entities until the 1970s, pharmacologists generally would have been extraordinarily surprised if told that it would be always impossible to visualize them or otherwise prove their actual existence.

The first steps toward such proof came in the early 1970s with the use of radio-labeled hormones, such as insulin, and radio-labeled drugs, in particular those which bound to catecholamine receptors.[45] Using radio-labeled drugs, it was possible to demonstrate binding to catecholamine receptors and to estimate both how many receptors a particular piece of membrane held and the affinity with which particular drugs bound to those receptors. The estimates derived from such experiments accorded very well with the predictions of classical receptor theory. Radio-labeled binding also led to the discovery of further drugs—a good indication that the technique had something going for it. From there it has proved possible to isolate the receptor proteins and indeed to go further than that and to clone them, so that what were once theoretical entities now have a very substantial existence.

The initial discoveries in catecholamine receptor binding were made by researchers investigating cardiac and respiratory biochemistry and physiology rather than psychiatric disorders, but there was no reason not to extend the new techniques into psychiatry. Just as this technology came on stream, in 1972, George Ashcroft, Donald Eccleston, and others at the MRC Brain Metabolism Unit in Edinburgh put forward the first receptor hypothesis in psychiatry—essentially reformulating the amine hypotheses in receptor terms.[46] Chasing the target Ashcroft and Eccleston offered, psychopharmacologists began using the new technology within a few years of its development—placing psychiatry, almost for the first time, not just within medicine, as the discovery of the psychotropic drugs had done, but at the forefront of medical developments.

The antidepressants were all found to down-regulate Beta-adrenergic receptors.[47] Even the ones that didn't block catecholamine reuptake acted in this way. This down-regulation took up to two weeks to appear, which appeared excitingly consistent with the peculiar delay there is between taking an antidepressant and improving. Fridolin Sulser and Jerzy Vetulani proposed, in 1976, that down-regulation of Beta receptors was precisely the mechanism of action of antidepressants. In the same year Steven Stahl and Herbert Meltzer in the *American Journal of Psychiatry* and others elsewhere were putting forward the dopamine receptor hypothesis of schizophrenia. Psychiatry had entered the receptor era.[48]

A considerable part of the appeal of receptors was that receptor function had, from the start, been portrayed in terms of keys fitting locks. This was very much the same language as the magic bullet terminology of therapeutic specificity. It seemed that progression to the receptor level must be right—it would inevitably yield the specific lesion that underlay depression, the mechanical fault. This was a very powerful lever when it came to getting grant monies from funding organizations. Searching for a very discrete and particular disturbance in a system "known" to be abnormal in the disease being investigated, with a technology designed to detect what one is looking for, is the perfect recipe for extracting grant monies from those who have them.

During the 1970s the major psychiatric disorders became defined as disorders of single neurotransmitter systems and their receptors, with depression being a catecholamine disorder, anxiety a 5HT disorder, dementia a cholinergic disorder, and schizophrenia a dopamine disorder. The evidence to support any of these proposals was never there, but this language powerfully supported psychiatry's transition from a discipline that understood itself in dimensional terms to one that concerned itself with categorical ones. This legitimized the rise of biological psychiatry, which in turn fostered a neo-Kraepelinian approach to diagnosis and classification, as embodied in DSM-III.

Receptor Retrospective

A number of points have emerged from two decades of receptor binding and grinding—as it is called. The first is that there are a great number of proteins that can be isolated and cloned that drugs bind to, but it is not clear that all of these are receptors. To be a receptor, the binding of drug to a protein should bring about a functional change of some sort. Often, however, it is not clear whether there is any relevant change, in which case the protein is just a binding site. A second point is that even when the proteins are not simply binding sites, our capacity to detect minute amounts of receptor and neurotransmitter proteins within the brain is now such that we can detect their presence everywhere in the brain, including places where they have little or no functional relevance. There have been claims that we now know of over a hundred different neurotransmitters and, given that each of these neurotransmitters will have a number of receptor types associated with it—anything from five to fifteen—the question of sorting out the biology of nervous transmission has become incredibly complex. An alternative view, however, is that a great number of these receptors and neurotransmitters, while not dismissable as simply binding sites or artifacts, are redundant.[49]

A great deal of the appeal of receptor models in the late 1970s lay in the impression that these were solid little proteins that only changed slowly. These changes, moreover, along with the delayed process of getting an antidepressant into the brain, were believed to coincide with therapeutic change. Researchers weren't interested in a receptor or its changes unless those changes happened within the right time frame—the implication being that receptor change somehow permitted recovery. In the mid-1980s it became clearer that receptor changes may be brought about as quickly as the blockage of amine reuptake.

This demonstration had little effect, but some recent studies on SSRIs and sexual function illustrate how wide of the mark 1970s ideas were. It seems clear that giving a 5HT reuptake inhibitor can delay orgasm. Alternatively a 5HT-1 agonist or 5HT-2 antagonist can bring it forward. In the case of clomipramine, for instance, 10–25 mg of the drug, which is less than one-sixth of the dose needed to treat depression, taken two to three hours before intercourse will produce a noticeable delay in ejaculation in men.[50] The significant point here is that it now appears that clomipramine does this by acting on brain receptors—that is, entry to the brain and receptor change can take place to a sufficient extent in two hours to bring about marked changes in behavior.

Far, then, from providing answers to the questions of how antidepressants work or to the question of what is actually wrong in the nervous systems of people who are depressed, the biological investigations involved have played a different role. In essence, they have provided biological justification for the new approaches that were taken up by psychiatry during the 1970s and 1980s. They have provided artistic verisimilitude by allowing psychiatrists, who talked about biology, to appear scientific. They have provided potent images that entered popular consciousness and replaced older notions of dynamic problems. To a greater or lesser extent many of these biological notions have provided badges of identity for particular groups of individuals within the psychiatric or neuroscientific communities. People identify themselves as catecholamine persons or 5HT persons and as biologically oriented rather than socially oriented. As a sociological phenomenon, the power of such ideas in disciplines such as psychiatry to command brand-name loyalty and the reassurance that such brand-name loyalty provides should not be underestimated.

As it transpired, Sulser was later to find that an intact 5HT system was necessary for adrenergic down-regulation to occur—as Brodie had once argued. Down-regulation had also been established in some strains of rat, but it was not clear that it applied to all rat strains (never mind other species of animal). There was never any evidence of supersensitive Beta receptors in depressed patients, which would have fitted with the proposed action for antidepressants. But Sulser's hypothesis didn't fall for any of these reasons. In 1979 Steven Peroutka and Solomon Snyder reported that they had cracked the problem of how to distinguish among the 5HT receptors. This coincided with the development of the SSRIs, and within a few years Beta receptor down-regulation was out of favor. It had never been disproved as an antidepressant principle. It had never even been properly understood as a phenomenon, which should have meant that "scientists" would keep on working to explain it. Its falling by the wayside seems to have more to do with the fact that 5HT became the politically correct neurotransmitter.

The 5HT Reuptake Inhibitor Story

While America went catecholamine, Europe had remained or had become even more interested in 5HT. In 1963 Alec Coppen at an MRC unit in Surrey had found that adding tryptophan, a 5HT precursor to an MAOI, appeared to boost the rate of recovery brought about by MAOIs. George Ashcroft and his colleagues at the MRC unit in Edinburgh had found that the 5HT metabolite, 5HIAA, was apparently low in the cerebrospinal fluid of depressed patients—the implication was there was an abnormality of 5HT turnover in depression. These findings were later replicated by Herman van Praag in Holland, who went on to link 5HT functioning to aspects of personality such as impulsivity and to symptoms such as irritability. In 1967 Coppen, reviewing the emerging field of biological research on depression, proposed a 5HT version of Schildkraut's catecholamine hypothesis. Because both 5HT and norepinephrine can be seen as monoamines, these hypotheses came collectively to be called the monoamine hypotheses; the 5HT hypothesis was a variation on a theme rather than a genuinely novel hypothesis.[51]

Meanwhile, as outlined above, it had been found that both desipramine and nortriptyline were much more potent inhibitors of norepinephrine than 5HT reuptake. The reasonable conclusion surely was that 5HT reuptake is rather peripheral to the mechanism of action of antidepressants. However, a number of clinicians, notably Paul Kielholz of Basel, felt that while no differences in the numbers of patients who responded to each antidepressant could be shown in large clinical populations, there were qualitative differences among the various compounds nevertheless. Kielholz's impression was that desipramine restored drive more clearly than imipramine but imipramine had greater mood-elevating properties. And indeed, while desipramine sold well in the United States, owing to the influence of Kline and Brodie, imipramine and amitriptyline remained much more popular in Europe. This was unexpected, for if they were only pro-drugs, they should have been associated with a greater delay in onset of action and possibly more side effects.[52]

Struck by this, Arvid Carlsson, the original champion of norepinephrine, reasoned that 5HT reuptake inhibition might not be so unimportant after all. He was virtually alone in the idea. In 1969, for instance, he visited Geigy and discussed the pharmacology of clomipramine with company representatives—pointing out that it was a more potent inhibitor of 5HT reuptake than any of the other drugs available. They weren't interested. At the time both Geigy and Ciba were working, independently, on maprotiline, a purer catecholamine reuptake inhibitor than anything else that was on the market. Moreover, even clomipramine broke down in the body to desmethylclomipramine, which was a preferential catecholamine reuptake inhibitor.

Back in Gothenburg, Carlsson, Peder Berntsson, and Hans Corrodi began to experiment with brompheniramine and chlorpheniramine, antihistamines that inhibited both norepinephrine and 5HT reuptake. They found that if a preferentially noradrenergic reuptake inhibitor was halogenated, it became a 5HT reuptake inhibitor—as, for instance, adding a chlorine atom to imipramine to get clomipramine. Halogenating brompheniramine led Hans Corrodi at Astra to produce zimelidine, the first specific 5HT reuptake inhibitor. It was a more potent 5HT reuptake inhibitor than clomipramine and, in contrast to clomipramine, didn't break down rapidly in the body to compounds that also inhibited norepinephrine reuptake. Carlsson, Corrodi, and Berntsson filed patents for it as an antidepressant in Britain, Sweden, and Belgium in April of 1971.[53] Astra took the compound into clinical trials, the results of which, confirming that it was antidepressant, were presented in 1980.[54] Zelmid was released for general use in Europe in 1982 and did well in terms of clinician satisfaction. It was due to be marketed in the United States by Merck and the license application had been submitted to the FDA when the cases of Guillain-Barré syndrome were reported, leading to its withdrawal before it appeared on the American market.

Carlsson also took the 5HT reuptake inhibition idea to Lundbeck in Copenhagen, where their chemist Klaus Bøgesø had many norepinephrine reuptake inhibitors. From one of these, Bøgesø produced citalopram, which was launched in Denmark in 1986. In the meantime Ciba and Geigy had merged and scientists there had begun to look at 5HT reuptake inhibitors. In the process, in June of 1973, they came up with a series of compounds containing some of the most potent 5HT reuptake inhibitors ever produced. One of these, CGP 6085, showed great promise in behavioral and biochemical screening tests indicative of potential antidepressant activity, but as outlined in Chapter 4, it was never developed further.[55]

While these developments were taking place European interest in 5HT was increasing. In 1976 Jouko Tuomisto, from Finland, reported that the take-up of 5HT by platelets from depressed patients was abnormal. This finding was quickly replicated by Coppen's group. A great number of other groups have also found that 5HT uptake into platelets is abnormal. Some groups have additionally found that it returns to normal with recovery, regardless of whether the antidepressant being used acts on the 5HT system or not. These findings have never really been pursued, however, because they were almost immediately superseded by a further development.[56]

In 1978 Steven Paul at NIMH and Solomon Langer in Paris found that it was possible to radio-label imipramine. The resulting radio-labeled compound bound to a site on platelets and nerve cells, which turned out to be the 5HT reuptake site. But even more interestingly, the number of imipramine-binding sites on platelets from depressed patients appeared to be reduced compared with normals. It seemed for one heady moment that a diagnostic test for depression was within reach—a test that would have picked out those with antidepressant-responsive depressive illness from those who had existential despair or neurotic misery. There followed almost a decade of intense work during which the Holy Grail slipped from the hand. It turned out that imipramine binding was not a sensitive tool. It bound to more than just the 5HT reuptake site. Binding levels varied with age, sex, season of the year, and time of day. Only about half of the groups who looked for it could find evidence of lowered binding in depression. Imipramine binding went out of fashion in the late 1980s.[57]

Another finding that caught the imagination came from Marie Åsberg and Lil Traskmann in Stockholm, who reported in 1976 that 5HIAA, the main metabolite of 5HT, was more likely to be lowered in the cerebrospinal fluid of depressed patients who were suicidal. As this finding was pursued by a number of groups, however, it too seemed to crumble in the hands of investigators. Åsberg and colleagues' studies suggested first that the finding occurred in depressives who were suicidal, then in depressives who had been or ever would be suicidal, then in people who were suicidal whether or not they were depressed, and finally in people who were impulsive, whether or not they were depressed or suicidal. The link between 5HT and impulsivity has remained, partly perhaps owing to its convenience for the marketing departments of companies with SSRIs, but probably also owing to the influence of Herman van Praag and the appeal of models of personality later developed by van Praag and others.[58]

Somewhat independently of all this, starting on May 8, 1972, in what they have since portrayed as a moonlighting series of experiments, David Wong and colleagues at Eli-Lilly, stimulated by Carlsson's proposals, produced fluoxetine.[59] They were clear they had produced a 5HT reuptake inhibitor, but they were less clear initially what it might be useful for.[60] After preliminary safety studies on animals indicated that the compound was likely to be safe in man, Lilly convened a series of meetings with clinical investigators to try and determine what market there might be for a 5HT reuptake inhibitor. At one of these in England, Alec Coppen suggested that they consider it for the treatment of depression, only to be told that of all the things it might ultimately be used for, depression was not likely to be one of them.[61]

Serotonin (5HT) was one of the first neurotransmitters to evolve. It is widely distributed throughout the body, being found in very large concentrations in the gut wall, in the walls of blood vessels, in blood platelets, and in the brain. In the brain, it is found in areas that control respiratory, cardiac, and appetitive functions rather than predominantly in areas associated with cognitive function. Drugs active on the 5HT system, therefore, can be expected to have a wide range of important actions on different body systems. In the 1970s there were strong indications, from work on rats, that 5HT drugs might prove effective antihypertensives—and in any drug company's estimation in the 1970s, the antihypertensive market was much larger and much more important that the antidepressant market. For all the developments with psychotropic drugs, psychiatry was still not part of real medicine. Accepting that one's drug would be developed as an antidepressant was very much settling for second best. It was only in the early 1980s, after the antidepressant credentials of zimelidine had become apparent and as perceptions of the size of the antidepressant market began to change, that Lilly began to speed up the development of fluoxetine as an antidepressant. The clinical trial evidence that it had antidepressant properties only became available in 1985.[81]

In the meantime a number of other 5HT-reuptake inhibitors were also in development. Duphar's fluvoxamine (Luvox/Faverin) first came on the market in Europe in 1983. Fluoxetine (Prozac) was licensed in the United States in 1987 and came on the market in European countries variably after that; it was licensed in Britain in 1989. Sertraline (Zoloft/Lustral) became available in the United Kingdom in 1990 and in the United States in 1992, while paroxetine (Paxil/Seroxat), which had been first developed in the 1970s, was brought onto the U.K. market in 1991 by SmithKline Beecham (SB) and onto the U.S. market in 1993. Its development had been delayed owing to SmithKline Beecham's uncertainty as to whether there was a market for this kind of compound and a general perception within companies that these new drugs were not as potent as the older tricyclics. Ending up somewhat behind the others, in an effort to distinguish their compound, SmithKline Beecham referred to it as a selective serotonin reuptake inhibitor (SSRI). The acronym was quickly generalized to the other 5HT reuptake inhibitors (and has been picked up by Wyeth, which has called venlafaxine (Effexor/Efexor), a compound that acts on both 5HT and norepinephrine reuptake, an SNRI—serotonin and norepinephrine reuptake inhibitor).

Compared with the tricyclic antidepressants, which were for the most part very similar molecules, the 5HT reuptake inhibitors are a very mixed bag of compounds. They are structurally quite dissimilar and their range of actions across a number of receptor systems is also somewhat dissimilar. Although all inhibit 5HT reuptake and are termed selective serotonin reuptake inhibitors by virtue of the fact that the parent molecules have little effects on catecholamine reuptake, they are not specific to the serotonergic system or specifically to 5HT reuptake blockade. They also differ in the potencies with which they block 5HT reuptake. Fluvoxamine is the most potent 5HT reuptake inhibitor of the currently marketed compounds, while citalopram (Cipramil) is the most specific to 5HT reuptake blockade.

The variation across compounds indicates that 5HT reuptake inhibition is neither necessary to nor sufficient for antidepressant efficacy. Drugs that block 5HT reuptake may be antidepressant, as may compounds that selectively block catecholamine reuptake. Indeed, there is a strong suggestion that in severe cases of depression, some of the older compounds that act on multiple systems may be more effective than the newer compounds. ECT is almost certainly the treatment that is least specific to a particular neurotransmitter system, but it is believed by many clinicians to be the most effective. What this suggests about depression is not that it is a disorder of one neurotransmitter or a particular receptor but rather that in depressive disorders a number of physiological systems are compromised or shut down or desynchronized in some way. Depression (at least the kind that responds to antidepressants) is a disease in the sense that it may involve a core defect, but it also involves a number of downstream disturbances. The action of antidepressants in correcting this may involve activity in a number of systems and the slow confluence of those effects into a therapeutic effect.[63]

Cosmetic Psychopharmacology

In complete contrast to the idea that depression might be a complex illness, whose management might require specifically nonspecific treatments, the selectivity of the SSRIs was used by Peter Kramer in *Listening to Prozac* to raise the possibility that we were at the dawn of an era when cosmetic psychopharmacology might become a possibility—a sculpting of personalities rather than just the treatment of illness. Far from being a new idea, however, this is actually an old idea with roots that go back to a time before the discovery of imipramine.[64]

In 1921 Carl Jung proposed that it would be useful to distinguish between extraverts and introverts. Extraverts, he suggested, handled their conflicts by acting out, leading to hysteria or psychopathy, whereas introverts were intropunitive and were more likely to be phobic. This idea was picked up by William MacDougal in 1929 and at much the same time by

Pavlov. Pavlov's interest in the idea stemmed from a flood in his laboratory in Leningrad, which had trapped some of his experimental dogs. Some of them afterward seemed more nervous than others. He elaborated these observations later into a theory of nervous types, based on differing amounts of cortical inhibition, which he thought might underlie different personality types making some people less susceptible to stress than others.[65]

Hans Eysenck put these various strands of thought together in a theory of personality in 1947. According to Eysenck, individuals differed on two intersecting dimensions—an introversion-extraversion and a neuroticism-stability dimension. This grid created four quadrants, or types, into which individuals could fall. These types, Eysenck noted, had some obvious similarities to the Hippocratic types, with the introverted and stable being sanguine, the introverted and neurotic being melancholic, the extraverted and neurotic being choleric, and the extraverted and stable being phlegmatic (see Figure 5.1). Where individuals fell could be determined by their answers on the Eysenck Personality Questionnaire (EPQ).[66]

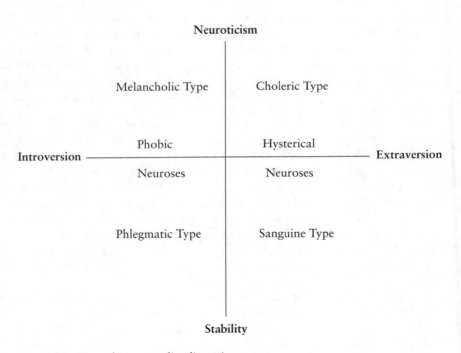

Figure 5.1   Eysenck's personality dimensions

This theory intersected with the psychopharmacological story in 1954, when Charles Shagass described the sedation threshold.[67] This refers to the amount of a barbiturate it takes to sedate an individual, which is characteristic for each person. Further work by Shagass in Montreal and by Gordon Claridge and Reg Herrington, working with Eysenck, showed that individuals' sensitivity to either sedatives or stimulants correlated very well with where they fell on the extraversion-introversion dimension on the basis of their answers to the EPQ. Hysterical subjects could be sedated with minimal amounts of barbiturates, whereas phobics and other neurotics often needed large amounts of sedative to produce any effect. The implications were that psychotropic drugs could be used to dissect out personality types or that the EPQ, for instance, could be used to predict responses to drugs. The emerging science of psychopharmacology for many looked like it would be more about such matters than about the therapy of supposedly specific diseases.[68]

For Eysenck, seemingly discrete disorders such as phobias were to be seen not so much as discrete entities as positions along a dimensional system. Illness was a matter of dimensions rather than categories. We all range along dimensions of nervous reactivity, just as we range along height and weight dimensions. Drug treatment could be used to shift us along a dimension by damping down or adding to the reactivity of an individual's nervous system. This was a research program that fit very well with the model psychoses of the high psychopathology of the early psychopharmacological period, described in Chapter 4, and Eysenck was a major contributor to early international symposia.

Since the initial formulation of his personality theory, Eysenck and a number of others have attempted to add a third dimension, loosely associated with aggression, to the system. Marvin Zuckerman and others have proposed a dimension associated with impulsivity or sensation seeking. Gordon Claridge and still others have argued for a dimension that is now termed schizotypy, which supposedly reflects a nervous system organization that grounds the propensity to schizophrenia.[69]

Much of this research program, however, came unstuck in the late 1960s, for a number of reasons. Above all, use of the prototypical drugs on which Eysenck's model was based—the amphetamines and the barbiturates—was proscribed or marginalized in the 1960s. Any research in this area, therefore, was going to look less relevant. More problematically, with imipramine, even more than with chlorpromazine, a generation of drugs emerged that could not be seen as falling on a spectrum between stimulant and sedative. Imipramine, though of benefit in states that Eysenck's theories suggested needed a stimulant, was paradoxically beneficial while having what appeared to be more sedative than stimulant properties.

A further factor involved the medical mood at the time, which was one of reaction against dimensional models. The psychoanalytic framework was inherently dimensional—everyone is ill to some extent and defensive reactions, in so far as they are all doing the same thing, merge into one another to some extent. While Eysenck was one of the fiercest critics of Freud and psychoanalysis, his dimensional ideas could not have been expected to take hold at a time when dimensional babies were being thrown out wholesale with the analytic bathwater.

The change in thinking can be illustrated by considering the fate of two review articles written by Gordon Claridge. One, published in 1972, outlined a theory of the schizophrenias as nervous types, very much in the Eysenckian tradition.[70] It was well received. An update in 1987, in contrast, incorporating emerging work on the genetics of schizophrenia and indicating how genetic research might be reconciled with a dimensional approach, received a very hostile response.[71] However, the more recent emergence of the concept of schizotypy, in particular, has brought this line of work slowly back to some kind of respectability in the past few years. One possibility now being reexamined is the idea of using drugs to test for vulnerability to mental disorders.

The dimensional approach has also seen another development, according to which the idea that Prozac might bring about personality changes is not so far-fetched. This has been the emergence of biochemical typologies, the best known of which have been proposed by Bob Cloninger and Herman van Praag.[72] In general, such schemes have also proposed three dimensions but have done so with one eye on what is known about neurotransmitters. Because the neuroleptics have been shown to reduce interest in behavioral goals and sensation seeking, there have been proposals that this dimension might be linked to the dynamics of dopamine in the brain. According to van Praag, 5HT regulates irritability and impulsivity, while norepinephrine regulates the level of reward experienced in activities. Drugs that are highly selective to one neurotransmitter might, if such models have anything to recommend them, be expected to bring about a shift in basic aspects of temperament or personality. The reports that a small number of subjects on SSRIs find themselves "better than well" might fit in with this—although it should be remembered that imipramine generated similar reports in the late 1950s.

An argument that most participants to the debate would accept is that even if something like this were true, the effects of Prozac on personality for most people would be nonexistent, because most people lie in the middle of the dimensional range. "Transforming" someone who lies near the middle of a biological dimension is much harder to do than transforming someone who lies closer to one end or the other. Another issue is whether these dimensions have to do with temperament rather than with personality. Temperament is that basic biological given, the nervous system with its characteristic level of reactivity that one has been born with; it is one of the inputs to personality, which is also shaped by environmental factors and social situations. In some people the temperamental input may be relatively large, but it is not clear that drugs with a specific effect on temperament would always have the same effect on all personalities or indeed have much effect for many personalities. These questions of temperament and personality were important in European psychiatry before World War II. In line with thinking first introduced by Esquirol and built upon by Kraepelin, disease was defined as something that cut across personality and its development or its reactions, but personality and temperament were never neglected to the extent that they have been since the introduction of psychotropic drugs. The 1962 amendments to the 1938 Food, Drug and Cosmetics Act, it should be noted, focused on diseases in a manner that assumed they could be dissected cleanly from their constitutional background, which is reasonable for bacterial disorders but much less so for other medical and for psychiatric disorders.

The more common response at present to claims that some people are "better than well" on Prozac, however, is not to invoke a dimensional model. Rather, observers account for such reports categorically on the basis that the widespread use of Prozac reveals that there are many people who don't know they are depressed, who can be called subsyndromally depressed, and who think of their depression as part of their personality. These individuals get better when treated properly. As the remainder of this book will emphasize, there are powerful forces stacked in favor of the categorical interpretation. It remains a moot point whether circumstances will in the near future permit a real reemergence of alternative dimensional possibilities.[73]

Life after the SSRIs?

The 245 years between La Mettrie's *Discours sur le bonheur* and Kramer's *Listening to Prozac* mark an epoch. There are similarities in the theses of both texts. Both advocate the importance of a medical input to the question of happiness. But whereas one was rejected violently in its day, the other became something of a cause célèbre. On this axis the experiences of Herman van Praag in the early 1970s seem to lie closer to those of La Mettrie than to those of Kramer. We have come a long way in recent years. What such a fundamental change in attitudes stems from or may signify in the longer term is unknown.

We have also come a long way in terms of our understanding of the workings of the brain. The 1980s saw the emergence of the neurosciences, which have been growing exponentially in the 1990s. These advances have been funded in great part because of the ability of simplistic hypotheses such as the catecholamine hypothesis to capture the imagination of the lay public, of the administrators of grant-giving bodies, of those scientists outside the field who are often responsible for directing research funds within fields adjacent to their own, and indeed of executives within the pharmaceutical companies themselves. This has led to the diversion of large amounts of funding toward basic neuroscientific research. But where once these basic sciences fed off clinical practice, developments in neuroscience now proceed independently of clinical practice.

The psychopharmacology associations were set up largely by clinicians, who were keen to interact with basic scientists who might be able to shed light on the new phenomena that were being revealed by the drugs. At meetings of these associations in the 1960s, the biological aspects of presentations were highly simplistic and served a symbolic function more than anything else. In the 1970s and 1980s, there was a more interactive partnership, with the Beta receptor hypothesis of depression and the dopamine receptor hypothesis of schizophrenia the expressions of a common credo. But in the 1990s, clinicians increasingly complain that neuroscientific developments appear to have marginal, if any, relevance.

There are increasing concerns among the clinical community that not only do neuroscientific developments not reveal anything about the nature of psychiatric disorders but in fact they distract from clinical research. The analogy of the drunk and the lamp post is cited—a policeman finds a drunk scrambling around at the base of a lamp post and asks him what's happened. The drunk replies that he has lost his keys and the policeman asks whether they're somewhere in the region of the lamp post. The drunk says, "No, they're several hundred yards away." On hearing this the policeman, somewhat surprised, asks why the drunk is searching underneath the lamp post, only to be told that there's no point hunting where there isn't any light.

There has been astonishing progress in the neurosciences but little or no progress in understanding depression. The fact that the SSRIs are no more effective than other antidepressants questions the idea that depression is the kind of target that a specific magic bullet will someday hit dead center. The fact that both specific norepinephrine reuptake inhibitors and specific 5HT reuptake inhibitors may cure it points strongly to the fact that it is simply not a single neurotransmitter disorder. A number of sexual dysfunctions or premenstrual tension, in contrast, may fall into the category of specific behavioral disturbances that respond to very specific interventions. Is there a difference between a disease and a dysfunction?

Should medicine be involved in rectifying dysfunctions in addition to diseases? In a number of important respects, the management of sexual dysfunctions or of obesity, for instance, would appear to differ from the question of cosmetic psychopharmacology. Dysfunctions, such as premature ejaculation or delayed orgasm, are relatively specific disorders that affect large numbers of people and that, in the case of premature ejaculation, for instance, appear to be constitutionally rather than environmentally determined. While obviously not the same thing as a disease, such a disorder would appear to fall into a different category from that of personality sculpting.

The failure of depression to respond more effectively to more specific interventions raises a further question. The era of receptor technology brought about a radical change in drug development programs. In the 1960s, companies used the reversal of reserpine's effects as a screening test for antidepressants. In the 1970s and early 1980s, in conjunction with behavioral biologists, they also developed a range of animal models—behaviors that were modified by antidepressants—the best of which, like the learned helplessness model, have convincing resemblances to human depression. While one use for such models was simply to screen compounds for possible antidepressant efficacy, another research interest lay in understanding the disease process, so that interventions might be more rationally tailored.

More recent developments in receptor technology, however, and in particular the emergence of an ability to clone receptors have permitted, indeed almost compelled, a different form of drug development. Specific compounds can be designed to target specific receptors, and this is how drugs are now produced. Large numbers of compounds can be developed inexpensively and much more rapidly than if they had to go through animal screening procedures. These new compounds are as selective as magic bullet mythology ever could have wished them to be, and this type of "rational" drug development promises an endless string of specific agents. But this is the chemists' rather than the biologists' vision. Chemists seek compounds that are selective in their effects to one or at the most two targets. Such compounds, however, in the case of both the antidepressants and the antipsychotics, seem if anything less effective than the older, "dirtier" drugs. This has led to something of a crisis in that the chemists cannot readily engineer compounds when more than two targets are indicated.

Despite this, the lure of molecular biology remains potent, and the philosophy that for every receptor there must be a corresponding illness seems to be matched at present by an increasing number of illness categories. DSM-IV has over 350, where DSM-IIIR had only 292 and DSM-II had 180. Today's classification systems make it possible to have many different illnesses at the same time—something that happens nowhere else in medicine. It would seem inevitable that there must be a collapse back toward larger disease categories at some point.

But if the SSRIs are no more potent than other antidepressants, there are a number of points that emerge from their development all the same. These drugs are different from the tricyclic antidepressants. They are active across a range of conditions other than just depression, such as obsessive-compulsive disorders and panic disorder. In some sense that has not yet been clarified, they are anxiolytic or psychic energizers. There are some hints that they break up associative chains of thought so that they cannot come to dominate the mind as readily—a very different kind of anxiolysis to that produced by the muscle-relaxing effects of benzodiazepines such as Librium or Xanax. At present, however, there would appear to be very little effort being devoted to establishing just what it is that the SSRIs do that appears to cut across clinical syndromes, that seems to lie midway between the effects of classical antidepressants and classical neuroleptics. Arguably, clinicians need to establish just what that something is, and why it is helpful, say, in obsessive-compulsive disorder or in depression, in order to be able to deploy their armamentarium rationally.

In the case of depression, it would seem that "anxiolysis" of this kind may be a therapeutic principle, as may the more energizing effects of the MAOIs or the tonic effects of the tricyclics. What does this tell us about depression? It is far from clear that establishing what that something is would be in the interests of the industry, as it would imply a relativity rather than specificity to therapeutic principles. But the very future of the psychiatric profession as an independent entity would appear to hinge on answering such questions. For whatever reason, there seems to be some failure of the profession to get control of its boat in the psychopharmacological rapids.[74] Until clinical scientists start chasing leads such as these and using what the drugs reveal to advance their own science, the field will be dominated instead by the pharmaceutical industry, which at present deploys clinical entities to advance its purposes, as we shall see in the next chapter.

There is one further lesson of the SSRI story. Whatever implications might be suggested by a theory or an animal model, the scientific community is at present locked into an industrial process: only when a compound gets a license from the FDA and is produced and marketed extensively is it possible to decide whether it is an antidepressant or not. Drugs that in theory or by virtue of effects on an animal model should be antidepressants cannot be treated as such until they are proved to be so, and the only party with the capacity to provide proof is the pharmaceutical industry. There is a real sense at present, then, that knowledge in psychopharmacology doesn't become knowledge unless it has a certain commercial value. The survival of concepts depends on the interests with which they coincide. This state of affairs should not be seen as the fault of the industry. It is a consequence as much of regulation and reimbursement concerns by insurance companies as it is of the industrial development of drugs. This issue will concern us in the next two chapters.

The End of an Era?

Within the larger story running from La Mettrie to Kramer, there has been another story outlined in this chapter. This is one that falls within the plot originally drawn up by Brodie, Axelrod, and Carlsson. There have been no radical departures from the script laid out by Brodie in 1956. The recently launched antidepressants venlafaxine, "the world's first SNRI" (serotonin and norepinephrine reuptake inhibitor), and mirtazepine (Remeron/Zispin) trade heavily on supposed interactions between the serotonergic and noradrenergic system—an idea first put forward by Brodie in response to the challenge from Carlsson and subsequently developed by Sulser, another of Brodie's protégés. The SSRIs are slugging it out trying to distinguish themselves from one another. Current marketing strategies have turned to focusing on the side effects of the different drugs. They have turned in particular to an enzyme system called cytochrome P450, which controls the breakdown of drugs in the body and therefore their likelihood to interact, possibly unpredictably, with other drugs a person may be taking. This is the enzyme system first discovered by Axelrod in 1955, which led to his leaving Brodie's lab.

However, the script may be coming to an end. It seems unlikely that compounds will be discovered that both act on monoamine systems and are substantially more effective than current antidepressants. In the current climate of cost containment, unless a new compound is markedly more effective than available treatments or has a substantially faster onset, it is unlikely that a company will be keen to market it. The price at which it would have to be sold would not adequately cover the development costs. It was thinking such as this that led Ciba to pull out of classical psychotropic development. Indeed, many companies in the late 1980s shut down the programs that were working on producing the next generation of antidepressants. Far from being a disaster, however, this may provide an opportunity for clinicians and others to take stock.[75]

Company efforts are being concentrated instead on understanding the processes of neurodegeneration and how to intervene there. At present there seems likely to be a much greater margin of return on anti-Alzheimer compounds than on antidepressants. In contrast to the situation in depression, there is already greater understanding of some of the processes underlying neurodegeneration than there is of the processes that regulate mood. There are also clear indications that some already existing agents, notably the MAO-B inhibitor deprenyl, have effects on some of these processes. The reversal of neurodegeneration, or the development of agents that are neuroprotective, will clearly be of benefit in cases of dementia, but they may also prove beneficial for individuals with recurrent depressive disorders or with schizophrenia. It is not possible to work on the brain without potentially shedding light on what happens in mental illness. Abandoning an exclusive focus on monoamines and efforts to develop drugs for mental illness within preset guidelines may paradoxically allow the emergence of what Schildkraut called a new "reductionist oversimplification" that might be "of heuristic value."

Alternatively, if the monoamines are the critical neurotransmitters, this suggests that while we may be able to improve therapeutic efficacy to some extent, we may soon approach a limit—after all ECT, the most effective treatment, can still take some weeks to work, doesn't cure everyone, and doesn't cure permanently. If this is the case any improvement in outcomes will only be obtained by combining antidepressant principles, some of which may be psychotherapeutic techniques. If the experience of depression makes someone more neurotic, it is not clear that resolving his depression with a drug will necessarily remove his newly acquired maladaptive responses. The scars of depression may be both biological and psychological, in which case treatment in some cases may require the development of combinations of pharmacotherapy and psychotherapy—but as we shall see in Chapter 7 the logic of current developments militates against such combinations.

Another possibility is that we will need to consider a return from categorical to more dimensional models. Given that he was pilloried for his biological views in the 1960s and 1970s, there is a certain irony in the fact that Herman van Praag has emerged as a champion of the dimensional approach in recent years, formulating amine models of psychopathology consistent with such a view.[76] Whatever the specific details of van Praag's models, the perceptible trend in neuroscientific and genetic research is toward complexity and biological diversity rather than any convergence on unitary mechanisms of disorder. When the human genome is fully mapped the consequences for the affective disorders seem more likely to be that a range of biological factors will be found to trigger these disorders or inhibit their resolution, with different factors operative in different people, rather than that one or two unique factors will be found to cause depression and rigidly determine its course. For instance in the case of the eye disorder retinitis pigmentosa, which had all the appearances of a disease likely to have a single cause, over eighty different genetic inputs to the condition have been determined.

If something similar holds for depression (which has never looked as homogeneous an entity as retinitis pigmentosa), two implications follow.

One is that a range of antidepressant principles will be needed for the management of the depressive disorders. The other related point is that biology, far from underpinning notions of specificity with market value, may act to restrict the applicability of and the market for any specific pharmacotherapeutic approach, in that simple blood tests may indicate that the number of depressed people likely to respond to a particular anti-depressant is dramatically smaller than the number of people known to be depressed. What this possibility reveals is that there is no necessary equivalence between biology and specificity despite the fact that the sell-ing of the concept of specificity has invariably been in terms of biological metaphors. At present, however, one hundred years of categorical models and arguably the interests of both the pharmaceutical industry and the medical profession, as the next chapter may demonstrate, stand in the way of any early realization of van Praag's vision or any use of biology to dismantle the corporate superstructure of the health business.

For security against robbers who snatch purses, rifle luggage and crack safes, one must fasten all property with ropes, lock it up with locks, bolt it with bolts. This—for property owners—is elementary good sense. But when a strong thief comes along he picks up the whole lot, puts it on his back, and goes on his way with only one fear—that ropes and locks and bolts may give way.
—Chuang Tzu, 323 B.C.[1]

6   The Luke Effect

Drugs are clearly commodities to be traded in the marketplace, but scientific ideas are also tradable commodities. In part their survival depends on the confidence that consumers have in their companies of origin—such as Yale, Oxford, Harvard, and Cambridge—the famous Matthew Effect, described by Robert Merton in 1968.[2] In part the survival of ideas depends on their effectiveness. But in addition to effectiveness and company of origin, there is the question of whose interests these ideas coincide with, the degree to which they have a clear commercial or practical application, and the extent to which they can be packaged in a manner that will achieve brand-name recognition.

The Matthew Effect gets its name from the parable of the talents in the Gospel According to Matthew, which ends with the moral that to him who has, more shall be given and from him who has not even, that which he has shall be taken away. The idea of a coincidence with commercial interests suggests another effect—a Luke Effect, taken from the parable of the sower in the Gospel According to Luke. The sower sows the seed; some falls on stony ground and withers, some falls on fertile ground but is choked by weeds, while some falls on fertile ground and yields a bountiful crop. This parable ends with an exhortation to those who have ears to listen, which I will argue is what drug companies do very successfully, as part of the business of bringing the development of their compounds to fruition.[3]

In psychiatry, this effect is particularly important. It has become more and more difficult to distinguish between the marketing of scientific ideas and the marketing of psychotropic compounds. The selection of scientific ideas according to their coincidence with commercial interests is becoming an increasingly important factor in shaping the academic marketplace. Put another way, drug companies obviously make drugs, but less obviously they make views of illnesses. They don't do so by minting new ideas in pharmaceutical laboratories, but they selectively reinforce certain possible views. And to return to the problem facing Herman van Praag at the end of the last chapter, in terms of brand-name recognition, this process is in many respects easier to accomplish with categories rather than with dimensions.

We have seen an example of this right at the start of the psychopharmacological era, when in order to market amitriptyline Merck marketed the concept of depression by buying and distributing 50,000 copies of Frank Ayd's book on recognizing and treating depression in general medical settings. In a very similar manner today, companies invest in or otherwise support campaigns on the part of the American Psychiatric Association or the British Royal College of Psychiatrists to sell the idea that depression is underrecognized in primary care and that it should be recognized and treated appropriately. There is no evidence as far as I am aware to prove Peter Breggin's suggestion in *Toxic Psychiatry* that undue influence is brought to bear on such psychiatric associations by pharmaceutical companies.[4] In general, such campaigns have come about in response to the collection of disinterested scientific information and indeed are likely to have been launched after lobbying from mental health professionals. The *disinterested but selective* support of campaigns such as these, however, favors particular evolutionary configurations of ideas, as will become clear.

The Antipsychotics and Kraepelin

At several points in the course of this book, it has been hinted that the emergence of antipsychotics and antidepressants selected Kraepelin's ideas on mental illness. There was nothing inevitable about this evolution, however. It didn't happen because chlorpromazine was obviously and only an antischizophrenic agent. The facts of the matter are that chlorpromazine was first used in cases of manic agitation; one of the first randomized trials it was put through was for anxiety states, where it was found to be useful, albeit at a cost in terms of side effects, particularly demotivation. Chlorpromazine's European brand name—Largactil—which had been given to it before the idea that it was an antischizophrenic agent took hold, reflects initial perceptions of its likely *large* range of *act*ion. It took some time before the idea that this group of drugs might be "antischizophrenic" emerged. In part this emergence may have come about because virtually every clinical problem that led to someone's being hospi-talized in America in the 1950s and 1960s was labeled schizophrenia. This, on the one hand, biased drug companies against developing antide-pressants because the apparent size of the market didn't seem to justify the development costs but equally, on the other hand, put a premium on views that saw schizophrenia as a common and significant mental illness.

Included in chlorpromazine's large range of action were beneficial ef-fects in many cases of depression. The subsequent discovery of more spe-cific "antidepressants" has led to something of a downplaying of early observations of the effects of neuroleptics in depression,[5] in a manner that has left many wondering whether the division into antidepressants and antipsychotics isn't itself something of a convenience for marketing pur-poses. This nihilistic view can be refuted, not least because of studies re-ported by Klein and Fink comparing imipramine with chlorpromazine, which are discussed further below,[6] but it remains the case that after the introduction of the tricyclic antidepressants, companies producing both antipsychotics and antidepressants had no incentive to sell the antidepres-sant effects of their antipsychotics and every incentive to distinguish be-tween them.

In taking this approach and dividing their compounds into two large groups, the antidepressants and the antipsychotics, the companies helped select a Kraepelinian model of mental illness and the classification of mental disorders into the two broad categories of manic-depressive insan-ity and dementia praecox (schizophrenia) that went with that. In the United States in the late 1950s, Kraepelin was a relatively obscure German, a figure from the prehistory of psychiatry, yet by the 1970s and 1980s, many American clinicians had come to see themselves as neo-Kraepelinian. It was this ideology that was to issue in the creation of DSM-III in 1980.[7] One of the factors that needs to be taken into account when attempting to explain such a dramatic change and the switch from the dimensional view of mental illness found in the works of Meyer and Freud to the categorical views of Kraepelin is the influence of the interests of both the pharmaceutical and the insurance industries.

This evolutionary path left a number of ecological niches, however, which hint at possible alternative developmental routes. In 1958, for exam-ple, P. V. Petersen and colleagues in Lundbeck, a Danish pharmaceutical house, synthesized chlorprothixene, the first thioxanthene neuroleptic.[8] From this came two antipsychotics in common use to this day, clopenthixol in 1959 and flupenthixol in 1963. Although these drugs enjoyed a steady use, Lundbeck was a relatively small company without the resources to pour into a lavish marketing campaign. The firm has been especially re-sourceful, instead, in producing a range of short-acting and long-acting intramuscular preparations of their compounds, which have been popular, especially in Europe.

For the purposes of our story, of greater interest is the fact that unlike other companies producing antipsychotics, Lundbeck has capitalized on the "antidepressant" possibilities of neuroleptics in the marketing of flupenthixol. A number of uncontrolled studies were reported from Denmark in the late 1960s in which depressed subjects showed a response to flupenthixol that was comparable to that shown by other antidepressants: that is, 60–70 percent responded.[9] This was followed up by a number of double-blind studies, in which flupenthixol was compared with and found to be superior to a placebo and "equivalent" in efficacy to nortriptyline and amitriptyline.[10] It was noted that this response was brought about by low doses of flupenthixol—lower than doses then in use for the management of schizophrenia. The idea emerged that flupenthixol might be antipsychotic in high doses and antidepressant in low doses. The marketing of this idea, by virtue of a halo effect, could have been expected to be good for the antipsychotic sales of flupenthixol also in that a proportion of subjects on other antipsychotics who seemed "depressed" or flat would be more likely to be changed to flupenthixol as a consequence.

There were efforts to account for the antidepressant profile of flupenthixol in terms of distinctive neurobiological effects. Specific effects on presynaptic dopamine autoreceptors and on D-1 receptors were touted as the basis for a difference between it and other neuroleptics, but the flurry of biochemical explanations ignored the basic question of whether flupenthixol was actually in any meaningful sense of the word an antidepressant—or if that is too loaded a term, whether it was in some respect a compound similar to imipramine or phenelzine. In general, studies with flupenthixol were conducted on patients with mixed anxiety-depressive states, of a milder degree of severity, rather than in groups of patients with more severe and clear-cut endogenomorphic depressions. The point here is that to be "imipramine-like" a compound should ideally show a response in severer cases and indeed a differential response, such that the differences between the active drug and the placebo increase with increasing levels of severity of the condition.

Furthermore, where an antidepressant effect of flupenthixol was present, it was clearly there from the first week of treatment and often within the first twenty-four hours. Such rapidity of onset is characteristic of the effect of an antipsychotic in anxiety disorders but not of an antidepressant. Whether one has schizophrenia, depression, or is simply a control subject, all antipsychotics produce a reduction of tension or agitation within a few hours. The fact that schizophrenic illnesses typically do not substantially improve for several weeks after therapy has been instituted has tended to obscure the immediate therapeutic effects of these compounds—which differ markedly from the short-term effects of the antidepressants. In fact, just as convincing "antidepressant" credentials, based on clinical reports and trial results, can be made for thioridazine, chlorpromazine, perphenazine, thiothixene, and sulpiride.[11] In the case of all of these compounds, it would seem that their clinical benefits are apparent within the first few days of treatment and that they are most apparent in groups of mixed anxiety-depressive states.[12]

Part of the problem is that it is relatively easy to construct antidepressant credentials for a major tranquilizer. In contrast to the way Kuhn set about detecting antidepressant effects, such effects can be discovered by someone who has never seen a depressed patient in his life. Anything which sufficiently reduces scores on the Hamilton Rating Scale for Depression (HRSD) in a substantial number of depressed subjects can be claimed to be an antidepressant. A great number of the questions and observations on the HRSD concern anxiety. A consequence of this is that any treatment or procedure which reduces anxiety or tension may bring about relatively large and relatively rapid changes in the overall HRSD scores and may therefore be an "antidepressant."

While on one level this would seem to amount to something of a statistical fiddle, what shouldn't be ignored is that, a priori, tension reduction would seem to have something going for it as an "antidepressant principle." One of the problems of current clinical trial methods is that distinctions among different kinds of antidepressant principles are not encouraged. Flupenthixol clearly seems to be doing something different from imipramine, but how like one another are the MAOIs, the SSRIs, and the tricyclics? When it comes to clinical trials, which are almost all run by the pharmaceutical industry, the methods employed seem designed to obscure any clinical differences among "antidepressants" that preclinically may be extraordinarily diverse pharmacologically.[13]

Another evolutionary pathway would appear to be opening with the development of a new generation of antipsychotics (sertindole, ziprasidone, and olanzapine). Following the success of clozapine, companies have sought to develop antipsychotics that block 5HT-2 and other receptors in addition to the traditional blockade of D-2 receptors. A number of companies are currently coming to the marketplace with what will probably be essentially similar compounds. They must compete with one another, with clozapine and risperidone which have got there first, as well as with an older generation of compounds which will be much cheaper. In such circumstances, one can predict that at least one of the companies will look for a strategy to distinguish its compound from its competitors.

In 1987 DSM-IIIR introduced a "new entity"—delusional disorder (DD)—that appeared perfectly suited for the purpose of distinguishing one compound from another. Delusional disorder had formerly been called paranoid psychosis or paranoia. It was not commonly diagnosed, possibly in part because these terms are so close to paranoid schizophrenia that they create the general impression that the paranoid psychoses are members of the schizophrenic family of disorders. Far from this being the case, the delusional disorders do not show the dementia of dementia praecox and as such could be expected to be particularly responsive to neuroleptics—many of the people who have responded well to these agents in the past may have been suffering from paranoia rather than from schizophrenia. Most important, the criteria for diagnosing DD and schizophrenia, at present, are such that there are a large number of people who could be diagnosed as one or the other.[14]

Seeking a license for delusional disorder would allow the successful company to sell the concept of DD rather than just its compound—that is, to engage in an educational program rather than mere marketing, as Merck did with depression in the 1960s. A number of consequences might flow from this. One would be the very limited effect of producing a distinctive profile of the compound being sold that would lead to its recognition in a competitive marketplace. However, another possibility is that many patients, currently diagnosed as schizophrenic, would be likely to be rediagnosed as having delusional disorder, and a proportion of these would have their treatments switched on redesignation. A further predictable outcome is that efforts could be put into increasing the size of the market. It is known that there are a substantial number of individuals with DD, living in the community, who never come to the attention of the mental health services. These individuals are encountered by the legal system because of their querulous litigiousness or by the general medical system because of their delusional beliefs that they have a physical illness. Many see cosmetic surgeons because of beliefs that some part of their body is misshapen or otherwise abnormal. An educational campaign on the nature of DD would lead to pressure to increase the recognition of such disorders and to treat them.

But could drug companies effect such a large change in our views of mental illness? It was one thing to do so in 1960, when the drugs were first being introduced, but could the same trick be pulled off again forty years later? A consideration of the development of the concepts of social phobia, panic disorder, and obsessive-compulsive disorder may shed some light on what is possible.

The Long March of the MAOIs

With the introduction of amitriptyline in 1961 and its subsequent success, the discovery of the "cheese effect" of MAOIs and the outcome of the

1965 Medical Research Council trial in which ECT, imipramine, phenelzine, and placebo were compared and phenelzine appeared to be no more effective than placebo, companies with MAOIs had a considerable marketing problem. One response to the situation would have been a rejection of the MRC findings. Subsequent research suggests that this might have been the best option, in that it now seems clear that the dose of phenelzine used (45 mg) was too low, the duration for which it was given (four weeks) was too short, and the use of the Hamilton scale was biased against the MAOIs.[15]

Another possibility was to accept the findings but to look for another way to market the compounds. Essentially this is what happened. Despite the MRC study, many clinicians were convinced that the compounds worked—so if they didn't work for straightforward depression, they must work for atypical depression. The idea that there were atypical forms of depression had been around for some time. Indeed one way to see the division of depression into endogenous and reactive proposed by Roth and colleagues was in terms of endogenous and nonendogenous depressions, rather than biological and psychological, with the nonendogenous group including many subjects who were biologically depressed in a manner that retained mood reactivity rather than neurotically disturbed in a manner that might militate against treatment with antidepressants.

As early as 1959, E. D. West and Peter Dally had reported on the benefits of the MAOIs for a version of atypical depression.[16] Subsequently a variety of atypical forms of depression were proposed by various authorities. John Pollit and John Young, in 1971, floated the idea that there might be depressions with reversed functional shift.[17] Where typical depressions showed a loss of sleep and appetite, reversed depressions had increased appetite and sleep—features that are now ascribed to seasonal affective disorders. The idea that there might also be forms of depression manifesting as anxiety or phobic states, uncomplicated by personality difficulties—somewhat similar to panic disorder—was also canvassed.[18]

There were two features common to these various syndromes. One was the eminence of their proposers, which by virtue of the Matthew Effect could be expected to improve the reception of the ideas. The other, however, was a supposed preferential response to MAOIs, which drew the interest of the MAOI-producing companies. The specific response to an MAOI of any of these syndromes has never been replicated. Nevertheless these concepts flourished during the 1960s and 1970s, at least in part because it was in the interests of certain companies for them to survive. The method of ensuring survival lay in advertising campaigns that presented images of phobic or otherwise neurotic depressions responding to MAOIs, as well as company representatives who provided copies of the various articles on atypical forms of depression and their responsiveness to MAOIs. Today in all branches of the biomedical enterprise, drug companies disseminate large amounts of scientific literature. It is probable that literature from such sources makes up a significant proportion of what is read by many clinicians. Not unreasonably, the material that is passed on relates favorably to a company's marketing concerns. In this manner many concepts that might otherwise be retired early to inhabit the back shelves of libraries are given an extended lease on life.

While a profile for the MAOIs in terms of their effects on anxiety-related states was being constructed, a corresponding impression developed in the late 1960s and early 1970s that tricyclic antidepressants were only effective for the endogenomorphic form of the illness. Kuhn's concept of the nature of real antidepressants was being selected rather Kline's. But while Kuhn argued that tricyclic antidepressants are most clearly efficacious in the classical endogenomorphic form of depression, he also claimed that many seemingly neurotic states might be expected to respond to them on the basis that vital depression was the kind of condition that could be expected to give rise to neurotic developments. Treating the underlying depression could in such cases be expected to lead to a resolution of the neurosis as well. As early as 1960 Frank Ayd and Doug Goldman were exploring and describing the effects of imipramine in a number of such neurotic states, and, as we shall see, by the late 1960s Don Klein was also claiming that panic disorder responded to imipramine.

Thus while companies are in the business of suppressing clinical differences among their compounds in order to first achieve a slice of the big market, depression, thereafter they are not reluctant to differentiate the profiles of their compounds, the marketing of which may even extend to the selling of concepts. The advent of DSM-III in 1980, however, made the continued maintenance of idiosyncratic concepts, such as reversed functional shift depression, untenable. But if DSM-III was closing one marketing door, it unlocked another one that, for the makers of the next generation of MAOIs, opened onto what were potentially far more lush pastures.

SOCIAL PHOBIA

The development of the reversible inhibitors of monoamine oxidase-A (RIMAs) brought the MAOI crisis to something of a head. The two best-known RIMAs at present are brofaromine, which was developed by Ciba-Geigy, and moclobemide, developed by Roche, as we saw in Chapter 4. The development trajectory of these compounds was such that they were always likely to come onto the market shortly after the SSRIs. This would inevitably pose a problem—because a number of the potential marketing ploys that were available to any new generation of compounds, such as claims for greater safety in overdose and for a cleaner profile of side effects than the older compounds, were going to get used up by the makers of whichever set of compounds got to the market first. Added to that was the lingering cloud of suspicion over the MAOI group. Despite an impressive clinical trial profile, therefore, with minimal side effects and a clear separation from placebo that became clearer with increasing severity of the depression being treated, when moclobemide was launched it was slow to take off.

The problem was how to make a market for these compounds. This is where DSM-III's recognition of social phobia comes in. It is said that if one searches the nineteenth-century German literature, every possible syndrome can be found there and that social phobia accordingly is not new. For all practical purposes in Western psychiatry, however, social phobia only crept into the literature in the late 1960s, when in a review of cases presenting with phobias to the Institute of Psychiatry, in London, Isaac Marks and Michael Gelder noted that a small number of the phobic caseload appeared to be primarily anxious about social situations. Around this time there were also one or two more passing references to social phobia by Marks and by others such as Malcolm Lader who were investigating the effects of conditioning in a range of phobias.[19]

There were no dramatic descriptions of the new syndrome. No credit has been claimed for its discovery. When asked about it, Isaac Marks shrugs and says that in some sense it was always recognized that there were people who were socially anxious or phobic about social situations.[20] While it is clearly the case that other people have the potential to be at least as anxiety provoking as snakes and that activities such as speaking in public or setting up a date have always been recognized as major sources of anxiety, nevertheless the syndrome had not been formally distinguished before the late sixties. Even then it took a further ten years before a small number of groups got around to following up this lead and providing initial estimates of the frequency of the disorder and its characteristic features.

Again despite what may now seem obvious—that social situations are among the primary sources of anxiety for most people—up to 1985, it was estimated that social phobia was uncommon, almost rare.[21] The studies that were conducted, however, did enough to provide a set of inclusion and exclusion criteria by which it could be diagnosed, and this led to its incorporation into DSM-III in 1980.[22] To be diagnosed as having social phobia, subjects had to demonstrate significant fear of social situations, such as speaking in public, urinating in front of others (if male), difficulties with eating in public while engaging in social exchanges or problems with the kind of exchanges with bank tellers that used to happen more often before the development of automated teller machines, and they had to actively avoid such situations. There are elements of performance anxiety about this form of social phobia.

Another form was generalized social phobia. One could be generally anxious about most social situations that involved imagined scrutiny by others. Almost invariably people with this condition have thoughts like "I'm boring" or "I am going to mess this up and make a fool of myself." There is a general fear of blushing, stammering, or otherwise having problems with social situations, and this can lead people to take evasive action when they see someone they know approaching on the street, for instance, or writing to someone rather than visiting them or calling them on the phone, or taking a drink or some other drug before social encounters.

DSM-III also described an even more generalized form of social phobia, which was termed avoidant personality disorder. The criteria for this include a pervasive sensitivity to interpersonal rejection, which results in the individual concerned having no close friends or confidants and extreme social reticence. The person steers clear of social or occupational activities that involve interpersonal contact.[23]

Given that individuals who meet criteria for either generalized social phobia or avoidant personality disorder are liable to be miserable and unhappy and to have repetitive thoughts about their worthlessness, overlaid on a primary anxiety disorder, it should be clear that many of these persons, before the publication of DSM-III, would for the most part have been diagnosed as having neurotic depression, mixed anxiety-depressive disorders, or even depressive personality disorder. The social phobia concept, however, potentially extends somewhat further into regions formerly occupied by the concept of shyness.[24]

The inclusion of social phobia in DSM-III quickened the pace of research on the condition only marginally. In 1985 Mike Liebowitz and Don Klein of Columbia University could still paint it as a neglected disorder. But today it is commonly discussed, symposia are devoted to it, journal supplements appear on it, a number of large research projects are devoted to it, and estimates of its frequency place it as affecting up to 10 percent of the population. The volume of work on social phobia is now such that one commentator has dubbed it the panic disorder of the 1990s.[25]

What has happened? Starting in late 1986, reports began to trickle out of the Columbia University group, in particular, that pointed to a response of social phobia to phenelzine.[26] Unlike the use of the MAOIs for the various atypical depressions, these responses have been replicable. Reports from some centers point to levels of response of up to 80 percent. This is distinctly superior to the responses to benzodiazepines, betablockers, or any other compounds. Of great interest is the fact that the MAOIs seem to be as useful if not more so for the more generalized types of social phobia and even for avoidant personality disorder—these compounds seem to strengthen the personality in some way. Considerable research is also being put in to the development of cognitive and behavioral packages for the treatment of social phobia, but it is proving a difficult condition to treat. Exposure to the feared situation is difficult to achieve, because social situations are often very transient. The self-devaluative attitude, moreover, is ever present but doesn't lead to habituation.

Since the initial reports with phenelzine, Roche has supported a number of studies with moclobemide for social phobia, and it has been shown to be of sufficient benefit to enable the company to pursue a license for it for this disorder (with the SSRI companies in keen pursuit).[27] A convenient reality for marketing purposes is that this condition is probably the least diagnosed of psychiatric conditions. Educational campaigns on depression are built around figures which have indicated that primary care physicians miss up to 50 percent of cases of depression, but for social phobia it is estimated that less than 3 percent of those who could be diagnosed are actually picked up. Even more conveniently, the reality has caught the imagination of patient groups. A number of groups, including the World Health Organization, are getting involved in the business of educating the public about the nature of social phobia and the potential for treatment; many of these efforts are supported by Roche. Obtaining a license will almost certainly lead the company to market social phobia rather than moclobemide, in the expectation that sales will follow.

Would this development have happened or would it have happened as quickly without pharmaceutical company involvement? The issues are complex. Although there are clearly marketing interests at stake, it would be a mistake to see this simply as a cynical marketing exercise. One reason for exercising caution in interpreting what is happening is that while social phobia has been an infrequently recognized condition in Western psychiatry, the picture worldwide is somewhat different. In Japan, a condition called Taijin-kyofu, which appears to be roughly comparable to social phobia, has for the better part of this century been regarded as the commonest of the neurotic disorders. It is thought, there, that social phobia may lead to a paranoid disorder, and recent work in the West would appear to confirm this; individuals with conditions such as delusional disorder and body dysmorphic disorder appear to have many features of social phobia also.[28]

Some of the dynamics that have conspired to keep social phobia out of view in the West, as well as some of the dynamics that go into the shaping of our notions of mental illness, can be glimpsed from a consideration of the possible overlap between social phobia and pathological forms of shyness. Clearly a lot of people are shy, but shyness has never been viewed as a possible medical condition in the West. Between East and West, therefore, there have been differences either in which difficulties in interacting socially have led to mental health consultations, or in which social expectations lead to difficulties in interacting sufficient to attract a diagnosis of mental illness rather than just shyness. Now, however, that the concept has taken root in the West, a great many people and enterprises other than Roche can be expected to have an interest in nurturing its development. Self-help books on the management of social phobia, which advertise themselves as pertinent to anyone who may be shy, have begun to appear, as have magazine articles, television programs, and advertisements for therapy.

The selling of social phobia could be seen as solely a marketing exercise, in which a number of unexpected bedfellows can be found together. Alternatively, the great increase in interest in this condition, and the almost inevitable increase in estimates of its prevalence that can be expected, may indicate that there are a number of different ways to slice the psychopathological cake. It is not clear, a priori, that there is a particular way to slice that cake that will come closest to some basic reality. It may even be that drug company involvement in selecting certain concepts is bringing us closer to a best-fit classification scheme, although at some point soon alarms about the medicalization of the neuroses entailed in the forthcoming mass diagnosis and treatment of social phobia seem certain to be raised, following as it does on similar developments regarding panic disorder and obsessive-compulsive disorder.

Panic Disorder

In 1962 Don Klein was working at Hillside Hospital in New York. He was the resident "druggist." This was at a time when most American psychiatrists believed analysis to be the only proper form of treatment for mental illness. In many institutions there was a grudging acceptance of the new drugs. It was felt that drug therapy somehow sidestepped important dynamic issues, papering over the cracks rather than properly curing anything. Involvement with such a therapy was not desirable, and therefore it followed that analysts should not prescribe—it was thought that the act of prescribing, the introduction of an escape route, would necessarily compromise the therapeutic relationship. This led to a situation in which one doctor in the hospital was commonly designated as the prescriber or druggist. He or she would institute drug therapy, leaving the analyst free to continue with psychotherapy.[29]

While working at Hillside, Klein and Max Fink instituted some of the most radical studies ever done with the new drugs. In one series of investigations they randomly compared chlorpromazine, imipramine and placebo, giving them to patients regardless of whether the diagnosis was schizophrenia, depression, mania, or anxiety. Although there are many studies showing that a great number of the neuroleptics may be of benefit in depression, theirs was one of the few randomized blind studies to show that imipramine was superior to a neuroleptic, confirming Kuhn's original observations that it truly was different.[30]

In the course of this study Klein noted that there appeared to be a form of periodic anxiety that also responded to imipramine. His attention had been drawn to this by the case of a particular patient. He had been asked his advice on a man who like almost everyone else in American hospitals at the time had been diagnosed as having schizophrenia, but who had responded poorly to chlorpromazine. The man had clear and circumscribed episodes of intense agitation but no evidence of delusions or hallucinations. During the episodes of anxiety, he would rush to the nurses' station, in fear for his life. The attacks would last a matter of minutes, only to recur several hours later. An atypical picture.

Klein suggested trying imipramine. The dose was titrated up over a few weeks, during which time the patient complained of side effects and noticed little or no benefits. After a month, the patient was still complaining that the treatment appeared to be of no help and his psychotherapist also complained that the drug was of no use, but the nursing staff noticed a difference. One nurse in particular picked out the fact that they were no longer seeing the rush to the station and the panic of a man who thought he was going to die. The patient still experienced a significant degree of generalized anxiety, and, to this day, Klein and others would argue that treatment in one sense only starts once the acute episodes of anxiety have been brought under control. There is still a great deal of unlearning of avoidance behavior and recovery from the debilitating effects of anxiety that needs to happen before the person can be said to be better.

Investigating this and similar responses to imipramine further, Klein came to the conclusion that the attacks of paroxysmal anxiety, which he was later to call panic attacks, came first, and that these were then overlaid with phobic reactions, generalized anxiety and depressive reaction patterns. The general reaction to this announcement was that he had got it wrong. Isaac Marks and the behavior therapy lobby argued that Klein was misdiagnosing cases of agoraphobia. The importance of the misdiagnosis was that according to Marks the hallmark of a phobia was the pattern of avoidance that went with it and that exposure of subjects to what they were avoiding was the proper therapy. Others felt that the response of these states to imipramine indicated that Klein had misdiagnosed cases of depression—had failed to spot that a vital depression, as Kuhn had suggested, could underlie many cases of what seemed to be just anxiety states.

The initial dismissal of his ideas was facilitated by the fact that Klein was operating from Hillside Hospital and from outside of the university system. It did not depend solely on this factor, however. Martin Roth, one of the most eminent British psychiatrists at the time, had described what he termed a phobic-anxiety-depersonalization syndrome, which bore many resemblances to Klein's panic disorder.[31] Roth and colleagues, however, suggested that the disorder responded preferentially to MAOIs—an idea that gained some currency in the United Kingdom as part of an effort to find a niche for the MAOIs. A multiplicity of names and a syndrome that fell outside the conventional diagnostic framework, therefore, probably also played parts in holding back the recognition of panic disorder.

Things began to change with the creation of a task force to draw up DSM-III. Klein was known to Bob Spitzer, the chairman of the task force, and he was included in the panel to review the anxiety disorders. Spitzer and a number of the others involved in drafting DSM-III were keen to get rid of the older term anxiety neurosis, because of its Freudian connotations. There was agreement that Klein's panic disorder should be included, partly because a clear set of inclusion and exclusion criteria could be set out for it but also because its inclusion chipped away at the anxiety neurosis monolith. Agoraphobia with and without concurrent panic disorder was also included, as was social phobia. However, it was clear that there was more to the anxiety disorders than simply panic disorders, agoraphobia, and social phobia. Floundering somewhat, members of the anxiety disorders subcommittee stumbled on the notion of generalized anxiety disorder (GAD), and consigned the greater part of the rest of the anxiety disorders to this category.

It was known that benzodiazepines could be of benefit in many anxiety states, and given the responsiveness of panic disorder to antidepressants, the expectations were that GAD would be responsive to benzodiazepines. In line with perceptions that the antidepressants were "stronger" medicines than the benzodiazepines, it was thought that GAD was a common and not so severe condition—one that might be responsive to both tranquilizers and talking therapies—while panic disorder sat at the apex of the anxiety disorder pyramid—a rarer and much more severe condition.

In the late 1970s, however, David Sheehan and James Ballenger, operating from Harvard, ran a study of imipramine and phenelzine against a placebo in panic disorder patients. They concluded that the ideas that Klein had been putting forward were correct—that there was a periodic form of paroxysmal anxiety that responded to imipramine.[32] But the fact that the study came from Harvard helped legitimize the concept of panic for many (another instance of the Matthew Effect). The data were presented at an American Psychopathological Association meeting in 1978, at which Sheehan suggested that companies should systematically test current and future antidepressants for antipanic activity.

Listening to the presentation were Jim Coleman and Bob Purpura from Upjohn, who approached Sheehan afterward to inquire about a new compound Upjohn had, a novel benzodiazepine, called alprazolam. The company had some evidence to indicate that alprazolam might have antidepressant activity. Did Sheehan think that it might be useful for panic disorder? Conventional wisdom was that a benzodiazepine should not have been of much benefit in panic disorder, but a novel benzodiazepine that had antidepressant activity—who knew? Sheehan suggested conducting a trial of alprazolam in patients attending his anxiety disorders clinic.[33] The trial was run, and the results for alprazolam were encouraging. A number of the patients who had previously been crippled with anxiety showed clear-cut responses to it.

Upjohn consulted Sheehan and others, in particular Gerry Klerman, then one of the most influential figures in U.S. psychiatry, on how to move forward from there. The company was persuaded that it would be worth its while to run a clinical trial program for the purpose of obtaining a license for the use of alprazolam in panic disorder. The strategy was to come onto the market with a drug that would on the one hand be perceived as being effective against the severest anxiety disorders while on the other hand would be found in practice to be as safe and convenient to use as the older benzodiazepines. The expectation was that this would lead to alprazolam's general displacement of other minor tranquilizers from the anxiety disorders market.

Achieving this required a number of studies that would demonstrate an efficacy for alprazolam in panic disorder. Upjohn funded two key studies. In the United States, a phase one set of investigations compared alprazolam with a placebo in panic disorder. These were published in the *Archives of General Psychiatry* in 1988. A second set of investigations, phase two, comparing imipramine, alprazolam, and placebo, involved study centers in Spain, Denmark, Germany, France, Sweden, Austria, Belgium, Mexico, Columbia, and Canada. The outcomes of the second phase were published in 1992 in the *British Journal of Psychiatry*. There was in addition a study that was eventually run out of London and Toronto, which had been in the first instance designed to look at the longer-term outcomes of treatment with alprazolam, imipramine, and behavior therapy. Isaac Marks had been approached several years previously about getting involved in such a study, and he had agreed to participate. There were difficulties, however, in coming to an agreed protocol, so that what, for some, began life as an effort to establish the effect of long-term treatment with alprazolam ended up being a study that looked at the long-term outcome of a shorter course of treatment with alprazo-lam. In the course of this research Upjohn withdrew its support, as we shall see.[34]

Before we look at the details of the investigations, one point needs to be made. In order to run a clinical trial, a set of workshops have to be or-ganized in which trialists are briefed on the details of the protocol, tested for inter-rater reliability, and generally engaged in discussion on the mer-its of the protocol and the possible place of the treatment being investi-gated in the management of the condition being looked at. During the trial there is often a need for follow-up workshops to assess rates of re-cruitment. After it there are meetings to outline the results and discuss how best to move forward. Depending on the approach of the company, such meetings can be held at expensive venues. The company may also fund speakers at symposia within meetings of the American Psychiatric Association, the Royal College of Psychiatrists, the British Association for Psychopharmacology, the American College for Neuropsychopharmacol-ogy, the CINP, and others. Senior figures in the field sympathetic to a concept such as panic disorder or to the company can arrange for sym-posia to focus on the question of panic disorder, for instance, or current treatments for it.

All these activities are legitimate. Upjohn, just like many other compa-nies before it, sought to influence the marketplace through such means. Whether it has done more, however, has been a matter of some debate. In 1991 Peter Breggin's *Toxic Psychiatry* was published, which at least im-plied more—that Daniel X. Freedman, the editor of *Archives of General Psychiatry,* and Gerry Klerman had been unduly influenced by the com-pany.[35] It is the case that the 1985 alprazolam studies were accorded a singular prominence in the *Archives,* which, given that the *Archives* is the most influential psychiatric journal in the world, must have been worth something to the company. Gerald Klerman, as lead coordinator of the studies, wrote an introductory article laying out the rationale for the stud-ies.[36] The implicit charge that a person's opinion has been corrupted by any association with a drug company is all but impossible to decide upon, in that a great number of senior clinicians will have been consulted by one or another company at some point. Many indeed will seek or actively take on a consultancy in order to use the funds provided for their services to support their broader research portfolio.

Does alprazolam work for panic disorder? It does. The points of con-tention have centered around the question of whether it works suffi-ciently well to constitute a therapeutic advance. Following the *Archives* articles, Isaac Marks and a number of others wrote a letter to the journal,[37] which Marks later claimed Freedman was slow to publish and then only published shorn of some of its central points and with a reply from Klerman and colleagues.[38] When the *British Journal of Psychiatry* reported the outcome of the phase two studies,[39] Marks and colleagues were given space to comment and did so vigorously,[40] to which Klerman replied.[41] Klerman had died before Marks and colleagues reported the London-Toronto outcomes, in 1993 in the *British Journal of Psychiatry*,[42] but close colleagues of his, Martin Roth, Myrna Weissman, and others, commented on this work[43] and Marks and others replied.[44] The various correspondences had a ferocity not seen since the debates on lithium. The correspondence on these issues rather dramatically pits Marks (from the Maudsley) and a few colleagues against many of the best-known names in the field, creating an impression of a few Greeks holding out at Thermopylae against the Persian hordes.

In their opening comments on the phase two studies Marks and colleagues comment that in their view "this study's elephantine labor has resulted in the delivery of a mouse." Among other things they strongly implied that any apparent efficacy of alprazolam resulted more from statistical manipulation on the part of Klerman and colleagues than from a genuine efficacy of alprazolam. In reply, Klerman noted that the study had been submitted four years earlier to the *British Journal of Psychiatry* for publication but had been held up owing to repeated critiques and requests for revision on the part of the referees. The fact that it had only finally been published with Marks's comments suggested an embarrassed trade-off with the editor of the journal, Hugh Freeman. Klerman noted that during the course of that four years other studies had appeared, and Marks was now complaining that the original study had not addressed the issues raised by subsequent studies—which on the face of it smacked of shifting goalposts. Marks, however, claimed that Klerman was well aware of the results emerging from these other studies as early as 1988.

There had been a series of increasingly acrimonious meetings between the two at this stage. The most public of these was in Geneva in March 1990, at a meeting to mark a decade of research on panic disorder.[45] At the gathering Klerman, who was by then wheelchair bound as a result of diabetic and renal problems, and Marks, who had formerly been a close friend of Klerman's, and who shared with him many common interests including a concern to demonstrate the efficacy of psychotherapeutic approaches to nervous disorders, angrily and publicly disagreed.[46]

The tension was heightened because the results of the London-Toronto study had begun to be available, and they appeared to show that behavior therapy provided longer-lasting effects than alprazolam. Richard Swinson, who uniquely had also participated in the initial studies on alprazolam, first presented these at the Carlsbad Golf and Country Club in La Jolla in November of 1990. He was surprised at the response. A significant proportion of the audience was clearly supportive and enthusiastic. But many others were hostile. He put the hostility down to the fact that, by 1990, a growing number of American psychiatrists were only reimbursed for treatment if that treatment involved prescribing. The study was submitted to the *New England Journal of Medicine,* which refused to accept it, for reasons which still surprise its authors.[47] It was finally taken by the *British Journal of Psychiatry.*

Following the critique of the London-Toronto studies by Roth, Weissman and others, Marks wrote that "monitoring and support [by Upjohn] stopped abruptly when the results became known. Thereafter, Upjohn's response was to invite professionals to critique the study they had nurtured so carefully before. The study is a classic demonstration of the hazards of research funded by industry." He added that the designs of the studies involving alprazolam had been dictated by Upjohn's interest in registering its drug with the FDA rather than the interests of science. He made a plea for a recognition that "pharmaceutical marketing and science are different disciplines." The imputation that the efforts of the industry are dangerous was being made in the inner sanctum of psychiatric debate.[48]

It is difficult not to feel some sympathy for Marks, who has championed the cause of making effective psychotherapies available for self-treatment to sufferers from psychological problems but who is faced with the problem that "with no major industry to advertise its superiority, however, exposure [therapy] may take a long time to become widely used." Yet as is usual with real life, the issues get murkier on closer inspection. Not only does Marks criticize Upjohn and its efforts to market alprazolam, but in a subsequent issue of the *British Journal of Psychiatry* he and colleagues drew attention to faults in a study of cognitive therapy in panic disorder. This critique draws attention in turn to the fact that the psychotherapies, in general, are very reluctant to engage in studies that would pit one modality against another in order to find out what are the effective elements common to both. There was also, in Marks's case, a history of struggle with the pharmaceutical industry, as the next section will bring out.

Still, quite apart from the question of whether alprazolam actually works other than in the short term for panic disorder or whether its effects are as good as behavior therapy or what these altercations may reveal about the disinterestedness of the scientific protagonists, there remains the fact that panic disorder has come from being virtually unrecognized in 1980 to being commonly diagnosed. It is even widely diagnosed in the United Kingdom, where prescriptions of alprazolam are not reimbursed under the national health service and despite the hostility of many prominent British psychiatrists to this American import. How has this happened? It happened because Klerman and others in designing the alprazolam studies were at least as interested to see how the new DSM and American concepts such as panic disorder internationalized as they were to see how the drug worked, and the clinical trial program was a potent tool to this end.

But in addition Upjohn, in the first instance, and subsequently other companies have found it in their interest to sell the concept of panic disorder. With the recognition of panic disorder has come awareness that the condition is quite common and that far from being of extraordinary severity it may respond to a wide range of interventions. This has meant that a number of companies have sought licenses to treat it, and it has been in all their interests to keep panic disorder on the symposium agenda, or as the subject of satellite meetings, or featured in television documentaries. Even if the documentary doesn't mention drug treatment, even if it gives the impression that behavior therapy is to be preferred, increasing the recognition of panic disorder will inevitably lead to an increase in the prescription of agents that have been associated with its treatment.

The seeking of a license to claim that alprazolam—or any other compound—is efficacious for panic disorder, furthermore, is ipso facto a potent marketing tool. In these times of increasing litigiousness, prescribers are taking increasing care to be seen to follow accepted rules. Many are uncertain about the exact meaning of a license. Many feel that they are on uncertain ground prescribing off-license—as one would be doing if for instance one were to prescribe imipramine for panic attacks, despite the fact that there is more data on the beneficial effects of imipramine for this condition than there is for any other compound. Few seem to realize that licenses are not intended to place a limit on what a prescriber can prescribe; they are rather intended to limit what a drug company can claim. Many clinicians forget this, however, and they are susceptible accordingly to advertising which claims, for instance, that alprazolam is the one and only treatment licensed for the treatment of panic disorder, as the early advertising for alprazolam did.

In this manner companies make their markets. It may often be far more effective to sell the indication than to focus on selling the treatment. It may be that the medicalization of the neuroses is proceeding too fast apace. It may be that a recognition of features of some anxiety states, such as hyperventilation and patterns of avoidance, and a recognition of what people can do for themselves are downgraded as a result of the focus on the periodic eruptions of paroxysmal panic, which suggest a biochemical disturbance appropriately managed by drug therapy. However, although the pendulum may have swung too far, it must be conceded that the concept of panic disorder has now been established and a classification system that includes it is surely more accurate than one that doesn't.

Clomipramine and Obsessive-Compulsive Disorder

One of the most compelling instances of a Luke Effect, indeed the paradigm that established the mold for the subsequent discoveries of social phobia and panic disorder and for the drawing of battle lines between Isaac Marks and the pharmaceutical industry, involved the marketing of clomipramine and subsequently the other 5HT reuptake inhibitors for obsessive-compulsive disorder (OCD).

As outlined in Chapter 2, imipramine took Geigy Pharmaceuticals and the rest of the pharmaceutical industry by surprise. Ordinarily, when a compound is produced it is produced as part of a series and when one is found to be active, others from the same series can be tested to see which compound produces the best trade-off between therapeutic activity and side effects. Sometimes a compound that appears to be part of a series shows an activity that is completely different from that of other members of the series and is, for reasons that were not initially appreciated, better characterized as belonging to a different series. So it was with imipramine.

The chemists at Geigy, Roche, and Merck, however, were slow to build on this. Roche and Merck, as we have seen, produced amitriptyline, which was almost identical to imipramine, but even so they expected it to be an antipsychotic. Geigy chlorinated imipramine, seeking to produce a more potent compound on the basis that a chlorination of promazine produced chlorpromazine which was markedly more potent than the parent compound (see Figure 6.1). The result was G34586, later to be called chlorimipramine or clomipramine. Despite the example of imipramine, G34586 was first given by Walter Pöldinger to ten schizophrenic patients, in what is surely a telling example of the fact that the antidepressant market had not yet been made. Like Kuhn, Pöldinger thought he could discern an antidepressant effect in some of the schizophrenic patients, and this led to the prescription of the compound for depression. But even though it was synthesized in 1958, it only reached the American market in 1990, under the trade name Anafranil and then for obsessive-compulsive disorder rather than for depression, even though by then many clinicians worldwide believed it to be the most potent antidepressant.

In Geigy's hands, clomipramine began to progress slowly toward the market, at a time when the FDA's regulations for registering a compound were progressively tightening, following the thalidomide disaster. Clomipramine posed a further problem because it was the kind of molecule the FDA was being accused of licensing too easily—a "me too" product incorporating a minor variation to its molecular structure designed to

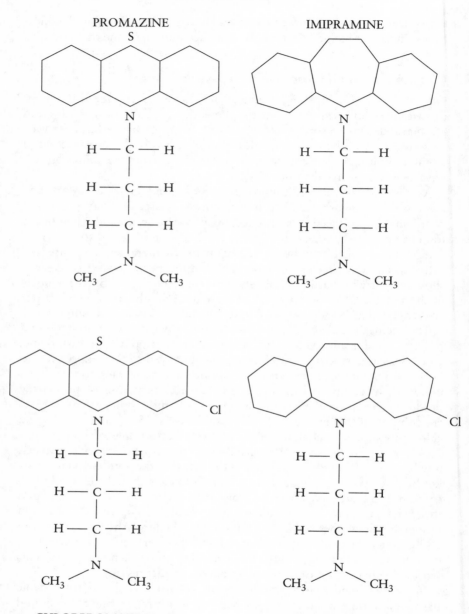

Figure 6.1   Promazine, imipramine, chlorpromazine, chlorimipramine

get around patent laws. There was, at the time, no clear evidence that it was more potent than imipramine, but there was some evidence that it was more toxic. While safer in overdose that most other tricyclics, during the 1970s clomipramine had a higher rate of unexplained sudden death. This may have owed something to its availability as an intravenous preparation.[49]

Very few antidepressants come in an intravenous preparation. Usually a company produces a tablet or capsule formulation first and moves on to a liquid or injectable preparation when the first form has begun to lose market share. Perhaps aware of the difficulties clomipramine would have in penetrating the marketplace, Geigy had produced an injectable form almost from the start. This had something of a vogue in specialist centers, with clomipramine units being established to treat difficult cases. A number of the leading figures in the field in the United Kingdom and Europe were enthusiastic advocates of this form of treatment. Whether it seemed better because it was more distinctive or whether it really did something different from the oral preparation is less certain. Given intravenously clomipramine is not metabolized as quickly by the liver, and more of the parent compound, which is more 5HT-reuptake inhibiting, gains access to the brain. The treatment is not without hazards, however. A small proportion of subjects have a slow metabolism of clomipramine and other tricyclics, and the infusion of large amounts of the drug intravenously can lead to a confusional state with lethal consequences.[50]

The job of preparing the regulatory submission in the United Kingdom fell to George Beaumont shortly after his appointment as medical director of Geigy UK. It was clear to Beaumont that clomipramine was not going to take the market by storm unless a distinctive profile could be established for it. Running through reports generated by its use in a number of European centers, he noticed that Jean Guyotat in France had claimed, in 1967, that it might be of some use in OCD.[51] He also became aware of a clinic run by Dr. Juan Lopez-Ibor in Madrid in which clomipramine was being given intravenously for a range of severe anxiety states, apparently with good results.[52]

Beaumont organized a group of British clinicians to go to Madrid to see Lopez-Ibor's procedures in practice. They were impressed with the results and came home happy to participate in a series of studies to replicate Lopez-Ibor's findings in a sample of British patients. Geigy then funded a number of studies in which clomipramine was given orally or intravenously to patients with a range of what were then thought of as anxiety neuroses—phobic disorder and OCD. In order to do anything systematic that could be used for registration purposes, Beaumont and his colleagues had to invent the methodology of drug trials for OCD. There were no established rating scales, because the condition had been thought

to be too rare to warrant systematic investigation. The Leyton Obsessional Inventory was one of the only scales that could be used. The studies, however, were largely open or single-blind.[53]

The results were supportive of what had been found in Madrid and
clear-cut enough to be used for registration purposes. If anything, the response of the phobic patients was somewhat better than that of the OCD
patients. These and other studies formed part of Ciba-Geigy's submission
to the Committee on the Safety of Medicines in the United Kingdom,
which led to a license for clomipramine, in 1975, for the treatment of depression and associated phobic and obsessional states.

Picking which way to market the compound, thereafter, posed certain
problems. The market for phobic disorders appeared to much larger than
that for OCDs, but the makers of MAOIs had successfully targeted that
market. Moreover, pushing clomipramine for phobic disorders would
have come up against clinical perceptions that the tricylics were not particularly useful for this group of disorders. The alternative was to go for
OCD. The problem with OCD in the mid-1970s was that, while it was
viewed as being difficult to treat and a treatment for it would accordingly
have been welcomed, it was also thought of as being rare.

Ciba-Geigy lost interest in the compound for a number of reasons.
First, it seemed that it would never get a license in the United States,
where the FDA had turned down submissions for registration. Second,
Geigy had been merged with CIBA, and clomipramine, as an "old" Geigy
drug, was not in favor. Third, with the development of the catecholamine
hypothesis, the laboratory focus was on drugs that would specifically inhibit norepinephrine reuptake. As chance would have it, at the time of the
merger, both Ciba and Geigy were working on maprotiline, which did
just that. When Arvid Carlsson called on Geigy, before the merger, to talk
about the possible 5HT-reuptake inhibiting properties of clomipramine,
he found even Geigy scientists uninterested.

The matter was left to George Beaumont. Clomipramine had been
taken up by British and European psychiatrists, who, possibly in part because the cachet of being able to give it intravenously, had begun to think
of it as a second-line antidepressant but perhaps the most potent. Its developing reputation led to a steady if unexciting penetration of the UK
market. A number of clinicians, picking up on its possible usefulness in
obsessive-compulsive disorder, gave it to patients who presented with any
obsessional features. At the time, obsessionality was seen by many as a
possible prodomal feature of schizophrenia. Thus if a young person presented with obsessional symptoms along with a family history suggestive
of possible schizophrenia, he or she was liable to treatment with
clomipramine, sometimes in very high doses (up to 1000 mg/day), on the
grounds that whatever the risk of side effects, it would be outweighed by

the benefits of forestalling the development of a possible schizophrenia. Clomipramine was also available in Canada, and its reputation led to its being smuggled across the border into the United States or to Americans picking up supplies of it when on visits to Britain or Europe.[54]

George Beaumont had by this time become increasingly interested in the obsessional disorders. Isaac Marks, in contrast, had become increasingly irritated with claims that clomipramine was an antiobsessional agent, and he suggested a study to tease out whether it or behavior therapy was the better treatment for OCD. Marks, working at the Maudsley Hospital, was one of the leading exponents of behavior therapy in the world. He was an advocate of behavior therapy for OCD for three reasons. One was that he had been involved in the development of a particular form of behavior therapy called exposure therapy, for which he campaigned widely, partly on the basis that it is a form of treatment that gives control of the illness to the sufferer. Second, although far from being an analyst, Marks shared the view that the neuroses were appropriately treated by nondrug means. In the case of OCD, behavior therapists were able to claim that there was a logical link between the theory of what was wrong and the practice of behavior therapy. Finally, there was the question of evidence; the evidence that behavior therapy worked was convincing according to Marks, while the evidence that clomipramine did anything other than what Kuhn suggested—treated an underlying depression—was inconclusive.

A protocol for the study was hammered out between Beaumont and his advisers and Marks, whereby OCD subjects would be entered into a blind trial in which both behavior therapy and clomipramine were delivered and the results compared. Depressed subjects were excluded. Everybody got behavior therapy, as Marks felt it would be unethical not to give a treatment that had been proved to be of benefit. As of 1977, Marks began to report the results at meetings, although the official paper did not appear until 1980. His claim was that behavior therapy clearly worked and that clomipramine had minimal effects. These effects had a slow onset, had not been shown to be sustained beyond treatment with the pills, and probably stemmed from an influence on depressed mood. This last point was a shade thin, if it is borne in mind that the study protocol had excluded subjects who were clearly depressed. It is all but impossible for subjects with OCD not to register scores on the Hamilton scale, which can then be divided into higher and lower scores, but whether the higher scorers should be seen as more "depressed" is another matter.[55]

Faced with Marks's interpretation of the findings, Stuart Montgomery from St. Mary's Hospital in London offered to run a straightforward comparison of clomipramine and placebo in OCD. Beaumont agreed and the trial was run. Clomipramine was substantially superior to placebo.[56]

At that time, no other drug had ever been demonstrated to be superior to placebo for OCD. Part of what was striking in fact was that, compared with studies in depression, the response of OCD to placebo was so poor. Recently, there has been a phenomenon called placebo drift, so that in clinical trials many patients with OCD now show some response to placebo. The reason for this may be that when earlier studies were conducted, the only visible patients with OCD were a small number of very disturbed individuals who had marked and obvious incapacities. With the increasing recognition of OCD, this has changed and a range of patients are now picked up who have milder forms of the disorder, some of whom may respond simply to enrollment in the therapeutic endeavor.

In the mid-1980s Marks and his group took up cudgels again. In response to criticism that the effects of clomipramine could not have been picked up in his original trial design, he ran a further study comparing clomipramine with a placebo in patients who were also receiving exposure or antiexposure clinical management instructions. As reported in 1989, his conclusions remained essentially the same.[57] He questioned the use of clomipramine for OCD given the availability of a superior and less hazardous treatment. He also noted that by then clomipramine had developed something of a reputation as a specific treatment for OCD.[58] Taking issue with this, he argued that while there were a number of studies supporting claims that clomipramine had some effects, the impressions of specificity seemed to arise more because there were so few studies of anything else. Objections from Ciba-Geigy that there were studies comparing clomipramine to other antidepressants such as desipramine[59] met the rejoinder, heard first in the lithium arguments, that the patients in these studies had not been properly randomized.

In this case, however, the debate did not follow the lithium route. There had been further developments. Clomipramine had moved into studies in the United States for OCD. In addition, following the rejection of the license application for depression for clomipramine by the FDA, with the emergence of the SSRIs American clinicians were able to explore the possibility that clomipramine's superiority in OCD compared with other antidepressants stemmed from the greater potency of its actions on the 5HT system. This led to trials, first of all of fluvoxamine, and thereafter with fluoxetine[60] and since then with both paroxetine and sertraline for OCD. In all cases these studies have demonstrated that drugs active on the 5HT system are of modest benefit, and all these compounds have received a license in the United States, the United Kingdom, and elsewhere for the treatment of obsessive-compulsive disorder. As mentioned, clomipramine was finally licensed by the FDA for use in the United States for OCD in 1990.

These trials have a number of striking features. One, as noted above, is the issue of placebo drift. Second, if I had been asked in the mid-1980s whether there were enough patients to support the number of multicenter trials that have since been run on OCD, or whether there would be sufficient patients to treat in the event that all the SSRIs were licensed for the condition, I would have answered no. I thought at the time that the market for even one SSRI in this area was limited. Epidemiological studies in the United States and elsewhere have since found, however, that far from being as uncommon as was previously thought, OCD may affect up to 3 percent of the population—up to 7 million people in the United States and 10 million in Europe, a substantial market. So where have all the patients come from?

In 1989 Judith Rapoport published a book called *The Boy Who Couldn't Stop Washing*.[61] It contains over twenty compelling clinical vignettes, which once read would crystallize obsessive-compulsive disorder in the mind of a sufferer or of a diagnosing clinician. The accounts brought home just how disabling the condition could be. One vignette tells the case of a person who, plagued by concerns to keep his apartment just right, resorted to cleaning out to the door, pulling it shut behind him, and then living on a park bench for two years. Clearly such a condition can be as disabling as schizophrenia. Common to all the cases described was an unwillingness to come forward and acknowledge the full extent of the problem for fear of being thought crazy.

Two other cases may be useful here. I was referred a young man, eighteen years old, whose primary care physician was concerned because he had just finished his secondary schooling and was due to go to university that autumn but had presented with a worrying complaint—he had apparently begun seeing things. The referring doctor was unclear just what his patient was seeing but was worried about early schizophrenia. Such a symptom, indeed, until quite recently would have very commonly led to a diagnosis of schizophrenia without much effort to dig deeper and establish the true nature of the condition. The physician wondered if he should advise his patient not to go on with his schooling.

When I saw the patient, however, it became clear that what was involved was not visual hallucinations. The young man had misled his physician to some extent, owing to his embarrassment at the nature of his problem, which was that when confronted by young children he often had an image of himself interfering with them sexually. This intrusive worry is not untypical of OCD, where sufferers are haunted by repetitive images of themselves engaged in activities such as harming their children or shouting out obscenities in church. Far from revealing the true nature of the person, arguably, these worries become the problem they do is in part because they are actually deeply repugnant to the individual. Such symptoms, however, are always likely to lead to guardedness on the part of the sufferer. Clinicians from Freud on have recognized that OCD is the most guarded neurosis—hence the failure to make the diagnosis.

A further case involves a twenty-eight-year-old woman, who was referred for a picture of mixed anxiety and depression. On her initial visit she was interviewed for close to ninety minutes and was on her way out the door when she let drop a comment that alerted me to a possible diagnosis of OCD I had missed. Further questioning established the diagnosis, although it seemed a mild case and perhaps not in need of aggressive treatment. She entered a trial of a compound for OCD and was interviewed every few weeks thereafter, concealing all the way through the extent of her problems. It was known that she had problems checking things, such as electrical plugs—she had agreed to this. But one day in clinic, in the presence of her husband who didn't normally attend, she said that she had improved as regards checking. He agreed, but prompted her to reveal that the previous evening he had come home from work, before her—unusually—and had been unable to make himself a cup of coffee. The reason was that not only did his wife check things, but to make doubly sure she removed the cords from any appliances they could be removed from and took them to work with her. This is not the kind of complaint that is concealed because the sufferer feels it to be reprehensible, but in her case, as with many others, there can be a reluctance to reveal what is perceived to be the full stupidity or lunacy of the problem.

The Boy Who Couldn't Stop Washing sold widely and Dr. Rapoport was featured on a number of television shows and on educational videos for clinicians. Her work and that of others has brought the existence of OCD to the attention of a much wider audience. Obsessive-compulsive sufferers' groups have been established and the condition is rapidly gaining increasing legitimacy in both clinical and nonclinical eyes.

Of particular importance is a new sensitivity to the fact that OCD can present without any obvious manifestations of compulsive behavior. It is now clear that subjects with intrusive imagery or impulses can attempt to neutralize these by purely mental rituals, leaving observers none the wiser. Renewed interest in both the obsessive and the compulsive aspects of the disorder has also led to a recognition of possible affinities between OCD and other disorders with intrusive cognitions, such as body dysmorphic disorder or delusional disorder, and disorders of compulsive behaviors such as Gilles de la Tourette syndrome.[62]

The question arises, then, to what extent has the revision of our views of OCD been shaped by the interests of the pharmaceutical industry? At this stage, it should be remembered that there are up to five companies with an interest in promoting their products for this indication. One way in which such shaping can happen is for the industry to sponsor conferences where OCD is discussed, to support the publication of journal supplements giving the results of studies on OCD, and to distribute such supplements or copies of books on OCD, such as *The Boy Who Couldn't Stop Washing,* as well as to support the foundation and continuing existence of patient support groups.

Although their occurrence does not mean that anything unethical or untoward has been done in the case of OCD, all the above methods of making a market have been employed in various areas of medicine and within psychiatry for various indications, and it would seem that the OCD story is not untouched by such influences. Indeed, far from considering this to be a problem, a good argument can be made that these efforts to rehabilitate OCD have redressed a balance. At the turn of the century, before the rise of manic-depressive insanity and schizophrenia, Pierre Janet, who had emerged as the leading figure in French psychiatry, saw OCD or the diathesis to obsessionality as the most common psychosyndrome. With the rise of schizophrenia and Freud's global view of anxiety neurosis, OCD along with hysteria was eclipsed.

The reemergence of OCD points to new ways to slice the psychopathological cake. It challenges our concepts of neurosis and psychosis in a way that no other condition does. If it is a neurosis, as it has been traditionally classified, it is a neurosis that is commonly as severe and disabling as a severe case of schizophrenia. Many clinicians do not see it as primarily an anxiety disorder—in which case, if it is a neurosis, there is a need to clarify what this term means, as hitherto it has meant a disorder predicated on a maladaptive handling of anxiety. Equally, however, the problem cannot be solved by redesignating OCD as a psychosis, at least as the term is used now, because it is not commonly accompanied by delusions or hallucinations.

Writing in 1989 before OCD had reemerged quite as fully as it now has and before the current craze for evidence-based medicine, Marks bemoaned the difficulties in getting services to take up treatments that have been shown to be effective. He was primarily arguing the case for behavior therapy, whether delivered by trained nurses or even delivered by self-help methods—bibliotherapy.[63] He wondered whether such approaches would ever be successful unless the marketing efforts of a large corporation were behind them. One way to solve this problem would be to get some of the larger pharmaceutical companies to diversify beyond pharmaceuticals and become health care delivery companies.[64] At this writing, the Pfizer pharmaceutical company has agreed to run a double-blind study of a computerized method for self-treatment of obsessive-compulsive disorder developed by Marks, John Greist, and David Baer.[65] The efficacy of the method will be researched in its own right and as a supplement to Pfizer's SSRI, sertraline, in the management of OCD. Its significance is that if proved effective, this piece of psychotherapy will be marketable as a product rather than as a service, which in due course may transform the mental health marketplace.

The Politics of the Orgasm

In the course of investigating the use of clomipramine, George Beaumont made what one hesitates to call a seminal observation. Visiting the Winwick Hospital in Warrington, where there was a clomipramine infusion unit, he met a patient being treated for depression, who observed that, when at home over the weekend, he couldn't ejaculate while having intercourse with his wife. He was treated intravenously with clomipramine in the unit during the week, but he was taking capsules when at home on weekends. Suspecting that the drug might be causing the problem, he experimented with varying the dose and found that he could control his ejaculations—the length of time to orgasm could be manipulated rather precisely.[66]

Reviewing the area, Beaumont found reports from as early as 1960 from Frank Ayd of patients' sexual difficulties while on antidepressants. Similar reports had begun to trickle in from elsewhere following clomipramine use, with a 1969 report from South Africa, for instance. Part of the problem was trying to judge what was caused by any particular drug and what was part of depression—which commonly leads to a loss of libido anyway. Another problem was trying to judge what part of the increased reporting stemmed from the changing climate regarding sex. Finally there was a question of disentangling troublesome side effects from a possible new therapeutic indication.

A second clinical observation opened the subject up further. Looking through routine reports filed by clinicians of side effects, Beaumont noticed a report from a primary care physician, Pulvertaft, near Peterborough, on a woman who claimed to be unable to have an orgasm while on clomipramine but who became orgasmic again following its discontinuation. This led to a recognition of the effects of clomipramine on female sexual functioning. (Indeed, in lectures on the preferability of behavior therapy to clomipramine for OCD, Isaac Marks has, on occasion, cited as an instance of one of clomipramine's disturbing side effects the case of a nun he was treating who was subject to three days of spontaneous orgasms upon withdrawing from clomipramine.) Beaumont then sponsored studies with Ray Goodman[67] at a clinic for sexual dysfunctions in Manchester and John Couper-Smartt in Liverpool.[68] It was found that in subjects with premature ejaculation, as little as 10 mg of clomipramine taken two to three hours before intercourse could be used to treat many cases of premature ejaculation in men and that orgasm could be similarly delayed in women.

These observations are of interest from a number of points of view. First there is the issue that such a small dose of clomipramine was needed, and second the rapidity of its action. The general prejudice was that the delay in onset of the effects of the antidepressants was owing to their slow entry into the brain and the resistance of brain receptors to rapid alteration. The findings with clomipramine appeared to blow all these ideas out of the water—if, that is, its effects were brought about centrally. The alternative was that these changes came about by a local action on the penis or vagina. The problems that these effects of clomipramine posed for theories on the mode of action of antidepressants were never addressed at the time, probably because of an unwillingness to address the issue of sex but also because of uncertainty regarding the location of clomipramine's effects on sex. It now seems clear that these are mediated through an action on the brain and that these observations pose a very considerable problem for any theory of antidepressant action.

As interesting or even more interesting than the failure to pursue the theoretical issues raised by clomipramine's effects on sexual functioning has been the failure to capitalize on these effects. Not understanding how a drug works has never stopped a company from marketing a drug for an indication if it was clear that it did work. It is now clear that all of the SSRIs have effects on sexual functioning comparable to those of clomipramine, affecting well over 50 percent of users and possibly up to 80 or 90 percent. Furthermore, company studies have indicated that up to one-third of men under the age of forty have problems with premature ejaculation. But at present, despite the initial observations with clomipramine that were made twenty-five years ago, none of the companies who are in a position to seek a license for the use of their drug for this indication have done so.

What is happening? Some idea of what is involved can be gleaned from reactions to Beaumont's initial observations. He presented the findings at a meeting in Jersey in April 1973. In the course of noting that these side effects might have a therapeutic role, he detailed an adverse events report from a primary care physician noting delayed ejaculation but commenting that "incidentally, his wife—a big film star type girl—wants him to stay on the drug as he can maintain an erection for considerably longer than ever before which is helping to resolve one of the causes of her upsets." A story about drugs to enhance sexual functioning appeared in the *Sunday Mirror* several days later—even though there had been no members of the press at the meeting. The managing director of Geigy was very upset and indicated that further publicity of this kind would not be welcomed. Although Beaumont subsequently funded a number of studies of the effects of clomipramine on sexual functioning, the Geigy profile in this area dropped and when Beaumont left Ciba-Geigy a few years later this line of work fell into abeyance.

This episode illustrates what is a general concern on the part of the pharmaceutical industry with its image—firms take into account the potential worldwide impact. This was most strikingly brought home in the case of the development of RU-486, the so-called abortion pill.[69] In developing this, Roussel encountered very substantial opposition from a range of different interest groups worldwide. It became clear to them that any monies they might stand to make through their development and marketing of this compound could lead to equal if not greater losses through an unwillingness of companies, clinicians, or patients around the globe to purchase or use any Roussel products.

Being perceived to develop pills to improve individuals' sex life was subject to the same risks and has been responsible for the reluctance to overtly develop this area—at least until the current use-patents for the SSRIs come close to expiring. In place of direct development, for this kind of indication, companies are prepared to wait for off-license prescribing by clinicians to lead to market development. There is nothing to stop any clinician from prescribing any SSRI or clomipramine for premature ejaculation, and an increasing number are doing so, influenced by a series of case reports of possible benefits and a number of studies. While such prescribing is off-license, in the event of any adverse reactions physicians could defend themselves by an appeal to the body of evidence that now exists to indicate that such a prescription has a rationale.

Off-license prescribing will be needed to establish other sexual indications such as the drug treatment of anorgasmia, impotence, and loss of libido. Work on the effects of SSRIs on sexual functioning in animals strongly suggests that drugs, such as buspirone, active on the 5HT-1A receptor, and trazodone, which blocks the 5HT-2 receptor, should bring orgasm forward. This is particularly likely to be of benefit in women or men with anorgasmia or loss of libido. Such compounds should be relatively reliably aphrodisiac.[70] Trazodone, for one, has developed something of a reputation in this respect. Following its pre-Prozac launch in the United States as an antidepressant, its sales were substantial, and it would seem that this owed something to its effects on sexual functioning. But even though there is clinical trial evidence that a very similar compound, yohimbine, has reliably aphrodisiac properties, the area was downplayed by Roussel.

The climate, however, seems to be slowly changing. Faced with the success of the SSRIs, the companies who have produced some of the post-SSRI compounds, Bristol-Myers Squibb with nefazodone (Serzone/Dutonin), Roche with moclobemide (Mannerix/Aurorix), and Burroughs-Wellcome with buproprion (Wellbutrin) have sought to highlight the issue—in terms of the sexual dysfunctions consequent on treatment with SSRIs. The marketing pitch is likely to increase awareness of the potential to prescribe for premature ejaculation and anorgasmia, even though none of the companies may ever seek a license for these indications.

Off-license prescribing will also be needed to establish indications such as obesity.[71] When compounds are developed that reliably cause weight loss, without an undue cost in terms of side effects, then they will probably be introduced on a license for use for weight loss before surgery, in the expectation that once their use is permitted for such an indication, use will leak outside that indication. Unless there are dramatic adverse developments, such use will establish the acceptability of treating obesity by medical means.

The issue of expanding off-license use has recently been the subject of some dispute in the case of Retin-A.[72] This drug was approved by the FDA in 1971 for use in acne. In the mid-1980s, there was some evidence that Retin-A might be of some benefit as a treatment for wrinkles. In 1988 Ortho, a division of Johnson and Johnson, held a news conference to outline the results of a study in the area. This almost inevitably led to appearances on talk shows by members of the medical profession. These tactics led the FDA in December 1990 to investigate whether Ortho's approach amounted to off-license marketing. In the course of the investigation, Ortho files and other material was destroyed by company employees. The company was subsequently charged in 1995 with obstructing justice but not with improper promotion. There would seem to be a very fine line between promotion and keeping the public abreast of what is genuinely newsworthy.

The New Medical State

The discoveries that "antidepressants" may be effective for OCD, social phobia, and panic disorder are of interest for a number of reasons. One is that they again raise the question of antidepressant discovery. Kuhn's claim that imipramine was a treatment for vital depression, in contrast to Kline's discovery of a psychic energizer which might be useful across a wide variety of neurotic states, appears less solid than it once did. In the 1960s, before imipramine achieved gold-standard status, there were a number of competing classification systems of the new compounds. Walter Pöldinger and Paul Kielholz saw imipramine as having mood-elevating or thymoleptic effects in contrast to the MAOIs which had drive-enhancing or thymeretic effects. Clomipramine fell mid-way between these poles, while amitriptyline veered off in a neuroleptic direction.[73] That there were differences between the various compounds was widely accepted.

The blurring of such distinctions was probably dictated primarily by marketing needs to gain access to the largest possible market—that is, depression—for which purpose the less complex the concept of depression and of drug effects the better.

The power of the pharmaceutical industry to shape psychiatric culture should, however, be seen as a power to shape rather than be dismissed as a corrupting influence. It could be argued that in many respects all that is happening is a redressing of an original corruption brought about by the private practice interests of the psychoanalytic establishment, which were dimensional in the extreme. In addition current psychiatric culture is not only shaped by the pharmaceutical industry. Considering the shifts in symptomatology that borderline personality patients have presented with over the past decade, Jerome Kroll has argued strongly that there is substantial evidence for a clear interplay between territorial and market forces in the mental health professions that acts to shape the clinical features that patients are described as presenting with. Psychotherapists no less than the pharmaceutical industry coach both their colleagues and the potentially treatable population as to the currently most appropriate clinical presentations.[74]

Given the many revisions of psychiatric nosology during the last thirty years, it is clearly a mistake to think that mental illnesses are something that have an established reality and that the role of a drug company is to find the key that fits a predetermined lock or the bullet that will hit an objective target. Although there are clearly psychobiological inputs to many psychiatric disorders, we are at present in a state where companies can not only seek to find the key to the lock but can dictate a great deal of the shape of the lock to which a key must fit. The situation is rather similar in some respects to that described by John Kenneth Galbraith in *The New Industrial State,* in which it was proposed that the marketplace, far from being shaped by the laws of supply and demand, is constructed by large corporations, whatever the branch of industry involved. Companies act to control not only the supply of products to a market but also the flow of ideas to the ideas market and to regulate the demands that come from the market.[75]

In purist circles there will be great resistance to the idea that one can create reality by marketing efforts. It can be argued, however, that it was ever thus. At the start of the psychopharmacological era, for instance, the distinctions drawn in Newcastle by Martin Roth, Leslie Kiloh, Roger Garside, and colleagues between endogenous and reactive depression, in contradistinction to the then dominant depressive spectrum views espoused by Aubrey Lewis and others, became extremely influential. But arguably these formulations owed much of their impact to their coincidence with the interests of the manufacturers of the first antidepressants, and it was the use of those antidepressants that internationalized the Newcastle formulations rather than any correspondence of drug responses with some preexisting truths. A similar argument can be used to account for the preeminence of the Hamilton Rating Scale, which was so well freighted to detect the effects of the tricyclic compounds.

Indeed in psychiatry, at least, a good case can be made that astute marketing extends right into the heart of academe itself. In psychopharmacology, the ideas that have caught on have done so because their originators have had a talent for coining a pithy title to describe a phenomenon— such as Type I and Type II schizophrenia or the amine hypothesis of depression. As Chapter 5 indicated, any consideration of the development of psychopharmacology makes it quite clear that good marketing of such ideas can capture a field, either before the evidence is in on an issue or even in the face of considerable contradictory evidence. Ideas, such as the dopamine hypothesis of schizophrenia or the amine hypotheses of depression, have functioned very much as brand names. They attract loyalty and are consumed to the extent that their proponents are effective promoters of their brand-name product and there is a body of people who feel some sense of community in endorsing particular brand names.

In the larger framework, the psychopharmacological era did a great deal to reestablish the Kraepelinian synthesis. There were two elements to this; one was a commitment to a distinction between disease categories and the other was a set of specific details about the disease categories— that is, that there were essentially two major mental diseases, manic-depressive insanity and schizophrenia. We shall see in the next chapter that the availability of modern psychotropic drugs led to the emergence of a neo-Kraepelinian school of psychiatry in the United States. This was responsible for the replacement of a dimensional diagnostic framework, DSM-II, with a much more categorically oriented one, DSM-III, whose concepts have yielded rich pickings for the pharmaceutical industry.

But if the psychotropic drugs support the categorical thrust of the Kraepelinian framework, more recently the efforts of the industry seem likely to modify or even subvert the specific details of Kraepelin's proposals. The developing interest in OCD, social phobia, panic disorder, and dysthymia has brought to the fore concepts that have played little or no part in the great game of psychopathology for almost a century. In addition, it would seem highly likely, with the development of atypical neuroleptics, active on both serotonergic and dopaminergic systems (the SDAs), that there will be increased interest in concepts such as delusional disorder in order to differentiate these antipsychotics from an older generation of compounds.

Companies are also in the business of disseminating information about relapse rates in depression and the need to consider potential lifelong treatment for some patients. Such suggestions will almost certainly run up against consumer resistance—from both prescribers and the takers of antidepressants. This has put a premium on the development of management packages to improve compliance (of both prescribers and takers). The pressure to develop in this way has increased as successive generations of compounds have appeared to be equi-effective, with no prospect of dramatic improvements in therapeutic responsiveness. Developments in the psychotherapies in recent years have been such that it becomes possible to conceive of creating such treatment packages—with a view to improving the image of the product with both prescribers and consumers and thus gaining a market advantage—and hence has led to company interest in working with packages such as that developed by Isaac Marks.

Finally, there is the issue of evidence-based medicine. As outlined in Chapter 3, this at first blush might seem an unimpeachably disinterested development, but the facts of the matter are that few groups in society have the resources needed to produce the kind of data that is increasingly becoming the currency of the health services. They who control the means of data production potentially control everything.

Control of the Media

The developments outlined in this chapter would appear to hinge on the issue of who controls the media. The media in question is the medical media, because owing to the prescription-only status of the drugs in question, the real consumers are the prescribers and not the takers. There are a number of ways in which company influence on the media operates to achieve what amounts at times to effective, though usually hands-off, control.

Traditionally there have been two areas of concern. There is advertising, the money from which has become necessary for the continuing operation of most psychiatric and medical journals—but we can ignore this, as "no one is ever influenced by advertisements." A much trickier issue has been the question of editorial boards of journals populated by individuals with company affiliations. This has raised concerns from the late 1950s regarding whether the independence of scientific journals is compromised as a consequence.[76] Concerns about affiliative links of this sort feature prominently in Breggin's *Toxic Psychiatry*. But the proliferation of journals is so great now that it should be possible for any author to circumvent problems of this sort.

More subtly, there are an increasing number of journal supplements which appear in the same format as their parent journals and which many readers may not appreciate are not peer reviewed in the same way as articles in the parent journal. Many journals are dependent on such supplements for their survival and in almost all instances it will only be the pharmaceutical industry which can afford to sponsor such supplements.[77] In addition to the sponsorship of supplements, however, there is the matter of dissemination. Where most independent researchers would feel themselves to be doing rather well if they have a few hundred reprint requests for an article, symposium supplements or favorable reprints may be bought in quantities from 10,000 to 50,000. There is nothing illegitimate or unethical about this, but which ideas will go further?

A similar influence can be detected in the development of consensus conferences. These will only be objective and yield results that are valid in so far as all points of view are represented at them. Consensus conferences, however, tend to be sponsored by pharmaceutical companies and to represent the utility of drug treatments for particular conditions, with little consideration of other treatments or effort to place treatments in a comparative perspective.[78]

In the field of postgraduate education it has become increasingly possible to link centers through optical networks or satellite or video links. There are clear advantages to this in that peripheral centers can now have access to quality presentations and the educational opportunities offered by centers of excellence. Such media at present have substantial sponsorship from the pharmaceutical industry, and accordingly it would seem likely that presentations through such media will feature speakers with a leaning toward biological treatments with the latest pharmaceutical compounds.

There are two developments that might significantly change the current pattern of development. One stems from efforts on the part of governments to exert some control over prescribing through purchasing consortia, whether managed care consortia in the United States or family health service authorities in the United Kingdom. Such developments have increased the importance of the psychiatric pharmacist. The significance of this is that companies have now begun to court pharmacists in a way that was not done before. If this courting is at the expense of efforts to influence clinicians, the delicate balance that has given rise to the possibility of selling compounds through a marketing of the medical model and of specific clinical concepts—a balance predicated on a congruence of interest between industry and physicians—may be lost.

Another development concerns the possible extension of over-the-counter status to include psychotropic compounds. During the course of 1994, in the United Kingdom, a number of compounds, notably the H2 antagonists, which were formerly prescription only, had their status changed to OTC. There would seem little reason why the SSRIs, for instance, could not similarly change status. Whether this is likely to happen or not may depend more on commercial considerations at time of patent expiry than on anything else. In a number of European countries, a range of hypnotics and tranquilizers that can only be gotten on prescription in the United Kingdom or the United States are available over the counter. It is accordingly rather difficult to argue in principle that there are reasons why such developments could not occur for other psychotropic compounds.

A major effect of a change to OTC status would be that company marketing efforts would switch to a greater emphasis on dimensional rather than categorical concepts, as lay views of mental disturbances are more generally dimensional. The most interesting scientific questions then would be how long it would take to dismantle the current categorical framework and what effect this would have on the psychiatric profession. These questions all touch on the nature of the disease concept and the extent to which society is prepared to medicalize its problems. Our interest in the question of health is becoming almost our central preoccupation, but as it does we seem to find ourselves ever more alienated from medical models of health and disease. These issues as they relate to psychiatry will be pursued in the next chapter and in the Postscript.

7 From Oedipus to Osheroff

The funding allocated by Congress in 1955 for the evaluation of psychotropic drugs led to the establishment of the Psychopharmacology Research Center.[1] Jonathan Cole, who as noted earlier served as the center's first head, in 1959 employed Gerry Klerman to coordinate the running of a nine-hospital study of chlorpromazine, which reported in 1964 and established the indisputable value of the antipsychotics in the management of schizophrenia.[2] Klerman, born in New York and trained in the Massachusetts Mental Center in its analytic heyday, was to become one of the most influential figures of the era. Working in the PRC he helped establish methods for the evaluation of drug treatments, liaising with figures such as Michael Shepherd, who was engaged in a similar exercise with the British Medical Research Council. He subsequently, as will become clear, participated in the creation of interpersonal psychotherapy, was a driving force behind the establishment of the epidemiological catchment area studies, moves to DSM-III, and the NIMH study on the evaluation of psychotherapy, and was the most important figure, after Osheroff himself, in the Osheroff case.

While planning the first trials of chlorpromazine, Klerman also inaugurated a series of studies of psychiatric orientations. These studies grew out of observations that psychiatrists had markedly different views on the question of whether the new drugs had contributed to the emptying of the hospitals, which had begun to become apparent around 1955–56. On the basis of a series of studies, Klerman drew distinctions among psychiatrists whom he termed socially oriented skeptics and three other groups—

psychodynamic skeptics, psychodynamic critics, and somatically oriented clinicians. He later fleshed these distinctions out into descriptions of social, psychotherapeutic, and somatic ideologies.[3]

This is a chapter about ideologies. As applied to beliefs about the nature and treatments of mental illness, ideologies can be distinguished by their judgments of what ought to be regardless of what actually exists. Evaluative judgments are their hallmark. A strong belief in one treatment for all illnesses would be evaluative in this sense, as would picking a treatment for humanitarian reasons rather than because of a proven ability to cure. Ideology is a term most commonly associated with politics. In politics, the protagonists of competing viewpoints have often been able to remain aloof from the issue of proof, and perhaps because of this, ideological divides in this arena have been able to persist over centuries. Scientific ideologies have also persisted over decades, especially in branches of science where critical experiments play a lesser role, as in the social sciences, history, psychology, and others.

This is all very well, as long as the science doesn't impinge on clinical practice, because no one expects the human sciences to be "hard" sciences. But once clinical practice comes into the frame, there is a potential problem, because the hallmark of a professional supposedly is that he or she will offer the best advice. Commitment to only one model of therapy would reduce a clinician to the level of an astrologer or a reflexologist. When it comes to treatment, therefore, it can be argued on the one hand that ideology should have little or no part to play. On the other hand, if enthusiasm for a treatment is an important part of therapy, as it may be in psychiatry even more than in other areas of health care, then disinterested assessment becomes much more problematic. Not believing passionately in the treatment being delivered or believing that it is only one of a range of possible options may well reduce response rates to that treatment. These were issues that were to recur for both Klerman and Shepherd throughout their careers.

Given his involvement in one of the first randomized control trials, his participation in some of the most influential multicenter trials of drug treatments, and his role as a founder and council member of the CINP, Michael Shepherd's credentials as a psychopharmacologist are impeccable. This notwithstanding, Klerman's socially oriented skeptical position has rarely been articulated so well by anyone else. Consider the following quote from Shepherd in 1960:

> In the light of current knowledge it would be difficult . . . to base the reputation of psychopharmacology on a large body of established knowledge about either the therapeutic value or the pharmacodynamic action of the psychotropic drugs. To what then can the high

status of the subject be attributed? . . . Status we may recall is a sociologic concept closely related to economic class . . . In retrospect it is clear that the newer pharmacological compounds were introduced to psychiatry at a time which favored their adoption: on the one hand a growing disillusion with psychological treatments had become increasingly evident even where these methods had been taken up with most enthusiasm; on the other hand the seams of the older physical methods of treatment had been exhausted and even their most ardent advocates were dissatisfied . . . The great majority of eclectic psychiatrists had long recognized the limitations of available methods of treatment and they have welcomed the clinical possibilities of the new drugs, finding great advantages in the relative flexibility of pharmacotherapy and less obviously in the employment of measures which are more in keeping with the traditional forms of medical care. Their eagerness to employ pharmacologic methods of treatment, stimulated by a vigorous campaign on the part of the pharmaceutical industry has played an important role in the widespread adoption of drug therapy in psychiatric institutions.[4]

Whence this skepticism? Shepherd was also a staff member at the Maudsley, which under the leadership of Aubrey Lewis had turned decisively toward social psychiatry in the 1950s. Lewis in London and David Henderson in Scotland had adopted the psychosocial models formulated by Adolf Meyer in America, which in due course gave rise to the community mental health movements in both countries. At the founding meeting of the CINP, Lewis was famously to say that "if we had to choose between abandoning the new psychotropic drugs and abandoning the industrial rehabilitation units and other social facilities available to us, there would be no hesitation about the choice: the drugs would go."[5]

Indeed Lewis put a social stamp on British psychiatry as the Maudsley Hospital became the training ground in the 1950s and 1960s for the first wave of psychiatrists to fill chairs in medical schools in Britain. Biological psychiatry, in contrast, developed outside the Maudsley and settings that owed allegiance to it. When it came to establishing a British Association for Psychopharmacology in 1974, none of those involved initially had any connection with the Maudsley, Oxford, or Cambridge.[6] Within Britain, there was some incomprehension on the part of those who saw themselves as biological psychiatrists at what they perceived as the Maudsley's lack of interest in the practical business of getting people well.[7]

This pattern wasn't unique to Britain. In America, psychopharmacology developed first in the state hospital system, in settings comparable to those in which it developed in Britain. The American College of Neuro-psychopharmacology (ACNP), which was founded in 1963, was initially an organization of scientists operating from outside the then analytically dominated centers of Harvard and Yale.[8] Similar divisions in German-speaking psychiatry can be seen in the confusion surrounding what Kuhn thought his developments signified. As in Britain, the high psychopatho-logical establishment in Germany was oriented to social psychiatry. These divisions underpinned with some feeling the debates on the merits of the MAOIs, the prophylactic effects of lithium, and other issues that have been covered in earlier chapters.

The social psychiatrists were skeptical of drug therapy and stressed the importance of environmental inputs to mental health. Their research fo-cused, for example, on the role that criticism by, hostility from, or the overinvolvement of relatives might play in precipitating relapses. They supported the delivery of community mental health services in order to avoid the secondary effects of institutionalization. Group therapy and family therapy were preferred to individual therapy. But above all the so-cial approach is associated with the view that the large hospitals were closing even before the introduction of the neuroleptics.[9] The new drugs at best may have accelerated the process. Extreme versions of this ap-proach found their way into the writings of the antipsychiatrists, in which illness was seen simply as a manifestation of social conflicts.

The most subtle version of the social critique was put forward by Shepherd in 1990, when he proposed that the success of psychotropic drugs might be accounted for in terms of an Oedipus Effect.[10] The Oedi-pus Effect was first described by Karl Popper in 1974; he contrasted it to Freud's Oedipus complex. The Oedipus Effect refers to way in which the pronouncements of an oracle can be self-fulfilling. Thus, in the case of Oedipus, the oracle predicted that Laius, king of Thebes, would be killed by his own son. At birth, therefore, the child was bound hand and foot, taken away from the court, and left on a remote mountainside to die. He was found, however, by a shepherd who took him home, looked after him, and gave him the name Oedipus because of his swollen feet. Twenty-odd years later, when traveling on the road toward Thebes, trying to stay away from Corinth and the man he believed to be his father, whom he also had learned from the oracle that he would kill, Oedipus met a car-riage that had halted on the road. The driver was whipping the horse in order to get it to move on. A fight ensued in which Oedipus ended up killing the passenger in the carriage, who was Laius. Clearly he might have otherwise killed him, but the implication is that had he known it was his father he would not have killed him there and then. Had Laius not paid heed to the oracle in the first place, moreover, the situation that brought about his death would not have arisen.

Following a similar dynamic, Popper argued, patients in therapy present dreams with Jungian themes to a Jungian analyst but dreams with Freudian themes to a psychoanalyst. The hopes of those who go to have their futures read are transmuted into predictions, which are then noticed by the person when they occur. The expectations of the oracle (therapist) condition the behavior of the person, so that what the oracle predicts is what comes to pass. Where drug therapy is concerned, Shepherd argued, the modern era was clearly shaped in great part by the introduction of neuroleptics, but it is a moot point whether the drugs got people out of hospital or whether clinicians, feeling that they had a safety net to fall back on, were more prepared to take risks than they previously had been. Once they had begun to decant patients, they then discovered the dangers of institutionalization. On this issue the profession had arguably followed the oracular pronouncements of an eminent few rather than the evidence.

While the views of Michael Shepherd are clearly skeptical, they do not reach an ideological extreme. That was the preserve of the antipsychiatry movement that developed in both America and Europe during the 1960s and 1970s. Its champions were Erving Goffman and Thomas Szasz in America, R. D. Laing in Britain, Michel Foucault in France, Franco Basaglia in Italy, and others. In its most extreme form, antipsychiatry saw biological psychiatry in particular as an instrument of oppression, so much so that Herman van Praag, as mentioned in Chapter 5, became a target of death threats and his lectures were picketed. For the antipsychiatrists, there was no such thing as a mental disease—there was only labeling by a society that was not prepared to tolerate deviant behavior—and the drug or ECT treatment of such behaviors was simply punitive.[11]

There was a certain Calvinism to the movement, in that certain principles were held to be sacrosanct and accordingly no compromise was countenanced, even in the face of evidence that a considerable number of people got well with ECT, antidepressants, or major tranquilizers. While he was in Utrecht, van Praag's department was the only one in Holland in which ECT was still given.[12] In America, ECT was banned in several states and has only made something of a comeback in the 1990s.[13] There was little tendency to be pragmatic, to question whether there was any form of practice that could be reconciled with the evidence of efficacy. Retrospectively, this was not so surprising, in that the idea that treatments should be evaluated other than by authoritative opinion had still not taken root even within mainstream psychiatry, and there was no agreed mechanism by which to decide on the merits of competing authoritative opinions. The evaluative style of the antipsychiatrists, however, was overlaid by a belief that biological psychiatry in some way threatened a spiritual view of man. If the guilt of a melancholic patient could be taken away by ECT, where did this leave morality in general or a religion like Christianity in particular?

In contrast to the social skeptics, moderate psychotherapy skeptics believed that if the drugs appeared to work they did so in great part by virtue of the enthusiasm of the prescriber rather than by virtue of anything intrinsic to the drug itself. A more hard-line or critical position was that the drugs didn't work. Moreover, by subverting the process of getting to understand the defenses a person typically employed, they blocked an individual's possibilities of ever getting really well. For the committed psychotherapist, such a view precluded involvement with drug prescribing. Because of this, as described earlier, arrangements were put in place in many U.S. hospitals and clinics during the 1950s, 1960s, and 1970s whereby one or two clinicians, the druggists, would be entrusted with the task of prescribing, leaving the therapists free from taint. This arrangement persists to this day—with a possible cost in terms of human life, as the Osheroff case may make clear.

From this ideological vantage point, psychotherapy is seen as the ideal treatment for what amount to ethical reasons—it shows the most respect to patients. Prescribing, if it is undertaken, is distinctly second-best. There were some grounds for such a bias in the early postwar years in that, in the pursuit of a cure for disease, patients had been subjected to a range of physical assaults, including the removal of tonsils, gonads, stomachs, intestines, spleens, uteri, and teeth; they had had their sinuses or any other sites of possible focal sepsis drained; they had had injections of colloidal calcium, inoculations with rat bite fever organisms; they had been rendered comatose with insulin, put to sleep for days, or made hypothermic; and a psychosurgery program was just beginning, which was to lead to 50,000 operations—the majority of which were almost certainly unjustified.[14] Clearly psychotherapy was likely to be less damaging than physical treatments such as these. But the bias remains for many even now, when physical treatments are clearly less destructive than they once were—we want a treatment such as cognitive therapy to work for what seem quasi-ethical reasons.

The psychotherapeutic approach became dominant in American psychiatry after World War II. Up until then American psychiatry, in broad orientation, had been more community oriented than European psychiatry. It was more prepared to see disorders such as neurasthenia, alcoholism, and substance abuse as being within the psychiatric remit. In terms of theoretical models it was poised between the disease model of Emil Kraepelin and the psychosocial model of Adolf Meyer. World War II led to an influx of clinicians into psychiatry, whose experience in the management of combat reactions led to a decisive switch in orientation from a disease model to Meyerian and analytic approaches.[15] This coincided with an expansion in the number of psychiatric departments, professorships, and residency programs. These positions all went to analysts. The expansion happened at just the time that effective somatic treatments were about to appear, setting the stage for a series of conflicts between an analytic establishment and a new breed of somaticists, conflicts that have spanned thirty years.

An ideological conviction about the place of psychoanalysis developed preeminently in America. This in itself raises a question that has never been fully answered. Why should America have so avidly embraced a treatment which never had the same level of acceptability elsewhere? Some have argued that Americans have felt a need to escape from what was perceived to be an overly fatalistic and deterministic view coming from Europe—that there was a discomfort with pessimism and that this was likely to lead to an enthusiasm for any approach that offered possibilities for radical intervention. If this argument that the adoption of psychoanalysis was a defense against pessimism is right, an equal and opposite swing to psychopharmacology would not be inconceivable.[16] The recent craze for Prozac may have been shaped in part by such dynamics—the need to find some vehicle for hope, some justification for optimism, some solution for complexity.[17]

In contrast to the social and psychological orientations, the traditional European approach had been largely somatic and bound by a classic disease concept. Kraepelin had been its foremost representative. This approach saw most mental illnesses as being brain determined. Physical means of treatment were therefore not only appropriate but also ethical in that only they offered any real prospects for recovery. It was this belief that drove "experiments" with a range of treatments such as the draining of sinuses, the creation of abscesses and their subsequent draining, insulin coma therapy, convulsive therapies, a range of shock treatments, and psychosurgical procedures.

There have been bitter arguments about the intentions of those who took this approach to treatment,[18] but it is clear that not all these experiments were conducted by secretive Mengele clones. No less a figure than Heinz Lehmann, who was the first to test both imipramine and chlorpromazine in North America and who has been a vigorous advocate of the combination of drug and psychotherapeutic modalities, had tried out many of these other methods first.[19] And while it is difficult to see how psychosurgery can have done much for the doctor-patient relationship,[20] Joel Braslow has rather convincingly documented the difference that malaria therapy for general paresis of the insane (GPI) made in the 1930s.[21] Before its introduction references to patients in case notes were largely derogatory and denigrating, but afterward, presumably because of an atmosphere of greater therapeutic optimism, patients were seen in a much more favorable light.

The story of another Montreal psychiatrist, Ewen Cameron, provides a representative case by which to judge the somatic approach and the emotions aroused. In the late 1950s and early 1960s Cameron, the director of the Allan Memorial Institute in Montreal, developed a form of therapy called psychic driving, which involved giving ECT repeatedly to patients in order to erase supposed faulty memory traces with a view to then reprogramming them. Patients underwent this procedure to the point of forgetting their own name and even becoming incontinent. There are some suggestions that the treatment was driven in part by Cameron's interest in discovering a Nobel Prize–winning procedure—something that seemed distinctly possible. The discovery that part of Cameron's funding, for the investigation of these procedures which were of relevance to the topic of brainwashing, came from the CIA has been used to call into question his motives. Similar concerns have been raised about the work of investigators such as Paul Hoch who were working on the effects of LSD and other hallucinogens. But in point of fact, Cameron's department was among the least ideological in North America. He had recruited to it both analysts and somaticists as well as the premier historian of psychiatry of the day—Henri Ellenberger. Cameron also introduced the day hospital to North America.[22]

Part of the issue in Cameron's case would appear to have been therapeutic zeal. Although proper medical practice is usually portrayed in terms of adherence to the conventions of a Hippocratic oath, it has more often than not been driven by a belief that while it is clearly unfortunate if therapeutic ministrations injure patients, it would be an even greater sin if someone were to suffer owing to a failure of physicians to do all in their power to help.[23] Unchecked zeal can lead a Benjamin Rush to bleed even his six-week-old child, but it also led to the development of anesthesia. While retrospectively regressive ECT can be seen to have been going too far, the theory behind it, if simplistic, was nevertheless coherent and logical. And in the room next door to the person recovering from his ECT, there were liable to be patients being spoon-fed as part of an effort to get them to regress to childhood in accordance with the dictates of the equally coherent and logical theories of psychoanalysis. There would seem little doubt that as many lives were mutilated and blighted by the zealous application of psychoanalysis as were ever destroyed by physical methods of treatment.[24]

The excesses of psychosurgery and regressive ECT are a testimony to how badly wrong things can go, as are the thousands of lives that were almost certainly harmed by hours each week of personality-destroying psychotherapies of one sort or another, but these historical facts also testify that the basic therapeutic instinct is to apply any new development and often to apply it heroically. Given that practitioners are not licensed because they, in particular, have been shown to have this instinct, it is a fair bet that similar instincts are shared by the population in general. Indeed, this would seem to be one of the concerns that led to the FDA's pushing for the power to designate certain drugs as prescription only.

The issue of therapeutic zeal clearly raises a number of questions. The traditional question concerns the role of evaluation. On the face of it one role for the controlled trial would be to contain the "furor therapeuticus." It was this that led Michael Shepherd into battle against Mogens Schou; Shepherd believed himself to be faced in Schou with someone who was, like Cameron, a "believer" in a particular therapeutic modality.[25] History reveals how complex the evaluative process is, in that while Schou was possibly initially driven more by belief than by replicated evidence, the majority opinion to this day is that he was essentially correct. Furthermore, leucotomy and regressive ECT were not replaced because the formal procedures of controlled trials showed them to be ineffective but rather because the majority of therapists were convinced by their daily practice that the newer drug treatments were superior. Thus expert evaluation clearly has a role, even if it is less clear what that precise role should be.

Another issue is the question of power. One of the features of therapeutic zeal is that powerful therapists, whether biologically or psychologically oriented, seem to think nothing of ordering patients, the foot soldiers in their campaigns, over the top of the trenches and into the firing line. Where drug treatments are concerned, the restriction of prescribing to physicians only is the mechanism of conscription to a campaign. Whether an over-the-counter strategy would make a difference to patient welfare is a moot point.

The introduction of chlorpromazine and imipramine was greeted enthusiastically by the somaticists as marking the beginning of new era. This enthusiasm stemmed from the clear efficacy of the new compounds, but it probably also stemmed in part from the fact that these treatments were clearly nonmutilative. Before chlorpromazine, patients could opt for some of the older procedures and therapists could get enthusiastic about giving them—but this was only against a background of what was being lost by not going ahead with treatment. Where substantial benefits are at stake, substantial risks will be taken.[26] Other than in its most severe melancholic forms, depression for the most part was not seen as warranting the serious risks entailed in the older treatments. But while the newer treatments have proved much "gentler" and have led to more patients being drawn into treatment, the heroic ethic has been very visible at times in some of the dose regimes that came into use. Where patients failed to respond to antipsychotics or antidepressants in the usual doses, the doses were pushed to extraordinary levels, with undoubted effects in terms of permanent side effects in many cases.

Pharmacological Calvinism

> The need for frequent chemical vacations from intolerable self-hood and repulsive surroundings will undoubtedly remain.
> —Aldous Huxley (1951)[27]

The publication in 1993 of *Listening to Prozac* followed on the heels of Prozac's appearance on the front covers of *New York* and *Newsweek* and an increasing number of discussions of its effects on television shows and in newspaper and magazine columns. Prozac had become a media event, a subject for public debate and for inclusion in screenplays. Far from being unique in this, however, it was following a trail blazed by the bromides, the barbiturates, Miltown, and Valium before it. The mythology wheeled out—that its creation was a rational process, that it was a designer drug—had echoes in the 1955 reception of Miltown as the penicillin for anxiety.[28]

Indeed the very possibility of making a media event of Prozac owed a great deal to concerns about the minor tranquilizers before it. These concerns had done much to make health a topic of interest to television programs and national newspapers. The widespread use of Valium, in particular, had been seen by many as something that was actively brought about by drug companies, from whose activities both patients and prescribers needed to be saved. In the case of the minor tranquilizers, the advertising strategies of companies, it was argued, were dangerously seductive and aimed at a medicalization of the neuroses, which were essentially problems of living. Concerns were expressed about the role of drug companies in providing medical education and their role in the censoring of information that appears in the scientific press and at scientific meetings.

Something of a view emerged that people taking tranquilizers were only doing so to avoid social or political conflict. There was a particular concern about the greater use of tranquilizers by women, which was interpreted by some as a clear effort to suppress women. In 1967 Stanley Yolles, then the director of NIMH, wondered, "To what extent would Western culture be altered by widespread use of tranquilizers? Would Yankee initiative disappear? Is the chemical deadening of anxiety harmful?"[29] Others felt that the turn to tranquilizers had occurred because people wanted things to come too easily, because they had become lazy and did not want to take the trouble to find their own center. They had become afraid to define their own existence and were prepared to let a pill do it for them.[30]

Such attitudes led Klerman, in 1970, to coin the term "pharmacological Calvinism." This refers to a belief system in which drug use is held to be bad and potentially even dangerous if it makes you feel good. A drug that makes a subject feel good either is somehow morally wrong or is going to be paid for with dependence, liver damage, chromosomal change, or some other form of secular theological retribution. Following up this lead, an NIMH study in the early 1970s found that many lay people viewed nervous problems as a sign of moral weakness and the use of something like tranquilizers for such difficulties as further evidence of weakness.[31]

The pharmaceutical industry was a particular target of suspicion, with eminent psychiatrists being prepared to voice worries about the "pharmaceutical juggernaut" and politicians citing the problem as one of the greatest menaces faced by society in peacetime. Drug companies became potent symbols of evil; they were portrayed as producing compounds that damaged health while yielding exorbitant levels of profitability. The issues lent themselves to media portrayal in terms of the suffering of vulnerable, albeit possibly gullible, patients as a consequence of rapacious exploitation. After affected individuals were innocently lured into addiction by corrupt or incompetent doctors, their struggle back to responsibility could be depicted in terms of a heroic encounter with inner and outer demons. Uniquely, in terms of the addictions, tranquilizer takers were seen as victims deserving of pity and support rather than as social outcasts. The social outcasts, rather, were the doctors and manufacturers.[32]

It was not uncommon to find articles that lumped together the practices of transnational corporations, in terms of tax evasion, price discounting, and resource transferring, with the evils of tranquilizers. The prescribing power in the hands of physicians and the use of science as an advertising strategy (if only in the sense of being able to parade genuine "scientific" evidence of efficacy in the shape of controlled trials), conjoined to the use of brand-name tactics in this sector of the market and the increasing role of the industry in educational initiatives, clearly marked a series of departures that deserved scrutiny. But while all of these issues may individually be problematic, their commingling may be something of a problem in its own right. In many of the pieces written, it becomes difficult to disentangle moral judgments made for political motives from genuine medical concerns.

A study by Mitch Balter and colleagues from NIMH[33] reported in 1974 that far from there being evidence for overprescribing, a majority of physicians and patients remained concerned about the dangers of psychotropic drugs and prescribed or used them rather sparingly. A further study by M. Pflanz and colleagues[34] indicated that minor tranquilizer users were more likely than nonusers to be health conscious and in the case of women more likely to be employed rather than confined at home. The clash between the results of these studies and more general perceptions fit well with the existence of Klerman's notion of pharmacological Calvinism.

Indeed, there has never been convincing evidence for overprescription. Although sales of Valium rose so that in the mid-1970s it became in many countries the most commonly prescribed prescription drug, it is much less clear that this was a newly minted tranquilization of misery. In the 1920s, the bromides had been among the most popular prescription drugs. Even before that both coca and a range of opiates had been widely used in the latter part of the nineteenth century and the early years of the twentieth century, largely as a consequence of over-the-counter sales. The bromides had been followed by the barbiturates and subsequently most notably by meprobamate, which, marketed as Miltown, had enjoyed all of the cult status in the media that Valium and Prozac were later to achieve.[35] In addition, there has been a constant use of alcohol and a range of herbal or other remedies. It is, therefore, by no means clear that there was an increase in drug consumption. An alternative explanation is that there was a concentration of anxiolytic prescriptions in one brand, owing in part perhaps to perceptions of benefits of that brand but also to the dictates of fashion. The current sales for Prozac almost certainly owe something to the same effect.

In terms of drug company involvement in education, which has been a concern since the late 1950s, there is an almost inevitable problem in that a great number of new drugs must of necessity be new to many of the practitioners called on to use them. They will not have been taught anything in medical school about the mechanisms of action of the new drug or the rationale for its use. Education must come from somewhere. While many countries are now establishing courses for continuing medical education, the sources best placed to provide information will remain the pharmaceutical companies. Indeed, the pharmaceutical industry did a great deal to set up some of the initial psychopharmacology organizations and sponsor many ongoing events, often without reference to the promotion of any particular compounds. It can be argued that simply increasing psychopharmacological knowledge will lead to an increase in prescribing and that therefore even disinterested investment in education of particular types will expand the psychotropic market. But against that there is the fact that before a new drug goes into humans, there are no independent experts and it takes time for independent expertise to be generated.

Where the ideological battles of the 1960s and 1970s were fought over the tranquilization of anxiety, the sales of Prozac and *Listening to Prozac* point to a shift in the terrain of battle. Depression has replaced anxiety, and this new battlefront has at least temporarily outflanked the defenders of the moral high grounds. The only surprise in this perhaps is that it took so long for this change to happen. As outlined in Chapter 2, as early as 1961 Frank Ayd was writing about the existence of depressive disorders in primary care and general hospital settings. In the debates about tranquilizer prescribing held before congressional committees in the 1970s, Nate Kline put forward the view that there was a great amount of hidden depression in the community and that this might account for neurotic complaints.[36] From this viewpoint, antidepressants, far from being overprescribed, were probably considerably underprescribed.

The work that persuaded companies that antidepressants were a better bet than anxiolytics was inaugurated by Michael Shepherd at the Maudsley. In the early 1960s he began to conduct a series of large-scale epidemiological studies of psychiatric morbidity in primary care, finding that depressive-type disorders in particular were extraordinarily common, frequently undiagnosed, and accordingly, it might be presumed, for the most part ineffectively treated. Shepherd comments: "What very quickly emerged was that far from wasting my time, I had stumbled on a horrifying fact, which was that psychiatrists knew very little about mental disorder."[37] This work was taken up by David Goldberg, Peter Huxley, and others in Britain and by a series of NIMH-sponsored epidemiological studies in the United States.[38] The result was a picture of depression very different from the one that had faced Kuhn in his hospitalized patients. It seemed quite possible that a large proportion of the seemingly psychoneurotic complaints presenting to primary care physicians might stem from unrecognized and untreated depressive illness. And indeed this different picture, it was argued, must in some way be the more accurate one, as studies began to reveal that depressions in the community outnumber those in hospital settings by a ratio of at least forty to one.

Initially the pharmaceutical industry paid little heed to these findings. The primary markets were in major and minor tranquilizers. But as the storm clouds gathered over the benzodiazepines in the late 1970s, it was notable that the pharmaceutical companies developed an increasing interest in the issues of the detection of depression in primary care and the indications that in many cases depressive disorders needed long-term treatment. The availability of this research probably played a significant part in company decisions to proceed with the development of SSRIs like Prozac when there was little evidence that they were any more efficacious than the older generation of compounds.

Shepherd's work on the psychiatric disorders of primary care led directly to campaigns by both the American Psychiatric Association (APA) and the British Royal College of Psychiatrists (RCP) to alert physicians to the diagnosis of depression.[39] These have had considerable support from the pharmaceutical industry, with the newer generation compounds being portrayed as particularly suitable for the treatment of the primary care patient and for long-term treatment.[40] Such developments have been noted by critics like Peter Breggin, who use the point to support an argument that the members of the medical profession are little but the runners for the drug bosses.[41] In contrast, the APA and the RCP are viewed by members of the American College of Neuropsychopharmacology (ACNP) and the British Association for Psychopharmacology (BAP) as being largely in the control of their psychotherapy or social psychiatry constituencies. The truth clearly lies at neither of these extremes.

There must inevitably be a struggle, or a dialectical process, to determine the meaning of physical symptoms and where the boundaries of health and disease lie. Medicine rather obviously medicalizes distress. This has almost always been true. Some of the distress of life has always been taken to medical practitioners in the form of "medical" problems. The presentations have typically fit the stereotypes of the day, whether these involve epilepsy or neurasthenia or, more recently, M.E. (myalgic encephalomyelitis). The history of the interplay between society and the expression of distress in physical symptoms by its members has been wonderfully told by Edward Shorter in *From Paralysis to Fatigue* and *From the Mind into the Body*.[42] Good practice, or better yet wise practice, should take this into account.

Behind this drama of the shifting language of physical symptoms a common theme has emerged, which is that a substantial proportion of the population—on the order of 15 percent—have persistent physical symptoms.[43] Good arguments can be offered for a genetic input to this. Doubtless it would also be possible to construct mechanisms by which the process of communicating distress is internalized in physical symptoms at an early age and develops some autonomy from there. The problem is that there is no agreement as to why these symptoms are present or on how best to treat them—and in the absence of such agreement, the *conviction* that seeking medical help for the problem is a form of brazen escapism indicates an ideological rather than a scientific stance. The escapist idea was clearly mooted in the 1970s in arguments about minor tranquilizer use—before child abuse had become the issue it now is. Today it seems quite possible that a proportion of those presenting with physical symptoms may have been the victims of child abuse of some sort. The minor tranquilizers, as it turns out, are among the most effective agents we have to reverse the dissociative states that can be associated with such abuse. While clearly in such cases tranquilizers are not the ultimate answer, it is unclear that prohibiting them is the best way forward or indeed that prohibiting them does not victimize the victims even further.

Drug taking, even when officially sanctioned, remains an ambivalent reality in a culture that increasingly emphasizes the need for personal and interpersonal competence. Paradoxically it can sometimes seem that taking drugs is especially problematic when it has been officially sanctioned. But if medicalizing problems is a problem, it is a moot point whether psychologizing them is any better. The psychotherapies of our day are for the most part intensely individualistic in their approach. Arguably, miseries that would formerly have been taken to stem from the hand of fate or from economic or social factors are now being "psycho"-pathologized and "treated" by methods that involve the handing over of autonomy to professionals, who are often more concerned with maintaining professional status and turf than with their client's welfare.

These issues have, however, receded into the background somewhat. At least temporarily, there is an impression that drugs like Prozac and therapies like cognitive therapy are treating specific disorders, such as depression, in contrast to Valium and psychoanalysis, which many thought were being used inappropriately to manage what were seen as existential problems. The change in perceptions, so that it is thought that real illness rather than distress is now being treated, owes a great deal to a revolution by committee, which was effected by the developers of a new classification for psychiatric disorders unveiled in 1980—DSM-III.

Carving by Feathers?

The initial steps toward a classification of diseases began in 1853, when the First International Statistical Congress charged William Farr of London and Marc d'Espine of Geneva with the task of preparing "a uniform nomenclature of causes of death applicable to all countries." Farr suggested classifying diseases as posttraumatic, developmental, epidemic, constitutional, and local, with the local diseases categorized by anatomical site. D'Espine's suggestion was for a more symptom-and sign-based approach. The type of classification wasn't resolved until 1893, when the International Statistical Institute, chaired by Jacques Bertillon, plumped for Farr's approach, creating the Bertillon Classification of Causes of Death, which in 1948 became the International Statistical Classification of Diseases, Injuries and Causes of Death (ICD) with the publication of ICD-6. This incorporated psychiatric diseases for the first time. ICD-6 was updated through ICD-7 and -8 to ICD-9 in 1978 and ICD-10 in 1992.

In America, in addition to the ICD, there was a parallel system of classification for mental disorders, later to become the *Diagnostic and Statistical Manual*. A first classification had been put together by the American Medico-Psychological Association in 1918 and called the *Statistical Manual for the Use of Institutions of the Insane*. The impact of the combat neuroses that were encountered by clinicians in World War II led in 1952 to the creation of DSM-I.[44] This was created largely at the behest of a new generation of clinicians, who entered psychiatry in the course of the war without going through an asylum training. The result was a system that was notably Freudian and Meyerian in approach—it focused on neurotic reactions rather than diseases. DSM-I was given a more explicitly analytic cast in DSM-II, which was brought out in 1968. This in turn was scheduled to be updated in the mid-1970s to produce a classification system that would be in line with that proposed for ICD-9.

The updating of DSM-II, however, took place against a series of developments that were to radically change the face of American classification and with it world psychiatry. There was a growing concern over the reliability of psychiatric diagnoses. A cross-national collaborative study on diagnosis coordinated by the World Health Organization (the International Pilot Study of Schizophrenia) had made it clear that schizophrenia was being diagnosed far more commonly in the United States and Soviet Union than in the rest of the world. American and British psychiatrists faced with the recordings of interviews of the same patients arrived at different diagnoses.[45] For the Americans essentially everything was schizophrenia, whereas many cases of American schizophrenia were seen as affective disorders by the British. The controversies over lithium which were running at the same time highlighted the fact that, whereas diagnosis had been relatively immaterial up until then, with the advent of the newer drug therapies a correct diagnosis might actually make a difference to the outcome. The promise that a further generation of psychiatric drugs would correct ever more specific disturbances raised the premium on correct diagnosis.

There was also an awareness that the boundaries of psychiatric illness needed to be better defined. This was symbolized best by a dispute over the status of homosexuality, which had been classified as an illness in DSM-I and -II. A decision to drop the "diagnosis" of homosexuality in 1973 can be taken as an indicator of a more profound underlying shift—a retreat from a commitment to the cure of society, which was implicit in psychoanalysis, to a more recognizably medical model of psychiatric disorders. Opinion leaders within American society began to voice the concern that the analytic agenda had become a crusade that had taken "psychiatrists on a mission to change the world, which had brought the profession to the edge of extinction."[46]

It was against this background that the emergence of a group Klerman dubbed the neo-Kraepelinians was to assume significance.[47] This was a group who believed in the importance in psychiatry of traditional medical diagnoses. Among the early voices raising this point of view were Eli Robins and Samuel Guze from Washington University in St. Louis, who were later joined by Paula Clayton, Bob Woodruff, George Winokur, and John Feighner. Klerman and another group of researchers based around New York, including Don Klein and Robert Spitzer, made up another wing of what has since been dubbed an invisible college, whose ideas came from nowhere in 1970 to a position of dominance by 1980. The neo-Kraepelinian approach held that psychiatry was a branch of medicine, that there was an identifiable boundary between the normal and the sick, that discrete and identifiable mental illnesses exist, and that psychiatry should treat these and not problems of living and unhappiness. Psychiatric research should, moreover, be geared to establishing the validity of diagnostic criteria. The creation of DSM-III was the Trojan horse by which they effected entry into the citadel of psychoanalysis.[48]

The opening shots in the campaign came at an NIMH-sponsored symposium at Williamsburg in Virginia at the start of May in 1969. The brief of the meeting was to establish the outlines of a research program to investigate the psychobiology of depression. Several of the discussants, notably Robins and Klerman, focused on the need for a standardization of psychiatric diagnoses before research of this type would be possible. This conference gave rise to the Psychobiology of Depression Collaborative Research Program, for which Klerman was one of the primary coordinators.[49] While the biological research component of this program proved something of a costly failure, Klerman had enlisted Robert Spitzer of the New York State Psychiatric Institute and others to come up with a set of standardized operational diagnoses, the research diagnostic criteria (RDC), which were to form the heart of DSM-III.[50]

Meanwhile, in 1973 the APA gave the go-ahead to the revision of DSM-II. In April 1974 Melvin Sabshin, the medical director of the APA, approached Spitzer with a view to having him head up the task force that would produce the new classification. Spitzer had all the qualifications for the post. He had been trained as an analyst. He had been involved in the drafting of DSM-II, and some continuity between the drafting of DSM-II and -III was clearly required. He had come to particular prominence in the debate over the status of homosexuality. He had a recognized interest in research on classification systems, having been involved in the International Pilot Study of Schizophrenia and being even then involved in the business of drawing up the RDC for the Psychobiology of Depression Collaborative Research Program. But, in addition, he had a very particular vision of what was needed, commenting, "whether we like it or not, the issue of defining the boundaries of mental and medical disorder cannot be ignored. Increasingly there is pressure for the medical profession and psychiatry in particular to define its area of prime responsibility."[51]

Fatefully, he was given carte blanche regarding whom to enlist as members of the task force. Not surprisingly, he turned to others with an interest in diagnosis, notably the Washington University Group at St. Louis under Robins and Guze, as well as Klein, and others from Columbia, and in general focused on individuals with an interest in diagnostic methodology rather than psychodynamics or therapy. He produced a task force as dominated by the neo-Kraepelinians as that brought together to formulate DSM-II had been by psychoanalysts. It was a group that, it was argued, was not representative of mainstream American psychiatry, a group "whose ideas are very clear, very publicly known and [whose] guns are pointed at [psychoanalysis]," as Otto Kernberg characterized them. Opposition to what was going on began to surface in 1976, when field trials of the proposed new categories led to a growing awareness among practitioners that DSM-III was going to be a far more radical break with the past than any of them had anticipated.[52]

Although clinicians of other orientations were subsequently pulled into the process, partly for reasons of political necessity, by the time the base was broadened in this way, the ground rules for DSM-III had been set. The only clinical features that would be used in making a diagnosis were those on which any observer, whether he or she be a behaviorist, an analyst, or a biological psychiatrist, would be likely to agree on. This, many felt, was a default process that could only favor biological psychiatry and behavior therapy.[53] Traditionally, the process of diagnosis in medicine has been characterized as an effort to carve nature at its joints, in the manner that one might disarticulate a turkey, for instance. The process of arriving at DSM-III, in contrast, was likened by Bob Michels to an agreement that, in the absence of consensus as to where the joints of the bird might be, it would be best to carve by the feathers. The concentration on surface features almost inevitably meant a downgrading of dynamic factors and principles.[54]

Even though few of those involved in the construction of DSM-III were associated with work on drug treatment or the biology of mental disorders, the biological default in what was proposed came about as one of the assumptions of the neo-Kraepelinians—that the core symptoms of mental disorders stemmed from aberrant brain functioning and that as a consequence accurate descriptions would reflect that functioning. For the neo-Kraepelinians the task became one of specifying discrete disorders by means of inclusion and exclusion criteria. They had no time for portmanteau terms like neurosis, which could not be defined by a good description. For the analysts, in contrast, the term neurosis referred to basic conflicts presumed to be operative in the production of symptoms, whether the symptoms be those of OCD or phobic disorders or hysteria. From this point of view, the accurate description of surface features was simply less important—"clerks rather than experts can make this kind of classification." Indeed, treating the surface features of a disorder would likely amount to malpractice, as a surface set of symptoms, such as the occurrence of panic attacks, was for the analysts equivalent to the fever that might go with an infection. Treating the fever without treating the infection was surely not right. It would only lead to the outbreak of other symptoms elsewhere.[55]

Clearly there was not room for two such radically different approaches to coexist. The crisis came to a head when the task force proposed to abolish the entire category of neuroses. This proposal proved a stumbling block for many, who "knew" that unconscious conflict was at the root of psychoneurotic disorders. How could those who were framing DSM-III possibly ignore almost a century of carefully detailed case histories and truths that had won general acceptance, unless, as Otto Kernberg put it, Spitzer, despite his apparent flexibility, had "extremely strong negative feelings about psychoanalysis."[56]

There followed a set of political skirmishes. These were political in that the pitches made by both sides were based on what would secure votes rather than by appeals to the evidence—at the end of the day Spitzer's document had to be voted into existence by the members of the APA. The various lobbies within the APA were mobilized in coalitions by the competing interest groups. Both groups targeted the shifting votes in the middle, the majority of practitioners who used a mixture of drugs and therapy without a strong ideological commitment. Ultimately the struggle focused on the question of neurotic depression—the bread and butter of U.S. private practice.

As outlined in Chapter 1, it has always seemed reasonable to many to divide depression into two forms, a drug-responsive form, which most have assumed is more severe and resembles traditional descriptions of melancholia, and another more psychological form, which may be less severe and for a long time was seen as more appropriately treated by psychotherapy, with drugs being used, if at all, in an adjunctive role. These different forms have been called endogenous and reactive depressions, or psychotic and neurotic depression.[57] The DSM-III task force wanted to code for bipolar, major, and chronic depressions only, eliminating the term neurotic depression. At the last moment, in 1979, it seemed quite possible that four years of work on the new system might be voted out because of the unwillingness of the psychodynamically oriented practitioners to countenance the removal of the category of neurotic depression.[58]

Spitzer sought to compromise. He pulled the concept of dysthymia, a concept that hadn't been in use for over half a century, out of the classificatory hat to replace chronic depression, but this was rejected. At this stage, under assault from within his own task force for losing his nerve and betraying his principles, and facing increasingly coherent opposition from without, he suggested a coding of "Dysthymia (neurotic depression)." The determinants of the argument at this stage would seem to have been perceptions on both sides of the number of votes that could be relied upon to support the differing formulations. The final successful compromise was "Dysthymia (or Neurotic Depression)"—both the "or" and the capitalized N and D of neurotic depression were felt to be critical to the passage of the entire classification.

While Klerman was not formally part of the DSM task force, his role behind the scenes can be gleaned from the fact that when it came to representing the significance of what had happened to the World Health Organization, he was the person selected to make the case. In addition in 1982 a debate was held at the annual meeting of the American Psychiatric Association on the significance of DSM-III, and Klerman was lead speaker in its favor. Reviewing the decision to adopt it, he argued that the "development of DSM-III represents a fateful turning point in the history of the American psychiatric profession . . . the decision of the APA first to develop DSM-III and then to promulgate its use represents a significant reaffirmation on the part of American psychiatry of its medical identity and its commitment to scientific medicine"[59]—a clear statement of the neo-Kraepelinian credo. From 1977 through 1980, Klerman had served as the director of the Alcohol, Drug Abuse and Mental Health Administration and on the Presidential Commission for Mental Health.[60] From this base he had helped inaugurate a series of epidemiological studies (the epidemiologic catchment area studies, or ECA)[61] that used the new criteria, which at one and the same time familiarized clinicians and researchers with the new diagnostic methods and cemented a series of new Kraepelinian diagnostic entities, such as panic disorder, social phobia, and others, in place.

I have alluded to some of the reasons why a need was perceived for a new classification, but why did this extensive revision eventually win passage, given the opposition and the extent to which DSM-III actually was unrepresentative of much that was happening in practice? A number of factors can be pointed to. One was simply the weight of bureaucratic inertia. Spitzer and the team members he selected, once in place, built up a head of steam, and it would have taken considerable organization to derail them. After four years of work and a vast expenditure on field trials, the document was not going to be thrown out lightly. Another reason had to do with a progressive shrinkage of research funding, a decline that had started around 1965. This was being blamed, in part, on the lack of clarity in the old DSM-I and DSM-II diagnostic system, which meant that a great deal of the research that had been done was unreproducible. The new system, DSM-III, also appealed to the pharmaceutical industry and to the FDA, because it introduced much greater clarity into the regulatory process and indeed meant that studies which had been conducted anywhere in the world that adhered to DSM-III criteria could be used as part of a regulatory submission. Finally, in a changing financial climate, the proposed new classification scheme met with favor. Health maintenance organizations and the idea of capitation methods of payment had been given the go ahead in Washington, and the feedback from such sources indicated that greater clarity regarding diagnosis and treatment would be required if payment for treatment was going to be forthcoming.

If these were the reasons for the acceptance of DSM-III, a number of important consequences were not initially foreseen. The creation of a discrete set of disorders, such as panic disorder, social phobia, obsessive-compulsive, and other disorders gave the pharmaceutical industry a set of targets at which to aim its compounds. In the process concepts such as panic disorder were internationalized and something of an American hegemony in psychiatry was created. World psychiatry, which had previously been largely dimensional in its concepts (even in places where the orientation was somatic rather than psychodynamic), became categorical. The problem was the number of categories proposed and their subsequent proliferation from 180 in DSM-II, to 292 in DSM-IIIR, and over 350 in DSM-IV, as noted in Chapter 5. Are there really so many distinct psychiatric disorders? This question ultimately links back to the issue of specific treatments and specific illnesses.

But in addition to support for DSM-III among regulators and the pharmaceutical industry, there was also support for it from within the psychotherapeutic establishment. The 1960s and 1970s had seen the emergence of briefer and more focused cognitive and behavior therapies. These clearly fitted the economic climate better than psychoanalysis. Previously, under the influence of Jerome Frank, there had been something of an understanding that while drug treatments might target symptoms it was the job of psychotherapy to focus on interpersonal functioning. In contrast, neither behavior nor cognitive therapy had a priori difficulties with what was proposed in DSM-III. Indeed behavior therapy appeared to do best when the various different neuroses were separated out from one another and treated by approaches targeted at the specific symptoms or surface features each showed. With the adoption of DSM-III, therefore, the stage was set for the first major showdown between the pharmacotherapies and psychotherapy on ground acceptable to both.

The Evaluation of Psychotherapy

As early as 1952, Hans Eysenck had argued that there was no evidence that dynamic therapy of any sort worked for any psychiatric condition.[62] At that time the question could be brushed aside, as evaluation as an issue had yet to develop. When it did, however, psychoanalytic practitioners were to prove extraordinarily resistant to the idea of evaluating the outcomes of their efforts. The drug evaluation studies that developed, with their rating scales and statistical averaging of data, always appeared to the analysts to be insufficiently sensitive to capture the "shadings, degrees, and grades of thinking relative to mental illness."[63]

It was only with the development of behavior therapy during the 1960s and cognitive therapy in the 1970s that data on "therapy" began to emerge. This information, combined with data from studies of briefer focused dynamic therapies, created by the mid-1970s a body of data of sufficient size for meta-analyses to be conducted—that is, sufficient for all trials using any therapeutic technique for any condition to be lumped together and effect sizes to be calculated. A number of such analyses were undertaken, with mixed results. Some claimed that therapy had been shown to work—that all had won and all deserved prizes. Eysenck and others took exception to this and argued that only cognitive and behavioral therapies had been shown to work and that the data from such studies were being hijacked by the analysts to support a case for therapy generally. Feelings ran high, with Eysenck suggesting that the field was ideologically riven and that it was "very doubtful if empirical data can settle the issue."

The situation was most interesting in the case of depression. The position at the end of the 1960s, when the psychotherapeutic high water mark had been passed, was that endogenous depression seemed to have been established as a disorder that stemmed from chemical abnormalities in the brain, which it would have been quixotic to even think about treating with psychotherapy. Yet by 1980 the psychotherapeutic tide was flowing again and a range of behavioral interventions, interpersonal psychotherapy, and cognitive therapy had all achieved some success with depressed patients. The renewed interest probably owed most to two developments, the appearance of both cognitive and interpersonal therapies.

Cognitive therapy had been introduced by Aaron Beck in works published in 1967 and 1976. These were eminently accessible to a lay readership.[64] The cognitive model of what goes wrong in depression—essentially that depression stems from reasoning errors made by people, that depressed individuals think themselves into depression—was easy to understand and not unreasonable. It also provided a coherent therapeutic rationale: challenge the faulty thinking. Doing this seemed to work. Sur-

prisingly, however, it seemed to work as well if not better for major depressive disorders of moderate severity than for the neurotic depressions
for which it had been first freighted. Through the 1970s and early 1980s
a series of studies appeared apparently confirming the equivalent effica-
cies of cognitive and drug therapies and indeed pointing to a possible
extra benefit of cognitive therapy in postponing or reducing the likelihood of relapse.[65] The timing was right for an independent comparative
study between cognitive therapy and antidepressants. The comparison
came in the form of a study run by the NIMH. But it was not simply between cognitive therapy and imipramine; a third treatment, interpersonal
therapy (IPT), was included.

One of the key studies in changing the way depression was viewed had
been carried out at Yale in the latter part of the 1960s by Klerman,
Eugene Paykel, Myrna Weissman, and others. This group reported that
when subjects who were depressed were investigated systematically for
the occurrence of life events prior to the onset of their depression, it was
found that, contrary to expectations, in depressions with endogenous features rates were raised compared with those of a general population control and were as common as in supposedly neurotic depressions.[66] A
range of studies subsequently confirmed this finding. The idea that there
were patients with unprecipitated depressions, who needed to be shipped
off to the hospital and treated with drugs, and those with precipitated depressions, who should not be given drug treatment, had begun to break
down.

Paykel had joined Klerman in 1965 to study the role of psychological
and pharmacological factors in relapse following antidepressant withdrawal; the life events findings were a by-product of work on classification associated with this. Findings such as these suggested that it might be
worth looking at the possible protective role of psychotherapy in relapse.
The appropriate therapy had to be invented. The Yale group looked to
the ideas of Harry Stack Sullivan with a view to modifying these for delivery by psychiatric social workers. Myrna Weissman, a social worker, was
recruited, and the outlines of a therapy focused on the here and now,
rather than on childhood antecedents of current problems, to be delivered
on a once-weekly individual basis, were drawn up. This interpersonal
therapy was then piloted and shown to improve the social adjustment and
interpersonal functioning of depressed patients and, in a subsequent
study, to produce significant effects on the acute symptoms of
depression.[67] Unlike cognitive therapy, IPT explicitly took on board a
medical model of depression. It recognized that many of the symptoms
might stem from biological dysfunction but argued that dealing with interpersonal difficulties was nevertheless likely to promote a resolution of
the disorder.[68]

In addition to his role in the development of IPT and in demonstrating the place of life events in the genesis of depressive disorders, Klerman also later played a part in the conception of the NIMH studies. As the director of the Alcohol, Drug and Mental Health Administration, he was the most senior psychiatric administrator in the United States, and this position gave him the clout to authorize the funding of the study and to argue for the inclusion of IPT in the study protocols. He had long believed that the proper evaluation of psychotherapy was one of the most important endeavors mental health professionals could undertake.

The NIMH studies generated enormous interest and deep passions. Many livelihoods were potentially at stake. Planning for the project began as early as 1977. The training of therapists began in 1980. The study proper began in 1982 under the coordination of Irene Elkin.[69] It was scheduled to close in 1986. The data, however, only began to emerge in 1989, and subsequent analyses and papers were strung over four years.[70] Before the first publication, in 1989, the chief authors and the journal editor felt compelled to issue a "press" release dissociating themselves from views expressed in the media as to what the data actually showed.[71] Some of the interpretations being offered came from those who jumped in, attempting to put their spin on the results. Some of the confusions arose because a number of those privy to early drafts of the 1989 paper went to press having caught sight of some analyses only to find that the figures weren't included in the final publication—presumably because there was less than full agreement with the statements being made and politics dictated a retreat. Following the 1989 paper there was lengthy and vigorous correspondence in the journals.[72] Clearly, as Poisson might have remarked, there must have been something of a challenge to transcendence in the data. What then did the studies show?—or what was actually published?

The first issue is to look at what was actually done. The studies had to tackle a critical design question—is there such a thing as a psychotherapy placebo? The placebo control of a drug treatment functions in several ways. One is to simply control for natural variations in symptom intensities and for spontaneous remissions. But very few experts think that the role of the placebo arm in depression studies is simply this. One of the consequences of a move to multicentered drug trials has been to highlight differences in response rates between centers and therapists. Although part of these differences may stem from variations in the patients being recruited at each center, another possibility is that the extras a therapist does at one center over and above simply giving drugs, the bedside manner as it were, may contribute to responses compared with what is done at another center. One therapist in some way is proving more adept at mobilizing a person's therapeutic resources. It is only when a specific treatment can be shown to bring about response rates that reliably exceed those that can be brought about by factors such as the empathy and charisma of the therapist and the belief of the patient in the package that is being sold that it is conventionally said to "work."

But what this statement conceals is the fact that factors that are not specific to the drug may also work. When Beecher's wartime casualties came out of shock and had sufficient relief of their pain in response to an injection of distilled water, something was working, as we saw in Chapter 3. This wasn't just spontaneous variation. In other words a "psycho" therapy of some sort may make a difference and indeed a powerful difference—and not just to something "soft" like depression, but to conditions such as Alzheimer's disease. If this is the case, the problem becomes one of trying to determine which components of therapy contribute most. Because of this issue, IPT, cognitive therapy (CT), and imipramine were therefore compared with an inert medicine. In the case of both the inert medicine and imipramine, however, treatment was supplemented by a clinical management package the components of which were specified in detail beforehand. The point was not to provide a completely inert standard against which the other treatments could be compared—no real-life encounter between two human beings can be inert—but rather to provide an optimal therapeutic encounter, lacking only an element of supposed specific effect.

The second issue was how to ensure that patients enrolled for IPT got IPT and those down for CT got CT. There is a voluminous literature showing that therapists from different orientations in practice often do much the same things, leading to the conclusion that effectiveness may stem more from the personality of the therapist and clinical pragmatism rather than from any adherence to an ideological line. To overcome this, all therapies, including the delivery of drugs, were standardized and committed to a manual, and the investigators were trained until there was a consensus that all were doing what the study protocol called for them to do.

Two hundred and fifty patients with depression of mild to moderate severity were included. There were in fact no clear differences *in the overall group* between imipramine, IPT, and CT. But there were also no significant differences between these and clinical management alone. As the severity of the depressions increased, however, both imipramine and interpersonal therapy could be shown to be significantly more effective than clinical management, whereas cognitive therapy couldn't. There was some evidence that those patients whose interpersonal skills were best to begin with did best with IPT, while those whose cognitions were less neurotic to begin with did best with CT. This was a result to please the cynic and confound the theorist.

The results were very dispiriting to the cognitive therapists, but somewhat more encouraging for Klerman and the advocates of interpersonal therapy. The outcome of a follow-up study, which offered some evidence that patients who had had cognitive therapy were less likely to relapse than those who had had imipramine, were seized upon by the CT lobby, but even this finding was less supportive than it might have first appeared. As regards relapse, the principal finding, as with the lithium studies, was that depressive disorders were very liable to relapse. Indeed, the patients who seemed least likely to relapse were those who had gotten better on clinical management alone. A possible interpretation of this was that recovery on clinical management alone selected for patients with a lower risk of relapse. Perhaps cognitive therapy was doing the same, in which case there was nothing that cognitive therapy specifically did that brought about the outcomes.

If we stand back from the issues of professional turf that were of such concern when the study was first reported, a number of points emerge from the effort and the dataset. One is the power of randomization to prevent specific enthusiasms from waylaying research. This said, the example of the 1965 MRC study should be borne in mind. A single powerful study like this can get things wrong. The cognitive therapy bandwagon has kept on rolling in the wake of the NIMH studies, in much the same way that many clinicians kept on prescribing MAOIs after 1965. At this juncture, it is all but impossible to decide why this should be the case. One reason may simply be that cognitive therapy, like the MAOIs, does work. Another may be that its use has become closely identified with the fortunes of clinical psychology. In contrast to interpersonal therapy and behavior therapy, it is argued that the effective delivery of cognitive therapy requires high-powered training, which in the nature of things is better done "specifically" by one profession to ensure "quality" outcomes—a convenient recipe for a profession searching for its niche.

Another point raised is the matter of antidepressant principles. Clearly in this study clinical management of a high standard is in its own right antidepressant. Adding cognitive- or interpersonal-type interventions appears to boost the response rate. Adding imipramine boosts it further. One of the features of drug treatment of depression is that as the severity of the condition increases, the differences that can be demonstrated between active and inert pills also increases (see Figure 7.1). It was just such a differential that led Linford Rees in 1960, as we saw in Chapter 2, to conclude that imipramine and iproniazid had their place in the treatment of depression but that they were less antidepressant than ECT. Are cognitive and interpersonal therapies antidepressant, but less so than imipramine, or just not antidepressant at all?

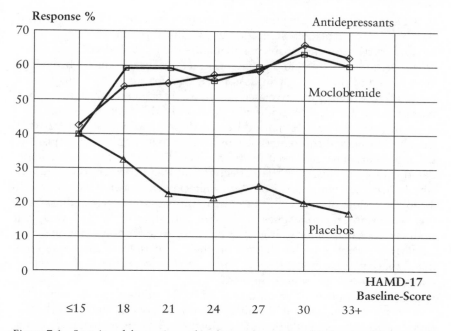

Figure 7.1   Severity of depression at baseline and response after four weeks' treatment with placebo and antidepressant. The graph presents the combined figures from a series of clinical trials comparing moclobemide, other antidepressants, and placebo. It shows a typical set of results—above a certain level of severity the remission rate on placebo becomes increasingly less likely. For the raw data, see J. Angst (1993), *Human Psychopharmacology* 8, 401–408.

The NIMH findings have been taken by some to show that neither cognitive nor interpersonal therapies are truly effective, at least not in the sense that antidepressants are. If they were effective, the reasoning goes, they shouldn't show a falloff with increasing severity as cognitive therapy in particular appears to do. A problem with this argument is that as severity increases yet further, the response to imipramine and other antidepressants also falls off, while that of ECT remains. The position that ECT is the only antidepressant, however, is unsustainable given the clear differences between imipramine and placebo at moderate severities of depression. We are forced back on an acceptance that there may be a number of antidepressant principles that may contribute independently to recovery.

What remains unclear from the NIMH study is what the nondrug principles might be. There are a number of potential antidepressant principles in the mix that is cognitive therapy. Its trademark is an effort to "correct" the thinking of depressed subjects, but in clinical practice elements of behavioral activation and problem solving are included. Indeed, common to cognitive, interpersonal, and behavioral therapies is a focus on behavioral activation and problem solving. As the Postscript suggests, perhaps these maneuvers, which could be classified almost as hygienic, are more important than the personality-modifying elements of the package.

As of 1995 no studies had addressed these possibilities. No less than the pharmaceutical companies, it would seem that cognitive therapists for the most part have not been in the business of doing studies that might demonstrate the existence of an effective nondrug treatment that could be simply delivered without the need for high-powered professional input. In 1996 the results of a first step to remedy this situation were published by Neil Jacobson and colleagues at Washington State University, and their findings appear to indicate that the behavioral activation components of cognitive therapy may be effective in their own right: the authors comment that these components "might be more accessible to less experienced or paraprofessional therapists."[73]

A further problematic finding of the NIMH study was that one of the three centers used had significantly worse results with cognitive therapy than the other two. While the CT results might have been better if all centers had performed equally, given the intense training and supervision that went into this study, another lesson might be that cognitive therapy as currently put together is too complex—unless of course the defense of a piece of professional turf is involved, in which case it is only just about complex enough. It takes much more time than interpersonal therapy to train in. As such it is at the opposite end of the antidepressant therapy spectrum to drug therapies, for which the prescribers are already trained for any conceivable future drug therapy at no extra cost. Interpersonal and behavioral approaches lie in a somewhat orthogonal position. They are relatively easy to train people in, so that primary care prescribers could deliver both drug and psychotherapy in one package, or alternatively these approaches can be formulated in a manner that would allow many people to help themselves.

In addition to contributing to an acute response to treatment, the idea that delivering something other than just drug therapy on its own might reduce relapse rates makes intuitive sense. The extra elements of the package may be no more than an education about depression and basic hygienic maneuvers, so that early recognition of a further episode may permit the individual to institute basic hygienic maneuvers at a time when they have the best chance of success.[74] Something more would seem possible in that the persistence of neurotic features after a response to antidepressants is one of the stronger predictors of relapse. Tackling this, then, should be a goal of early treatment in that the prevention of the development of maladaptive attitudes is likely to be more effective than efforts to remedy the problem after they have developed.

There has been another influential study involving interpersonal therapy that has shed some light on this question. This study, noted in Chapter 4, was conducted in Pittsburgh and looked at the role of imipramine and interpersonal therapy in the prevention of relapses and recurrences in recurrent depression. Maintenance IPT in this study appeared to confer benefits even in depressions characterized by biological dysfunctions.[75]

Transposing the debate to the field of ulcer therapies may shed further light on the salient points. Both lifestyle modification and drug therapy may offer antiulcer principles. Different behavioral interventions may each contribute something distinct, as may different drug treatments. Drug therapy alone would seem to be more effective than lifestyle modification alone, particularly as the condition becomes more severe, but a combination of both is likely to be of greater benefit over the longer haul than reliance on one alone. As regards relapse and recurrence, the question arises whether there are psychic or physiological factors in the depressive equation that play similar roles to that played by helicobacter pylori in the case of ulcers. If there are, treatments aiming at the prevention of recurrences may need to involve quite different principles to those that bring about initial responses. Where a therapy like imipramine may bring about the resolution of uncomplicated depressive disorders, it may be simply symptomatic when additional factors are operative. We need to continue hunting for what other principles need to be added to current "antidepressants" to eliminate the propensity to further episodes of depression rather than to simply forestall recurrences.

At present, however, there seem to be few forces acting to combine drug and nondrug therapeutic principles into integrated management packages. Patients increasingly seem to get one or another treatment modality, commonly drug treatment alone as it is so relatively inexpensive.[76] Yet there is no evidence that drug therapies are pushing a treatment like cognitive therapy out of the market. Therapies such as behavior therapy and interpersonal therapy seem much more likely to vanish. Is this because there are no large corporations or professional bodies behind them? It is time to consider a case directly relevant to these issues, which laid bare the state of ideological conflict in American psychiatry circa 1990.

The Osheroff Exchanges

In 1990 the *American Journal of Psychiatry* carried a series of exchanges by two of the grandees of the field, Klerman and Alan Stone, a professor of law and psychiatry at Harvard and a former president of the American Psychiatric Association, on the implications of a lawsuit that had been instituted in 1982 and was only brought to a close in an out-of-court settlement in 1987. The pieces attempted to assess the impact of the Osheroff case on the treatment of depression—which many thought might be considerable. The exchanges rehearsed many of the issues that have emerged in this and previous chapters.

The suit had been filed by a Dr. Rafael Osheroff against Chestnut Lodge. He claimed negligence because the hospital had failed to institute drug treatment and had persisted in the use of individual psychodynamic psychotherapy as the sole treatment for his severe depression. This delay in instituting effective treatment had allegedly cost him his marriage, his practice, and his profession. Dr. Osheroff had had episodes of anxiety and depression before being admitted to Chestnut Lodge in 1979. During this time he had been treated by Nate Kline, as an outpatient, and prescribed an antidepressant. Kline's notes suggested that the treatment had been of some benefit, but Osheroff had not continued with it.[77] He got worse and was admitted to Chestnut Lodge, where he was to remain for seven months. He was diagnosed as being depressed against the background of a narcissistic personality disorder. During this time, he apparently lost up to 40 pounds, had very poor sleep, and experienced frequent periods of marked agitation. Nevertheless, psychotherapy remained his only treatment.

The wisdom of this was questioned by an outside psychiatrist, who had been approached by the family. In response to this inquiry, the team members involved in the case at Chestnut Lodge reviewed their management but decided to continue with what they had been doing. Osheroff left Chestnut Lodge and was admitted to Silver Hill Foundation in Connecticut. He was diagnosed as having a depressive illness and was treated with a phenothiazine and a tricyclic antidepressant. Improvement was apparently noted within three weeks, and he was discharged within three months. His clinician at Silver Hill did not agree that he had a narcissistic personality disorder. He subsequently remained well for the following decade.[78] The exchanges between Klerman and Stone on this case are illuminating.

In his article entitled "The Psychiatric Patient's Right to Effective Treatment," Klerman pointed out that even Chestnut Lodge accepted that Dr. Osheroff had had a depressive disorder.[79] The difficulty lay with the fact that a supposed personality disorder was chosen as a target for treatment.[80] There were two problems with this: first, there were no treatments that were demonstrably effective for narcissistic personality disorder, and second, there was the failure to treat with therapies that were known to be effective for the depressive disorder that was also present.

Chestnut Lodge at the time maintained a split between psychotherapists and "druggists," alluded to in previous chapters, thus seeking to avoid sullying the hands of the therapist with drug treatments.

The case raised in acute form the question of proof of efficacy in psychiatry. Klerman argued that the most "scientifically valid" and most convincing evidence comes from randomized control trials. There was no such evidence of this nature for any form of treatment for narcissistic personality disorder, and, while there was some RCT evidence for efficacy for cognitive and interpersonal psychotherapies, there was none for psychodynamic therapies—the form of therapy in which Chestnut Lodge specialized—in the treatment of a major depressive disorder.

He pointed out that drug treatments must be scrutinized by the FDA prior to license and must be demonstrably safe and efficacious, but that this was not the case for the psychotherapies. It might be argued that if they could not be shown to do any good, then they couldn't be doing any harm either. The Osheroff case challenged that assumption, however, in that there appeared to be clear injury to his reputation and career and, one could presume, an increased likelihood of further depressive episodes, by virtue of the delay in instituting effective treatment. It might be further argued that if patients are willing to pay for an ineffective treatment, the government shouldn't intervene to stop them—but in Osheroff's case, although he may not have been entirely without blame in perpetuating his own condition, he was arguably in no fit condition to protest against the "experts." In such circumstances, doesn't the government have a duty to intervene?

The question of the efficacy and regulation of psychotherapy had been a matter on which Klerman had written for a decade prior to these exchanges. In a piece in the *American Psychologist* in 1983, he had noted that psychotherapy was in a position similar to that of drug therapies prior to the 1962 amendments discussed in Chapter 1. By 1976, for example, there were 164 different therapies available through 50 different therapy institutes in New York alone. Psychotherapy was consuming ever larger amounts of the health budget, but with little evidence that it worked for anything in particular.[81] The claims that could be made on behalf of different therapies were unregulated, and it seemed that therapies were likely to continue to proliferate without anyone ever having to provide proof of efficacy. The implication was that there were charlatans in the market place and that a number of practices were almost certainly harmful. In the absence of regulation, there was an onus on the profession, he argued, to provide treatment guidelines.

In 1990 Klerman's argument was that the conflict was not one of biological versus psychological treatment so much as one of prejudice versus evidence. In medicine, it has always been true that while clinicians might be held responsible for any adverse consequences of the delivery of an idiosyncratic therapy unsupported by evidence of efficacy, they would not be held responsible for the prescription of a therapy that a respectable minority of other therapists provided. On this basis, one might argue that analytic therapy can be justified. But, Klerman asked, should there be an obligation to inform patients of the possible options and to seek their consent for what is being proposed?

In reply, Stone accepted that Chestnut Lodge might have erred in persisting with its form of treatment for so long in the face of Osheroff's obvious deterioration, but thereafter he parted company with Klerman.[82] He suggested that the reason no consensus on therapeutic approach existed was because there were no compellingly effective treatments in psychiatry that could be delivered with minimal side effects—that there was a considerable gap between clinical trial data and clinical practice outcomes. (The outcome studies of antidepressant usage that were noted in Chapter 3 offer some support for this point.) If drug therapy for depression was clearly effective, Stone went on, then everyone would use it and his debate with Klerman wouldn't be taking place. The fact that not everybody was using drug therapy implied that there was more evidence to weigh in the balance than just the data that came out of clinical trials. This point is more problematic, in that the obvious efficacy of antibiotics, for instance, hasn't led everyone to jump on that therapeutic bandwagon; so much is this the case that many people might understandably feel that regulation to prevent other forms of therapy from claiming efficacy for the treatment of infections is still warranted.

Stone expressed alarm that Klerman's article might be used to promote more uniform standards of treatment in psychiatry based on one view of science and clinical practice. Linking the Osheroff case and the issues of efficacy in practice and public policy, he argued, would be a mistake because the science of evaluation was still, in 1990, not good enough to dictate policy or standards of clinical care. Indeed, as we saw in Chapter 3, at the time Osheroff was hospitalized the methodology being employed to assess antidepressant efficacy was seriously flawed (possibly even more so than Stone realized).

Stone also argued that psychoanalysts are in no different a position than surgeons—neither have to provide evidence from RCTs for what they do. In addition, quite apart from efficacy data, there is a larger question involving the issue of outcomes. An antidepressant may be shown to be efficacious in an RCT but because of side effects, or something to do with the shape or color of the capsule, or the style of those who prescribe it, few people may get well on it. Conceivably if the question were asked how many get better on a particular antidepressant compared with subjects receiving psychoanalysis, the answer may indicate that similar numbers get well with both or indeed even more may be improved by analysis. In terms of outcomes, Stone argued that it has never been shown that psychoanalysis is at a disadvantage.

In the actual case, Klerman, Don Klein, and Frank Ayd testified on Osheroff's behalf. Chestnut Lodge settled out of court without admitting liability, and accordingly no legal precedent was set. Klerman alone attempted to define the issues in public debate. He and Stone followed up their initial articles in the correspondence columns of the *American Journal of Psychiatry* but without reaching closure.[83] It seems likely that Klerman would have agreed with Eysenck who, revisiting in 1994 the debate his 1952 article had triggered, concluded that "the simple truth is that the supply of psychotherapeutic help nowadays is still determined by the personal preferences and interests of . . . therapists. There appears to be no way in which the facts unearthed by scientific research can be made to bear on the actual practice of therapists."[84]

But Klerman never had a chance to agree—he had died two years earlier.[85] He died with a letter on his desk from Isaac Marks criticizing the panic disorder studies he had coordinated, and he had spent the previous day composing a reply. The exchanges on this had been bitter and dramatic. At a meeting in Geneva in 1990 organized by Klerman to review progress on the study of panic disorder, critics of Marks from the floor had inveighed against the pharmacological Calvinism of those opposed to the drug treatment of panic—seemingly unaware of the significance of Geneva in this context and also unaware that the concept of pharmacological Calvinism had come from Klerman in the first place. A number of those closest to Klerman felt strongly that this dispute contributed to an earlier death than might otherwise have been the case. The intensity of feeling that went into the dispute can perhaps be better understood against the background of the NIMH studies and the Osheroff case. Klerman was no simple drug therapist. Both Marks and he had developed psychotherapies. Both had come up with treatment packages that were simple enough to "give away." Both argued for the need for public health policy to follow the evidence in the days before the discovery of evidence-based medicine. They differed mainly in their commitment to diagnostic categories. Klerman was a neo-Kraepelinian, Marks wasn't. As Klerman put it, everything seemed to hinge on "the validity of separating panic anxiety from other forms of anxiety and designating panic disorder as a diagnostic category that is separate from anxiety neurosis."[86]

Both had a common problem, however. In the debate with Stone, Klerman raised the question of who should be responsible for providing evidence of efficacy of treatments and offered his opinion that it should be those making claims of efficacy. The problem with this, which is the situation that applies to the pharmaceutical industry, is that it forces "corporate" development on therapeutic schools, and this prevents the development of therapies that might be given away, such as interpersonal or behavior therapy. Cognitive therapy is at present underpinned by a semicorporate structure in the profession of clinical psychology, and this has played no small part in its viability—at a cost, however, of further fracturing the therapeutic domain into rival exclusivities. As Marks has framed the issues, citing George Bernard Shaw, every profession is a conspiracy against the public.[87]

Klerman died a day after meeting with Michael Shepherd, who was visiting New York to lecture on the placebo. Talking to Shepherd, Klerman had outlined his role in the Osheroff case and in particular the subsequent debate and confided that he felt let down by the lack of public support from other members of the profession.[88] Shepherd had agreed to participate with him in a debate the following day in front of a Westchester Hospital audience on the role of specific and nonspecific factors in treatment.

Having pioneered the evaluation of psychotropic drugs, Shepherd believed that the search for specific efficacy was increasingly leading the profession down a blind alley. Evaluation must continue, but arguably the most important thing that efforts to date had shown was that the evidence of efficacy for nonspecific elements in treatment was much greater than that for the supposedly more specific elements.[89] As noted above, Stone had argued that the position of analysts was no different from that of surgeons. This didn't appeal to Shepherd, in that when a surgeon removes a gallbladder from a patient with confirmed biliary colic, he has the gallbladder to show for it and the patient appears well. In such circumstances, an RCT to demonstrate efficacy is not called for. In contrast, in surgical conditions where the outcome is "softer," as in the case of the relief of anginal pain, RCTs are called for and when done have often produced surprising results. In the case of psychiatry, such specific efficacy can never be demonstrated clearly enough to warrant not doing an RCT.

But the randomized trial was for Shepherd a means of tempering enthusiasm rather than a method of discovery, and the interpretation of the results of RCTs were what mattered. Stone was on grounds more to Shepherd's liking when he pointed out that a license from the FDA is not a guarantee of good practice. Good clinical practice is more often a matter of ameliorating secondary disabilities than it is of tackling a primary pathogenetic process. Entire therapies can be geared to this end: behavior therapy, for example, is predicated on the management of secondary disabilities. While Klerman and Marks had fallen out over panic disorder in part because the one was interested in the primary process and the other in correcting the secondary disabilities, interpersonal therapy was also first conceived as a therapy to correct secondary disabilities.[90] Besides which, where depression was concerned, the facts of the matter, as was well illustrated by the NIMH study, were that sensible clinical management was a potent therapy for depression—anything else was at best marginally more potent for most depressions.

Shepherd blamed Kraepelin for the current intense focus on specificity in psychiatry. His last published piece, which appeared three months before he died in 1995, was an extraordinary attack on Kraepelin, whose influence he saw as a potential danger to society.[91] Klerman, Shepherd felt, as one of the fashion-setters in U.S. psychiatry in the late 1960s, had engaged in Kraepelinism. Shepherd saw Klerman as flexible enough, however, to see the problems with this approach and ultimately to transcend it, but he was less sanguine about the rest of the profession.

Coda

In his 1983 piece on the question of efficacy in psychotherapy, Klerman drew attention to a vast expansion in the mental health industry. In 1955, 1 percent of the American population had contact with a mental health practitioner. By 1980 this had risen to 10 percent.[92] In 1896, the year Kraepelin formulated the modern concept of manic-depressive insanity, in North Wales there were approximately 200 admissions per million people to the local asylum.[93] This was a marginal increase over the 1850s figures, the decade after the asylum was opened. It was marginally lower than the 1930s figures (averaging 250 per million per year), the prewar decade. But after 1945 these figures began an inexorable rise. The same catchment population now has rates of admission on the order of 3,000 per million per year—a figure that could be doubled if those who were assessed and rejected for admission were included, and one that does not take into account all those being seen by community mental health teams or treated by primary care physicians.

In 1896, 20 percent of admissions in North Wales were diagnosed as having melancholia, 40 per million, most of which would now be diagnosed as major depressive disorder with psychotic features, a majority of which would probably now be given ECT. The comparable figures for depression now are 948 per million. Of these 948, 268 per million are diagnosed with severe melancholic or psychotic depressions. The figures for referrals for depression to community mental health teams will greatly exceed these and these figures in turn will be dwarfed by the figures for those being seen and treated, commonly with antidepressants, in primary health care settings. Although many of the key studies referred to in these pages convincingly indicate that depression can be a chronically recurring disorder, in 1896 the mean length of stay for the psychotic depressives was five months. They went home apparently well—without any of the benefits of modern therapy.

This huge increase in the diagnosis of depression may reflect some absolute increase in the incidence of the affective disorders. Diseases do rise and fall in frequency—neurosyphilis did so dramatically during the nineteenth and early twentieth century. The population is aging, and melancholic subtypes of depression, which have been thought to be more drug responsive, are more common in the elderly. But against this must be set the vast expansion of all psychiatric diagnoses in modern times and the fact that these at present are underpinned by depression and the use of antidepressants. This is not to say that the modern mental health industry has been created solely by the pharmaceutical industry marketing its products, although in general increased rates of referral and consultation in many areas of medicine occurred around the time of the introduction of the new treatments. It would seem important to try to understand why this should have been the case: with the secularization of society, is health becoming a focus for our concerns in a way that it hasn't been before, and does this explain the great increase in health-seeking behaviors generally? Whatever the answer to this question, an escalating use of medical resources needs legitimization, and in the field of psychiatry this has come from the opportune discovery of both antidepressants and depression. Depression has been the cornerstone on which the neuroses, carved out and shaped for construction purposes, have been laid in successive layers.

Commenting on these points two months before he died, Michael Shepherd offered the view that "these topics had spin-offs—like money, reputation, fraud, publicity and 101 other things which you can take out of the issues and make your own without getting involved in the scientific issues. That I think is where the conflict takes place. It cannot be resolved. It's got nothing to do with whether drug X affects condition Y."[94] Shepherd here would seem to be suggesting that in an ideal world science would be a disinterestedly rational exercise. Although there is very little in the antidepressant story that would suggest that we have been in the business of doing science, equally there is little in the story that would seem to support Shepherd in this view. Where the field has attempted a disinterested rationality, the results have been confounding rather than illuminative. At present the area is such that it almost seems we have to let whatever dynamics govern the field take us to extremes and back in the hope that successive traverses of the scanner will eventually reveal the shape of what lies beneath the surface.

Although the antidepressants have underpinned the medicalization of psychiatry, the antidepressant story in turn has been underpinned by the notion of specificity. As was outlined in Chapter 3, the specificity of modern medicine is not defined simply by the statistical methods that allow us to probe behind superficial symptoms to see if a treatment is working. It is defined by its relationship to a modern discovery—the power of non-specific therapeutic principles. Among other things, the antidepressant story illustrates the power of forces in modern society to capitalize on the notions of specificity and the benefits that can stem from this, the most telling example of which is the current rush to patent the human genome. But the antidepressant story also illustrates the problems that stem from the failure of modern medicine to capitalize on the discovery of the non-specific.

The woman in the Introduction was a good case in point. In psychiatry at present there is an almost exclusive focus on specific mechanisms (whether biological or psychological) that might be giving rise to her experience and how these might be appropriately adjusted, with little attention being given to the aspects of her situation that led her to seek help—the why of her problem. Has she been abused in the past and does she feel that her depression is evidence of the corpse rising to the surface? Is she unable to explain what's happening and as a consequence has she become terrified she is going mad?—could things be improved simply by telling her she's not? Is she afraid that this will go on for the rest of her life and is she struggling with the meaning of such a life? Is she afraid of the effects of her state on her marriage or her child?[95] For a number of years, coming mainly from a background in anthropology, there have been voices arguing for a reintegration of personal narratives into medical therapeutics—arguing that these *why* questions are at least of equal importance to the *how* questions.[96] Picking up on the narrative aspects of therapy is not something that needs to be left to the wisdom of therapists. At least to some extent, it can be standardized and taught. The problem rather is in seeing who might have an interest in fostering such an approach, as the possibilities for making money out of something that in principle could never be shown to be specifically effective would at present seem to be rather limited.

The mission for neuroscience and psychopharmacology has been and is to chase the question of *how* even if the idea of a soul appears to be banished in the process. There is both good and ill interwoven here together. Answers to the question *why* that are wrong may be tyrannies from which we need to be liberated. The scientific pursuit is good in so far as it liberates us and hands us control of our destinies. The radical materialism that seems to be the result of modern neuroscience and psychopharmacology should not necessarily lead to despair on the question of *why*: such a viewpoint is not incompatible with a life of high moral purpose or indeed with notions of transcendence.[97] But if all vision of *why* is lost within medicine or even temporarily suspended, in the vacuum that remains the question of whose interests are being served in medical interactions will inevitably be thrown into greater relief. We will have confidence in such a science only in so far as we have confidence in the fact that sectional interests don't dominate it and that the intentions of the key stakeholders are not duplicitous. It is difficult, perhaps increasingly difficult, to have such confidence at present.

As with many other aspects of the modern marketplace, therapeutics (both psychotherapy and pharmacotherapy) at present seems to be leading to an atomization of distress. Just as the ideal market arrangements would have everyone living in a single's apartment, each complete with washing machine, dishwasher, and refrigerator, so also treatment development in practice has disconnected individuals from their social milieu. For example, in 1982 Sanjeebit Jachuk and his colleagues in Newcastle looked at the perceptions of doctors, patients, and relatives of the effects of antihypertensive medication. They found that the doctors reported 100 percent efficacy—blood pressure was reduced. The patients were mixed in their views, some reporting benefits and others reporting problems with the medication. Three-quarters of the relatives, however, reported that treatment was having an adverse effect—patients were now complaining of side effects and the process of diagnosis and treatment had made them hypochondriacal.[98] Whose view is most ecologically valid? If treatment of a disease, which wasn't affecting a family up until then, has more effects on a family than the disease itself, who should be making the risk-benefit assessments and what should be put in the equation?

All these issues, the increase in utilization of medical services, the questions of specificity and risk-benefit assessment, all lead back to the 1962 amendments. Service utilization began its rise when patients were forced to go to medical practitioners to obtain treatments by virtue of the fact that only medics could prescribe. This seems a rather obvious point, but one that is overlooked by virtually every commentator in the field. Where before illnesses might be managed by turning to relatives, friends, or neighbors or by self-management, all these efforts have since 1962 increasingly been funneled through the medical services. Where once management might have aimed at improving functioning, since 1962 it has aimed at extirpating diseases—an approach immortalized in the quip that the treatment was successful but the patient died. In addition, the focus on disease entities in the 1962 amendments was, where psychiatry was concerned, essentially a Kraepelinian focus.

After 1962, if Kraepelin hadn't existed he would have had to be invented—which in one sense he was. The Kraepelin of neo-Kraepelinism involves a selective abstraction from the range of almost mutually contradictory views the real man appears to have held in the course of his career. If one is in the business of challenging established views in any field, invoking an esteemed predecessor, preferably one who is dead and cannot argue back, is always a good tactic. As Voltaire put it, history is a trick the living play on the dead. The portrait of a scientist whose empirical orientation transcended ideological division arguably comes closer to being a portrait of its creator, Klerman, than anyone else.[99]

When viewed from a wider perspective, the era since 1962 has been highly atypical. It has been an era when access to drugs, though not to therapists, has been restricted even though in the larger scheme of things it would seem that people have by and large had more confidence in their drugs than in their therapists. The dike erected by the 1962 amendments, however, seems to be slowly springing leaks. Pharmaceutical companies have begun to advertise prescription-only compounds directly to consumers, and an increasing number of prescription-only medicines are being reclassified for over-the-counter use. It might take little more than a minor stroke of a politician's pen, modifying slightly the current regulatory framework, to trigger the scanner into a further traverse—but this time in the opposite direction to the one in which it has been traveling for the past thirty years.

Klerman, Kline, Kuhn, Shepherd, Marks, Brodie, Axelrod, Carlsson, and a handful of others were the key figures of the age. It is difficult to know whether their individual and collective impact on the culture of our times has been greater than any contributions to whatever "science" there may be in psychopharmacology. Axelrod and Carlsson received their share of honors, but the others weren't even department heads. The opportunity for Nobel Prizes now seems to slipping away. For the clinicians it is perhaps gone entirely, in that no one person can discover a new psychotropic drug anymore and even theoretical developments in psychopathology stem now from the deliberation of committees rather than from individual genius. The grail may also have passed by the psychopharmacologists working within the neurosciences, in that neuroscience is no longer constrained by the clinical domain and future discoveries will be of primary relevance to neuroscience rather than to psychopharmacology. As regards the question of specificity and nonspecificity, after the completion of the human genome project, the elixir of specificity will be consumed to the last drop and it will then be apparent whether there are specific defects that determine the course of a depression or whether the role of biological factors is to contribute to the course of some of the depressive disorders. If the latter is the case, the therapeutic challenge will involve a combining of antidepressant principles, and the field will not be well served if it has become too fragmented.

Postscript

The discovery of the psychotropic drugs was as monumental a breakthrough as the discovery of psychodynamics. The histories of the two therapies also show a number of striking similarities. In both cases, there has been a focus on treatment specificity, a focus that has acted as a source of division into orthodoxy and heresy. In both cases, an increasing emphasis on long-term treatment and on possible difficulties with withdrawal recalls Karl Kraus's quip about analysis to the effect that the treatment had become the illness it purported to cure.

In addition, just as efforts to criticize psychoanalysis constructively or otherwise became a problem, with critics having their criticism dismissed as irrational and as evidence of a personal neuroticism, psychopharmacotherapy risks developing a similar problem and style, with an insistence on evidence-based medicine as the only basis for rational discourse functioning as the barrier to open debate. At present the dominant forces in the health marketplace endorse such appeals even though they are palpably wrong. But it is difficult to point out what is wrong without risking incoherence.

At times the current debate about the merits and proper place of pharmacotherapy appears to lack genuineness. But getting to the heart of the problem, which is that the majority of people who meet diagnostic criteria for depression or anxiety and who seek treatment because of how they feel have a condition or conditions that in principle cannot be shown to benefit *specifically* from available treatments, risks throwing open the gates of the citadel to a host of charlatans and mountebanks. Getting to the nub of an issue, however, is one thing; offering a solution is quite another.

The chain of events outlined over the last seven chapters would not have been history if things could not have gone in a quite different way. There are a number of points at which the story could have taken a different course, but I will pick one—the tenth of October, 1962, when the FDA was permitted to continue to designate certain drugs as prescription only, to insist on randomized control trials, and to encourage the pharmaceutical industry to develop compounds for indications that experts thought were appropriate.

This framework was put in place for a number of legitimate reasons. Consumers of medicines had shown an inveterate preference for drugs to treat halitosis or tonics and other agents aimed at the symptoms of nervous fatigue.[1] Putting a premium on the medical model was supposed to force the industry into developing drugs for medical diseases rather than for broader indications such as the reduction of tension or the provision of a tonic. Where once the production of a tonic would have been the goal and any clearing up of tuberculosis would have been seen as a useful side effect, now the cure of tuberculosis was to be the goal and if the drug also had tonic properties, these were considered side effects. The aim was to increase the risk-benefit ratio of treatment by reducing the side effects of treatments or by confining treatments to individuals who clearly had something to benefit from them.

One of the problems with this approach is that essentially the FDA, the medical profession, and the pharmaceutical industry signed on to a bacteriological model of disease. This was not perceived to be a problem at the time because the most conspicuously successful drugs that had arrived on the market thus far were the antibiotics, and other treatments such as insulin for diabetes appeared to have an almost comparable specificity. Diphtheria or staphylococcal infections, however, are not disorders with a significant superstructure of social constructions. Nor does the use of diphtheria antitoxin or the penicillin treatment of staphylococci require the incorporation of much in the way of nonspecific factors into the therapeutic act. Antitoxin in diphtheria will work regardless of the bedside manner of the physician.

The medico-pharmaceutical and regulatory complex was setting its eyes on the development of treatments for all other diseases that were comparably specific. But as may perhaps now be clear, the available psychotropic drugs do not have this kind of specificity. The antipsychotics and the antidepressants are neither specific in the sense of being specific to one disease nor specific in the sense of working regardless of the nonspecific milieu in which they are delivered. It is worth examining some of the consequences of this.

In the case of the antipsychotics, for instance, a premium was put on developing drugs to treat schizophrenia. These drugs quickly came to be seen as antischizophrenic agents, despite the evidence of benefits for a range of other conditions. When one generation of compounds failed to cure schizophrenia, the fact that a large number of individuals remained actively disturbed or suffering from residual symptoms has provided the justification for the development of subsequent generations of largely similar compounds. It has also provided the justification for biological psychiatry to capitalize itself by arguing for the necessity of PET and MRI scans to find out what is happening in the brains of those who fail to respond.

If, in contrast, chlorpromazine had from the start been available over the counter, and if it remained, as it clearly appeared in the first instance, a nonspecific treatment that provides a certain kind of tension reduction, events might have unfolded quite differently. Tension reduction would have been seen as an antipsychotic principle. The nature of this effect would have been more keenly characterized. If tension reduction with chlorpromazine failed to work, there would have been some justification for developing a number of variations on the neuroleptic theme on the basis that people are different and some people may respond dysphorically to chlorpromazine but not to haloperidol, for instance. But if tension reduction as a principle failed to work, there would not have been the market for or the pressure to develop an almost endless string of similar compounds.[2] Instead, faced with an unresponsive patient, the challenge to the clinician would have been to answer the question why it was that this particular condition was not responding to tension reduction as an antipsychotic principle. This arguably would have had two consequences, one being the addition of other therapeutic principles, whether pharmacological agents of a quite different type or behavioral treatments, hygienic elements, other social maneuvers, or some mix of all of these to improve the situation. The other consequence would have been that a greater premium would have been placed on theoretical efforts to model and understand psychotic and affective disorders.

Could chlorpromazine have remained available over the counter? Most probably it could. The hazards of chlorpromazine therapy in recent years have been owing to its prescription by medical personnel rather than to the intrinsic properties of the drug. In the pursuit of the chimera of specificity, the failure of an individual to respond to treatment has often meant that doses of these agents were escalated during the course of the 1960s and 1970s to heroic levels—levels that no patients would have contemplated giving themselves. Another notable hazard has stemmed from the drug's intramuscular injection in large amounts, which appears on occasion to lead to cardiovascular collapse and death—something the average patient would seem less likely to do to himself. While it would clearly be naive to suppose that the average consumer is the repository of all wisdom on these matters, it is worth noting that over-the-counter consumption of chlorpromazine or haloperidol would almost certainly have settled fairly quickly on the doses that forty years later the psychiatric profession is coming to accept are optimal.

Do the same arguments apply to the antidepressants? Consider the case of cyproheptadine. This tricyclic compound, developed in 1960, acts on the 5HT system. It stimulates appetite, it enhances sleep, it improves drive and libido. It has been widely used off-label as a tonic for many years by primary care physicians in Europe and elsewhere. A compound with a very similar profile of action on 5HT receptors, nefazodone (Serzone/Dutonin), has recently been released. It enhances appetite, improves sleep, and increases libido and drive, but it is marketed as an antidepressant. It is interesting to speculate what the consequences of marketing it as a tonic might have been. Arguably tonics would have had much greater acceptability with the public at large than "antidepressants." The notion of a tonic does not convey the same sense of constitutional or moral failing, nor does it require compliance with a medical model in quite the same way.

In many respects cyproheptadine, imipramine, and agents such as nefazodone would have fit the bill as the ultimate "humoral" remedies. They promote sleep and appetite. In high doses they produce perspiration, diarrhea, and an increase in nervous tone. In this older sense, as described in Chapter 1, they obviously work. In contrast, the current antidepressants are at present all but misbranded as antidepressants. They are effective for a wide range of "neurotic" conditions. Kline's term, psychic energizer, seems much more apposite—at least as applied to the MAOIs and SSRIs. Retrospectively, however, it is clear that the notion of a psychic energizer could never have caught on in the 1960s, because it lacked the suggestion of specificity that became politically correct in Western cultures following 1962. Worldwide, however, the most recent WHO study reveals an entirely different picture, with cases of depression more likely to receive a tonic or a sedative than an antidepressant and cases of anxiety at least as likely to receive an "antidepressant" as cases of depression.[3]

If imipramine had been developed as a tonic, the problem of treatment resistance would have required a range of other tonics of different classes in order to ensure that the person was not simply reacting adversely to a drug from a particular class. But thereafter, where clinicians now look to the possibility of cocktails of antidepressants (surely an indictment of specificity if ever there was one), there would have been a much greater premium on looking at other types of antidepressant principles. These might include psychic energizers, the prescription of routine or activity, and a range of other hygienic measures along with engagement in the solution of interpersonal difficulties or dysfunctional attitudes, as indicated in each individual case.

A recourse to such approaches is not blocked by the lack of a theoretical model to accommodate them. For instance, there are circadian rhythm models that offer scope to marry cognitive, interpersonal, and biological elements in one picture, a picture that is consistent with current epidemiological and physiological evidence on depression. Circadian rhythm hypotheses stand in contrast to both the amine hypotheses of depression, which are strictly biological and offer little if any accommodation for psychological factors in the genesis, presentation, or maintenance of a depressive disorder, and the cognitive hypotheses, which are equally neglectful of biological inputs to depressive disorders. Both cognitive and amine hypotheses are predicated on a primacy of specific therapeutic interventions; current circadian hypotheses are not.

As put forward by a number of groups, the most plausible hypotheses at present envisage that circadian rhythms are knocked out of synchronization by dislocations to social routines.[4] These dislocations would qualify as the life events that have been observed to trigger depressive episodes. Social dislocation gives rise to a disruption of the circadian system that is noticeable but not disabling—rather like a dose of flu or, more pertinently, like a case of jet lag or the dysphoric state that can result from shift work. More pertinently again the state has strong resemblances to the dominant presentation of "depression" in non-Western settings (that is, the majority of people who meet criteria for "depression" at any one point in time worldwide), which is of a neurasthenia.[5]

Whether such a state becomes a depressive "case" would then depend on a number of other factors, some of which may be biological and others social or personality based. There is some evidence that certain individuals are constitutionally programmed to adapt less readily to the dislocating effects of social cue disruption than others. In the case of shift work, the way to handle this is to avoid doing it. In the case of depressive disorders, such avoidance may not be feasible, in which case those with a constitutional predisposition may find themselves suffering more severely and at greater length from the dislocations caused by life events.

A depressive case would also be more likely to emerge in dysrhythmic individuals performing below par who then were faced with a challenge that required them to perform optimally. In such circumstances, performing less well than they expected, they would seek to account for their poor performance and, in the absence of a clearly physical explanation such as an infection, would be liable to blame themselves. Self-blame leads to demoralization and a consequent downward spiral that would potentially aggravate and maintain the dysrhythmic state, because the resolution of such a state requires motivated activity. Those who are jet-lagged who have the motivation to get out and onto local time get over their dysrhythmia much more quickly than those who wait to recover in their hotel rooms. Similarly shift workers with the motivation (because of mortgage payments or whatever) to work the system may do relatively well as their motivated activity facilitates a reentrainment of the clock, but those who haven't a similar motivation may end up suffering from what may become relatively permanent psychosomatic dysfunction.

Within such a perspective, a number of points follow. First, many affective disorders would be seen to be mild and self-limiting. It would seem highly likely that most if not all of us at some point in our lives would have episodes of mild biological depression of this type. Second, there would be no single genetic cause for depression. The role of genes would rather be to contribute a variety of constitutional factors that may predispose one to depression or (more probably) militate against a speedy resolution or remission. Third, the role of drug treatments would be to facilitate the resolution of a disorder rather than to specifically cure it. This role would be rather similar to the role of pharmacotherapy in the treatment of ulcers—a variety of different principles might facilitate a cure. The fact that major and minor tranquilizers, as well as tricyclics, MAOIs, SSRIs, and a range of activity-oriented "psycho"therapy packages may all facilitate recovery is entirely consistent with such a proposal.

In terms of nonpharmacological management strategies, the generation and maintenance of routines and involvement in a certain amount of routine activity would be a matter of basic hygiene. In the course of an acute episode, behavioral activation would also be a therapeutic principle, as would the removal of blocks to spontaneous self-healing by tackling, where indicated, the overlay of dysfunctional cognitions that a depressive disorder generates. There is also some evidence that interpersonal dynamics are particularly important in the entrainment of rhythms. A further implication of such a model is that biological disruption would generate depressive and other neuroses. These might persist once the underlying biological disorder resolved, in which case the role of drug treatment in the management of the resulting condition would be very limited.

In addition to the above, considering the narrative aspects of the presentation—why is this a problem now—will be important. In order to address this, I routinely write to people following consultations to offer them a view on how things appear to me to hang together.[6] I also give detailed handouts on drug treatments—their indications and side effects.[7] As part of a general strategy of embedding therapeutic acts in a network of optimal human encounters, I would also encourage all depressed or anxious subjects to participate in Rogerian-style encounter groups.

One of the problems with this model is that it requires research across boundaries. It requires some appreciation of biological factors from cognitive researchers and vice versa and an appreciation of social constructions from both. It is also difficult to capitalize upon quite as readily as the alternative models.[8] Adopting this model clearly commits a therapist to conceding that there are a range of therapeutic options and that therapy may require a judicious mix of different principles. Such an approach might seem like good practice. In practice, however, there are growing difficulties in taking such an approach. There are health schemes in the United States (and probably elsewhere in the West in the near future) in which the only act of a psychiatrist that is reimbursed is the act of prescribing. Lengthy amounts of time put into managing nonspecific aspects of care count for nothing.

The FDA and the 1962 amendments didn't create the idea of specificity, but they arguably put an enhanced premium on it. This has had side effects in its own right, one being a decline in medical confidence in the management of nonspecific factors of the therapeutic encounter. What relation this has to the seeming increasing alienation of the public from orthodox medicine is difficult to determine, but in the case of psychiatry a bacteriological model of depression as applied to the psychoneurotic complaints of the population at large is found by consumers to be unpersuasive.[9] In the face of this the act of therapy has been fractured with specific therapists on the one side, included among which are both drug and cognitive therapists, and on the other a cadre of therapists delivering nonspecific therapies—aromatherapists, reflexologists, physiotherapists, psychologists, and others, whose entry into orthodox health care is blocked.

In the alternative health movement, health is conceived in terms of an inner balance and harmony that is presumed to exist naturally, unless disturbed by the forces of modern civilization. This idea of health being a matter of balance and harmony is, of course, straight from Galen. The twist that the greatest threat to our health comes from the products of industrialization, whether they be food additives or the drugs of the chemical industry, is a reworking of Rousseau's myth of the noble savage. The idea that the natural state is a virtuous one and that the deals of civilization are chains that bind us was powerfully articulated in 1962 by Rachel Carson in *Silent Spring* with its proclamation that "for the first time in the history of the world, every human being is now subjected to contact with dangerous chemicals from the moment of conception until death."[10]

Some of the problems that regulation throws up have been framed by Paul Leber of the FDA in terms of delegated narcissism, which he has illustrated with a vignette of a child standing at the edge of a cliff with her father watching the sun go down. The sky is suffused with pinks. She has never seen anything like this before. After the sun disappears, she turns to her father and says, "Gee, Dad, that was wonderful, can you do it again?" We want the miracle of safe and effective cures and have, through government, put a law in place supposedly to prevent unsafe and ineffective treatments, notwithstanding the facts that drugs cannot be either entirely safe or entirely effective and that our methods for evaluating both safety and effectiveness are evolving and at present remain of limited utility. We ignore the fact that the funds allocated for independent evaluation are almost nonexistent. Added to this is the problem that, as with all political deals, politicians have oversold the protections that are available, and this is compounded by the fact that the pharmaceutical industry has sold both the safety and the efficacy of its products to an extent that has been unwarranted.

Against this backdrop, the health-related behavior of many of us seems to be increasingly shaped by dynamic defense mechanisms. Everything has become a hazard. There are any number of programs on television or stories in the newspapers about the range of risks from the air we breathe to the food we eat—saturated or unsaturated or transfatty, with additives or not, hormone treated or not, free range or not. This month we are phobic about eating butter. Next month we have the amalgam in our teeth ripped out. We jog or exercise compulsively, quite possibly shortening our lives in the process, driven by intrusive images of ideal body shapes. We obsessively ruminate on the adequacy of the expert evaluations we are offered, never getting closure on the issues. By this year's standards the evaluations of five years ago seem hopelessly flawed. And where at one time personal calamities or accidents were likely to be put down to fate or God, now there appears to be something of a belief that we are born with a warranty and an expectation that the origins of problems can be determined, that guilty parties can be identified, and that recompense will be offered.

When it all gets too complex there is a retreat to a paranoid response, the essence of which is to deny complexity. Because the FDA is charged with ensuring the safety and efficacy of drugs, the occurrence of any untoward event, for many, indicates some complicity between the agency and the industry. Why else would Prozac or Halcion have been left on the market after all those people who were taking these drugs assaulted their relatives or committed murder? The Scientology magazine *Freedom* features the FDA commissioner on its front page beside a "Killer Drug" headline, promising a story on how the FDA approved, whitewashed, and continues to protect the most deadly prescription drug in history. Peter Breggin cites Paul Leber's involvement in the licensing of Prozac as a case of his conspicuously greater concern for the interests of the pharmaceutical industry rather than those of consumers—notwithstanding the fact that the success of Prozac probably owes something to FDA stringency in restricting the availability of other new antidepressants on the U.S. market.

How can the situation be remedied? Should the conspiracy against the public that is prescribing rights be extended to nonmedical health professionals such as clinical psychologists? The extension of a conspiracy against the public would seem on the face of it unlikely to render that conspiracy any less of a conspiracy. If the element of prescribing rights is to be looked at again, then the more radical solution would be to do away with the category of prescription-only drugs and to make these compounds available over the counter. An alternative might be to retain new drugs on prescription-only status for a limited period of time, perhaps three to five years, to encourage expert involvement in their evaluation in clinical settings. Either of these options would transform the matrix of relations among the public, the medical profession, and the pharmaceutical industry. Putting some control back in people's hands would force medical personnel to rediscover therapeutic virtues other than the prescription of specific agents.[11]

Empowerment of the individual in this manner might also reverse some of the alienation many consumers of medical services now feel. For instance, as a prescriber I enjoy considerable advantages that are denied to others. I can take a minor tranquilizer before a public address or interview if I wish, without having to display my "weakness" to another. I can treat my tendency to dry skin with a steroid ointment of my choice—and is a tendency of this sort, aggravated by stress but wonderfully responsive to steroids, an illness or a disease? If I had premature ejaculation, I could take care of it without anyone knowing. If I had AIDS or some other fatal disease, I could prescribe myself the latest drugs or combination of drugs that the most recent research had suggested was useful or that from a theoretical point of view looked promising.[12] I don't have to go through a system that was put in place for addicts. I don't have to listen to someone who I know isn't listening to me advise me about drugs that he has never taken and probably never would.

A common response to proposals of this type is that patients are simply not sufficiently well informed of the possible benefits and risks of treatments to exercise choices of this kind. The tenor of such criticisms has overtones of criticisms made a century ago in response to demands from women that they be given the vote. The disasters predicted following universal suffrage never came to pass. It is not clear that a relaxation of these aspects of the current regulatory arrangements would bring about disasters. Patients would continue to consult medical practitioners, though possibly on a more equal footing; they would consult for advice rather than a prescription (in the sense of an order). The power relations would have been changed.[13] Indeed they might be restored to something closer to the position between the wars, which appears retrospectively as something of a golden period in medicine, when clinicians were more skeptical of therapeutic developments and appeared to have a much greater sensitivity to the social factors that might lead to a consultation.[14]

It would almost certainly be a mistake, however, to think that the story would have been entirely different if the 1962 amendments had not been in place. The development of the antidepressant story is a side effect of developments elsewhere in society. The 1962 amendments acted to channel in a particular direction forces that would have still been present—forces that called forth regulation and that make for corporatism. The imperative for the pharmaceutical industry to make money would have been present, whatever the framework in which it had to operate. The public's inveterate willingness to consume pharmaceutical compounds would also have been present. The technical development of modern medicine would no doubt have still progressed, facing us with recurring ethical crises just as the development of anesthesia once did. The story would have been different but not necessarily a better one.

Current Major Physical Treatments for Depression

1. Tricyclic Antidepressants

Drug Name	U.S. Trade Name	UK Trade Name
Imipramine	Tofranil	Tofranil
Amitriptyline	Elavil/Endep	Tryptizol/Lentizol
Desipramine	Pertofrane/Norpramin	Pertofran/Norpramin
Nortriptyline	Aventyl Pamelor	Allegron; also in Motival/Motripress
Clomipramine	Anafranil	Anafranil
Protriptyline	Vivactil	Concordin
Doxepin	Adapin/Sinequan	Sinequan
Trimipramine	Surmontil	Surmontil
Dothiepin	n/a	Prothiaden
Lofepramine	n/a	Gamanil

2. The Monoamine Oxidase Inhibitors (MAOIs)

Phenelzine	Nardil	Nardil
Tranylcypromine	Parnate; also in Parstelin	Parnate

3. Reversible Inhibitors of Monoamine Oxidase (RIMAs)

Moclobemide	n/a	Mannerix/Aurorix

4. 5-HT Reuptake Inhibitors

Citalopram	n/a	Cipramil
Fluvoxamine	Luvox	Faverin
Fluoxetine	Prozac	Prozac
Paroxetine	Paxil	Seroxat
Sertraline	Zoloft	Lustral
Venlafaxine	Effexor	Efexor

5. Other Antidepressants

Mianserin	n/a	Bolvidon/Tolvon
Mirtazapine	Remeron	Zispin
Trazodone	Desyrel	Molipaxin
Nefazodone	Serzone	Dutonin
Maprotiline	Ludiomil	Ludiomil
Viloxazine	n/a	Vivalan
Buproprion	Welbutrin	n/a
L-Tryptophan	Trofan	Optimax

6. Treatments for Bipolar Disorders or Prophylaxis of Recurrent Disorders

Lithium Carbonate	Eskalith/Lithobid	Camcolith/Priadel
Carbamazepine	Tegretol	Tegretol
Sodium valproate/ Valproic acid	Depakene	Epilim

Notes

Introduction

1. By mythical, in this sense, I mean a heuristic orientation to the unknown—not necessarily a bad thing.

1. Of Illness, Disease, and Remedies

1. Hippocrates cited in O. Diethelm (1971), *Medical Dissertations of Psychiatric Interest before 1750* (Basel: S. Karger), p. 16.
2. C. E. Rosenberg (1979), "The Therapeutic Revolution: Medicine, Meaning, and Social Change in Nineteenth-Century America," in M. J. Vogel and C. E. Rosenberg, eds. (1979), *The Therapeutic Revolution* (Philadelphia: University of Pennsylvania Press), pp. 3–25; W. Sneader (1985), *Drug Discovery: The Evolution of Modern Medicines* (Chichester: John Wiley & Sons); W. Sneader (1990), "The Pre-history of Psychotherapeutic Agents," *Journal of Psychopharmacology* 4:115–119; G. E. R. Lloyd (1983), "Introduction," in *Hippocratic Writings*, trans. J. Chadwick and W. M. Mann (Harmondsworth: Penguin Books).
3. P. Brain (1986), *Galen on Bloodletting: A Study of the Origins, Development, and Validity of His Opinions with a Translation of the Three Works* (Cambridge: Cambridge University Press); L. I. Conrad, M. Neve, V. Nutton, R. Porter, and A. Wear (1995), *The Western Medical Tradition* (Cambridge: Cambridge University Press); T. S. Hall (1976), *History of General Physiology*, vol. 1 (Chicago: Chicago University Press).
4. M. S. Pernick (1985), *A Calculus of Suffering: Pain, Professionalism, and Anaesthesia in Nineteenth-Century America* (New York: Columbia University Press).

5. Rosenberg (1979).

6. W. F. Bynum (1994), *Science and Practice of Medicine in the Nineteenth Century* (Cambridge: Cambridge University Press); E. Pellegrino (1979), "The Socio-Cultural Impact of Twentieth Century Therapeutics," in Vogel and Rosenberg (1979).

7. Nor were there the statistical methods to track the course of a disease through the social body; see Chapter 3.

8. L. M. Lawson (1849), cited in Rosenberg (1979), p. 20. J. McManners (1981), *Death and the Enlightenment* (Oxford: Oxford University Press).

9. Pellegrino (1979); Bynum (1994).

10. D. Healy (1993), *Images of Trauma: From Hysteria to Post-Traumatic Stress Disorder* (London: Faber & Faber), chap. 1.

11. Ibid. See also F. Sulloway (1994), *Freud: Biologist of the Mind* (Cambridge: Harvard University Press).

12. Paracelsus (1979), *Selected Writings*, ed. J. Jacobi (Princeton: Princeton University Press), p. 84.

13. Ibid., p. 95.

14. Cited in A. G. Debus (1976), "The Pharmaceutical Revolution of the Renaissance," *Clio Medica* 11:309.

15. I. Hacking (1975), *The Emergence of Probability* (Cambridge: Cambridge University Press). See also Debus (1976).

16. C. C. Mann and M. L. Plummer (1991), *The Aspirin Wars* (New York: Alfred Knopf).

17. R. Porter and D. Porter (1989), "The Rise of the English Drugs Industry: The Role of Thomas Corbyn," *Medical History* 33:277–295;

18. J. Liebenau (1987), *Medical Science and Medical Industry* (Basingstoke: Macmillan Press).

19. J. H. Young (1992), *The Medical Messiahs: The Social History of Health Quackery in Twentieth-Century America* (Princeton: Princeton University Press).

20. Liebenau (1987).

21. C. C. Hopkins (1927), *My Life in Advertising,* cited in Young (1992), p. 21.

22. L. K. Altman (1987), *Who Goes First: The Story of Self-Experimentation in Medicine* (New York: Random House).

23. Ibid.

24. J. Libenau (1987), *Medical Science and Medical Industry* (Basingstoke: Macmillan Press).

25. K. Menzi (1983), *Geschichte der Chemie in Basel: Schweizerische Zeitschift fur die Chemische Industrie* 5:15–30.

26. C. A. Russell (1971), *The History of Valency* (Oxford: Leicester University Press).

27. Tens of thousands of organic chemicals have been synthesized since 1860, and at present several hundred enter commerce each year. Petrol rather than coal is now the primary base for production, leading to the term *petrochemical industry*—a term that for many conjures up images of pollution and carcinogenesis.

28. Hoffmann-La Roche, in contrast, began in 1894 as a specifically pharmaceutical business.

29. Menzi (1983).

30. Mann and Plummer (1991).

31. J. P. Swann (1988), *Academic Scientists and the Pharmaceutical Industry: Co-operative Research in Twentieth-Century America* (Baltimore: Johns Hopkins University Press); M. Bliss (1982), *The Discovery of Insulin* (Toronto: McClelland and Stewart).

32. R. Bolt (1962), *A Man for All Seasons* (New York: Vintage Books), p. 38.

33. Liebenau (1987).

34. Ibid., p. 94.

35. P. Temin (1980), *Taking Your Medicine: Drug Regulation in the United States* (Cambridge: Harvard University Press).

36. Most other Western countries followed suit during the 1950s, sometimes in a rather piecemeal way, by successively amending existing dangerous-drug legislation to cover first of all cortisone and penicillin and then other compounds as they became available.

37. E. Shorter (1987), *The Health Century* (New York: Doubleday).

38. R. Kanigel (1983), *Apprentice to Genius: The Making of a Scientific Dynasty* (Baltimore: Johns Hopkins Press).

39. J. O. Cole and D. Healy (1996), "The Evaluation of Psychotropic Drugs," in D. Healy, ed. (1996), *The Psychopharmacologists* (London: Chapman & Hall), pp. 239–263.

40. D. Eisenhower, cited in D. Dickson (1988), *The New Politics of Science* (Chicago: University of Chicago Press), p. 59.

41. M. N. G. Dukes and B. Swartz (1988), *Responsibility for Drug-Induced Injury* (Amsterdam: Elsevier Press).

42. J. Braithwaite (1986), *Corporate Crime in the Pharmaceutical Industry* (London: Routledge & Keegan Paul); F. J. Ayd and D. Healy (1996), "The Discovery of Antidepressants," in Healy (1996), pp. 81–110.

43. L. Lasagna (1989), "Congress, the FDA, and New Drug Development: Before and after 1962," *Perspectives in Biology and Medicine* 32:322–343.

44. Cole and Healy (1996), pp. 239–263.

45. P. Knightley, H. Evans, E. Potter, and M. Wallace (1979), *Suffer the Children: The Story of Thalidomide* (London: Andrew Deutsch). See also Dukes and Swartz (1988); Temin (1980).

46. Lasagna (1989).

47. Temin (1980).

48. Diethelm (1971).

49. G. E. Berrios (1991), "The Two Manias," *British Journal of Psychiatry* 129:258–259; C. Fear, H. Sharp, and D. Healy (1995), "Obsessive-Compulsive and Delusional Disorders: Notes on Their History, Nosology, and Interface," *Journal of Serotonin Research* 1 (suppl. 1):1–18.

50. In fact, the French word *délire* is the root of both the modern term *delirium* and the word *delusion;* see Fear, Sharp, and Healy (1995).

51. N. Moore and D. Healy (1994), "Interview with Norman Moore," *Irish Journal of Psychological Medicine, Bulletin* 1:1–8.
52. R. Napier, cited in M. MacDonald (1991), *Mystical Bedlam: Anxiety and Healing in Seventeenth-Century England* (Cambridge: Cambridge University Press), p. 154.
53. H. Maudsley (1979), *The Pathology of Mind: A Study of Its Distempers, Deformities, and Disorder* (London: Julian Friedman), pp. 170–171.
54. E. Clarke and L. S. Jacyna (1987), *Nineteenth-Century Origins of Neuroscientific Concepts* (Berkeley: University of California Press). See also Healy (1993).
55. McManners (1981).
56. J. E. D. Esquirol (1838), *Des maladies mentales considerées sous les rapports médical hygiénique et médico-légal* (Paris: Baillière); trans. E. K. Hunt (1845) as *Mental Maladies: A Treatise on Insanity* (Philadelphia: Lea & Blanchard).
57. G. N. Grob (1994), *The Mad among Us* (Cambridge: Harvard University Press).
58. Esquirol (1838).
59. P. Kramer (1993), *Listening to Prozac* (New York: Viking Press).
60. Fear, Sharp, and Healy (1995).
61. Healy (1993).
62. P. Pichot (1983), *A History of Psychiatry* (Paris: Èditions Roger DaCosta); Pichot (1995), "The Birth of the Bipolar Disorder," *European Psychiatry* 10:1–10.
63. E. Kraepelin (1896), *Psychiatrie: Ein Lehrbuch fur Studierende und Aerzte,* 5th ed. (Leipzig: J. A. Barth); P. Hoff (1995), "Kraepelin," in G. E. Berrios and R. Porter, eds. (1995), *A History of Clinical Psychiatry: The Origin and History of Psychiatric Disorders* (London: Athlone Press), pp. 261–279; G. E. Berrios and R. Hauser (1995), "Kraepelin," in Berrios and Porter (1995), pp. 280–291.
64. E. Bleuler (1950 [1911]), *Dementia Praecox or the Group of Schizophrenias,* trans. Joseph Zinkin (New York: International University Press).
65. E. Kraepelin (1921), *Manic-Depressive Insanity and Paranoia* (Edinburgh: E. & S. Livingstone).
66. D. Healy (1990), *The Suspended Revolution* (London: Faber & Faber), chap. 2.
67. H. Hippius and D. Healy (1996), "The Foundation of the CINP and the Discovery of Clozapine," in Healy (1996), pp. 187–213.
68. Extraordinary as it may seem, therefore, it would have been effectively impossible to call what we now call depression "depression" at any time in history other than this century.
69. Figures taken from admissions to the North Wales Hospital during the period 1848–1948; the data have been derived from a study funded by the Wellcome Trust. I am grateful to Mervyn Jones, Pam Michael, and David Hirst for access to the records and to Marie Savage for working on them with me.

70. G. Beard (1880), *A Practical Treatise on Nervous Exhaustion (Neurasthenia)* (New York: William Wood); E. Shorter (1992), *From Paralysis to Fatigue: A History of Psychosomatic Illness in the Modern Era* (New York: Free Press).

71. G. N. Grob (1991), "Origins of DSM I: A Study in Appearance and Reality," *American Journal of Psychiatry* 148:421–431; S. W. Jackson (1986), *Melancholia and Depression: From Hippocratic Times to Modern Times* (New Haven: Yale University Press).

72. M. D. Beer (1995), "Psychosis: From Mental Disorder to Disease Concept, *History of Psychiatry* 6:177–200.

73. Healy (1990), chap. 2.

74. As will become clear in Chapters 5 and 7, however, many would consider that the question of the relative roles of biological and psychological factors in the etiology and pathogenesis of the neuroses is once again up for grabs.

75. Healy (1993), chap. 1.

76. Healy (1990), chap. 2.

77. J. Kassanin (1933), "The Acute Schizo-Affective Psychoses," *American Journal of Psychiatry* 113:97–126.

78. E. Mapother (1926), "Discussion on Manic-Depressive Psychosis," *British Medical Journal* 2:872–876; A. Lewis (1934), "Melancholia: A Clinical Survey of Depressive States," *Journal of Mental Science* 80:277–378.

2. The Discovery of Antidepressants

1. This group of drugs are also called neuroleptics or major tranquilizers.

2. J. P. Swazey (1974), *Chlorpromazine: A Study in Therapeutic Innovation* (Cambridge: MIT Press). The possibility for an earlier breakthrough had been there since 1899, stemming from observations of P. Bodoni, "Dell'-azione sedativea del bleu di metilene in varie forme di psicosi," *Clin. Med. Ital.* 38:217–222.

3. J. Hammon, J. Paraire, and J. Velluz (1952), "Remarques sur l'action du 4560 R.P. sur l'agitation maniaque," *Annales Médicale Psychologie* 110:331–335.

4. J. Delay, P. Deniker, and J. M. Harl (1952), "Utilisation en thérapeutique psychiatrique d'une phénothiazine d'action centrale élective," *Annales Médicale Psychologie* 110:112–131.

5. P. Pichot and D. Healy (1996), "The Discovery of Chlorpromazine and the Place of Psychopharmacology in the History of Psychiatry," in D. Healy, ed. (1996), *The Psychopharmacologists* (London: Chapman & Hall), pp. 1–27.

6. Swazey (1974); L. Cook and D. Healy (1998), "Chlorpromazine, Behavioural Pharmacology, and Drug Development," in D. Healy, ed. (1998), *The Psychopharmacologists*, vol. 2 (London: Chapman & Hall).

7. This is now known as the Douglas Hospital.

8. H. E. Lehmann and D. Healy (1996), "Psychopharmacotherapy," in Healy (1996), pp. 159–186; H. E. Lehmann and G. E. Hanrahan (1954), "Chlorpromazine: A New Inhibiting Agent for Psychomotor Excitement and Manic States," *AMA Archives of Neurology and Psychiatry* 71:227–237; J. Brady

and D. Healy (1998), "The Origins of Behavioural Pharmacology," in Healy (1998).

9. Lehmann and Healy (1996). See also L. E. Hollister (1994), "Review of International Colloquium on Chlorpromazine," in T. A. Ban and H. Hippius, eds. (1994), *Towards CINP* (Brentwood, Tenn.: J. M. Productions).

10. F. Ayd and D. Healy (1996), "The Discovery of Antidepressants," in Healy (1996), pp. 81–110.

11. Lehmann and Healy (1996).

12. Pichot and Healy (1996).

13. J. Delay, R. Laine, and J.-F. Buisson (1952), "Note concernant l'action de l'isonicotynil-hydrazide dans le traitment des états dépressifs," *Annales Médico-psychologiques* 110:689–692; J. Delay and J.-F. Buisson (1958), "Psychic Action of Isoniazid in the Treatment of Depressive States," *Journal of Clinical and Experimental Psychopathology and Quarterly Review of Psychiatry and Neurology* 19, 2 (supp.):51–55.

14. J. Delay, P. Denniker, J.-F. Buisson, and A. Heim (1959), "Le Traitement des états dépressifs par les dérivés de l'acide iso-nicotinique, isoniazide, and iproniazide," *Annales Médico-psychologiques* 117, 125–138; Pichot and Healy (1996); P. Deniker (1970), "Introduction of Neuroleptic Chemotherapy into Psychiatry," in F. Ayd and B. Blackwell, eds. (1970), *Discoveries in Biological Psychiatry* (Philadelphia: Lippincott), pp. 155–164; P. Deniker (1983), "Discovery of the Clinical Use of Neuroleptics," in M. J. Parnham and J. Bruinvels, eds. (1983), *Discoveries in Pharmacology*, vol. 1: *Psycho- and Neuropharmacology* (Amsterdam: Elsevier), pp. 163–180.

15. A. Broadhurst and D. Healy (1996), "Before and After Imipramine," in Healy (1996), pp. 111–134.

16. J. Thiele and O. Holzinger (1899), "Ueber O-Diaminodibenzyl," *Annales Chemische Liebiegs* 305:96–102.

17. W. Schindler and F. Haefliger (1954), "Derivatives of Iminodibenzyl," *Helvetica Chemica Acta* 37:4.

18. Acetyl-choline (Ach) was the first chemical shown to be a neurotransmitter (see Chapter 5). Drugs that antagonized its action are termed anticholinergic. At this stage there was no agreement that the Ach that had been detected in the brain had any behavioral role there; the accepted anticholinergic effects were a dry mouth, difficulty with micturition, constipation, and possible blurred vision.

19. Biographical information provided by R. Kuhn; see also R. Kuhn and D. Healy (1998), "From Imipramine to Levoprotiline: The Discovery of Antidepressants," in Healy (1998).

20. R. Kuhn (1970), "The Discovery of Imipramine," in Ayd and Blackwell (1970), pp. 205–217.

21. Ibid. See also R. Kuhn (1957), "Ueber die Behandlung de depressiver zustaemde mit einem iminodibenzylderivative (G22355)," *Schweizisch Medisina Wochenschrift* 87:1135–1140; R. Kuhn (1958), "The Treatment of Depressive States with G22355 (Imipramine Hydrochloride)," *American Journal of Psychiatry* 115:495–464; R. Kuhn (1987), "Some Questions and

Consequences Following the Discovery of a Specific Antidepressant Drug for Scientific Research and Practical Use," in B. E. Leonard and P. Turner, eds. (1987), *Thirty Years after Imipramine* (Oxford: Rapid Communications); R. Kuhn (1990), "Artistic Imagination and the Discovery of Antidepressants," *Journal of Psychopharmacology* 4:127–130; Broadhurst and Healy (1996); R. Kuhn (1977), in L. J. Pongratz, ed. (1977), *Psychiatrie in Selbstdarstellungen* (Bern: Verlag Hans Huber), pp. 219–257; R. Kuhn and D. Healy (1998), "From Imipramine to Levoprotiline: The Discovery of Antidepressants," in Healy (1998).

22. Broadhurst and Healy (1996); Ayd and Healy (1996); Lehmann and Healy (1996). An analogy with the Last Supper here is compelling: an audience of twelve, few of whom perhaps knew what was really being said, but the good news was later to spread and make the reputations and careers of those who listened to it.

23. Ayd and Healy (1996).

24. Kuhn (1958).

25. It can be noted that whereas Kuhn is directly in the tradition of Esquirol, Kramer (as will become clearer in subsequent chapters) takes a pre-Esquirol or post-Kuhnian line.

26. W. L. Rees (1960), "Treatment of Depression by Drugs and Other Means," *Nature* 186:114–120.

27. Swazey (1974).

28. H. Hippius and D. Healy (1996), "The Founding of the CINP and the Discovery of Clozapine," in Healy (1996), pp. 187–213.

29. When an international meeting was held in 1987 to mark the thirtieth anniversary of imipramine, Kuhn remarkably was not on the initial list of invited participants.

30. F. Ayd and B. Blackwell, eds. (1970), *Discoveries in Biological Psychiatry* (Philadelphia: Lippincott).

31. Broadhurst and Healy (1996).

32. W. Janzarik (1976), "Die Krise in Der Psychopathologie," *Nervenarzt* 47:73–80.

33. Lehmann and Healy (1996).

34. H. E. Lehmann, C. H. Cahn, and R. L. de Verteuil (1958), "The Treatment of Depressive Conditions with Imipramine (G22355)," *Canadian Psychiatric Association Journal* 3:155–164.

35. S. Garattini and D. Healy (1996), "The Role of Independent Science in Psychopharmacology," in Healy (1996), pp. 135–157.

36. Hippius and Healy (1996).

37. Ayd and Healy (1996).

38. P. B. Bradley, P. Deniker, and C. Radouco-Thomas (1959), "Neuropsychopharmacology," *Proceedings of the First International Congress,* Rome, September 1958 (Amsterdam: Elsevier).

39. O. Schmiedeberg (1877), "Ueber das Verhaltnis des Ammoniaks und der primaren Mono-aminbasen zur Hamstoffbildung im Thierkorper," *Naunyn-Schmiedebergs Arch. Pharmakol. Exp. Pathol.* 8:1–14.

40. O. Minkowski (1883), "Ueber Spaltungen im Thierkorper," *Arch. Exp. Pathol. Pharmakol.* 17:445–465.
41. M. L. C. Hare (1928), "Tyramine Oxidase: A New Enzyme System in the Liver," *Biochemical Journal* 22:968–979.
42. See note 18.
43. D. Richter (1989), *Life in Research* (London: Stuart Phillips).
44. D. Richter and D. Healy (1995), "The Origins of Mental Health Neurosciences in Britain," *Journal of Psychopharmacology* 9:392–399.
45. Noradrenaline, dopamine, and 5-hydroxytryptamine (5HT) are the only other neurotransmitters that will be introduced in this book. They are generally described as amines or monoamines. The discovery of their role in behavior is described in Chapter 5.
46. S. Kety (1976), "Introduction," in Symposium MAOI, CIBA Foundation Symposium Series, London, vol. 39, pp. 1–4.
47. C. E. Pugh and J. H. Questal (1937), "Oxidase of Aliphatic Amines by Brain and Other Tissues," *Biochemical Journal* 31:286–291.
48. H. Blaschko, D. Richter, and H. Scholassmann (1937), "The Inactivation of Adrenaline," *Journal of Physiology* 90:1–19.
49. E. A. Zeller (1938), "Ueber des Enzymatischen Abbau von Histamin und Diaminen," *Helvetica Chemika Acta* 21:881–890.
50. H. Meyer and J. Mally (1912), "Hydrazine Derivatives of Pyridinecarboxylic Acids," *Monatsschr Psychiatrie und Neurologie* 33:400.
51. H. H. Fox and J. T. Gibas (1953), "Synthetic Tuberculostats: VII Monoalkyl Derivatives of Isonicotinylhydrazine," *Journal of Organic Chemistry* 18:994–1002; M. Sandler (1990), "Monoamine Oxidase Inhibitors in Depression: History and Mythology," *Journal of Psychopharmacology* 4:136–139.
52. E. A. Zeller, J. Barsky, E. R. Berman, and J. R. Fouls (1952), "Action of Isonicotinic Acid Hydrazide and Related Compounds on Enzymes of Brain and Other Tissues," *Journal of Laboratory and Clinical Medicine* 14:965–966; E. A. Zeller (1983), "Monoamine Oxidase and Its Inhibitors in Relation to Anti-Depressive Activity," in Parnham and Bruinvels (1983). See also Sandler (1990).
53. Ibid. See also D. M. Bosworth, J. W. Fielding, L. Demarst, and M. Bonaquist (1955), "Toxicity to Iproniazid (Marsilid) as It Affects Osseous Tuberculosis," *Quarterly Bulletin Sea View* 16:134–140; *New York Times,* July 5, 1952, p. 1:2.
54. J. A. Smith (1953), "The Use of the Isopropyl Derivative of Isonicotinylhydrazide (Marsilid) in the Treatment of Mental Disease: A Preliminary Report," *American Practitioner* 4:519–520.
55. G. E. Crane (1956a), "Further Studies on Iproniazid Phosphate: Isonicotinal-Isopropylhydrazine Phosphate. Marsilid," *Journal of Mental Disease* 124:322–331; G. E. Crane (1956b), "The Psychiatric Side Effects of Iproniazid," *American Journal of Psychiatry* 112:494–501.
56. F. J. Ayd (1957), "A Preliminary Report on Marsilid," *American Journal of Psychiatry* 114:459.

57. Obituaries and biographical notes; G. Simpson and D. Healy (1998), "Early Clinical Psychopharmacology," in Healy (1998); M. Sheperd and D. Healy, "Psychopharmacotherapy: From Specific to Non-Specific," in Healy (1998).

58. N. S. Kline (1954), "Use of Rauwolfia Serpenthina Benth. in Neuropsychiatric Conditions," *Annals of New York Academy of Sciences* 59:107–132.

59. H. J. Bein (1970), "Biological Research in the Pharmaceutical Industry with Reserpine," in Ayd and Blackwell (1970), pp. 142–154.

60. In 1957 the Lasker Prize was given to Kline, Vakil, and Noce for work on Reserpine, and to Lehmann, Bovet, Laborit, and Deniker for work with chlorpromazine.

61. A. Carlsson and D. Healy (1996), "The Rise of Neuropsychopharmacology: Impact on Basic and Clinical Neuroscience," in Healy (1996), pp. 51–80.

62. F. J. Ayd (1991), "The Early History of Modern Psychopharmacology," *Neuropsychopharmacology* 5:71–84.

63. F. M. Berger (1970), "Anxiety and the Discovery of the Tranquillisers," in Ayd and Blackwell (1970), pp. 115–129.

64. T. A. Ban and D. Healy (1996), "They Used to Call It Psychiatry," in Healy (1996), pp. 587–620.

65. Ayd (1991).

66. Swazey (1974).

67. N. S. Kline (1970), "Monoamine Oxidase Inhibitors: An Unfinished Picaresque Tale," in Ayd and Blackwell (1970), pp. 194–204.

68. M. Chessin, B. Dubnick, E. R. Kramer, and C. C. Scott (1956), "Modification of Pharmacology of Reserpine and Serotonin by Iproniazid," *Federation Proceedings* 15:409.

69. H. P. Loomer, J. C. Saunders, and N. S. Kline (1957), "A Clinical and Pharmaco-dynamic Evaluation of Iproniazid as a Psychic Energiser," *Psychiatric Research Reports* 8:129–141; *New York Times,* April 7, 1957, p. 82.

70. G. E. Crane (1957), "Iproniazid (Marsilid) Phosphate: A Therapeutic Agent for Mental Disorders Anti-debilitating Diseases," *Psychiatric Research Reports* 8:142–154.

71. *New York Times*, April 7, 1957, "Science Notes: Mental Drug Shows Promise," p. 86.

72. *Newsweek*, November 23, 1964, p. 101; *Time,* November 24, 1964, "Medicine—Awards: A Lift from Depression," p. 106; N. S. Kline (1965), "The Practical Management of Depression," *Journal of the American Medical Association* 190:732–740; J. C. Saunders (1965), "Lasker Award Priority Claim," *Journal of the American Medical Association* 191:865; N. S. Kline (1965), "Dr. Kline Replies," *Journal of the American Medical Association* 191: 865–866.

73. J. C. Saunders (1958), "Discussion and Remarks," *Journal of Clinical and Experimental Psychopathology* 19 (suppl. 1):84–85; J. C. Saunders (1959). "Discussion and Remarks," *Annals of the New York Academy of Science* 80:719–725; J. C. Saunders (1963), "Treatment of Hospitalized Depressed and Schizophrenic Patients with Monoamine Oxidase Inhibitors: Including Reflections on Pargyline," *Annals of the New York Academy of Sciences*

107:1081–1089; J. C. Saunders, D. Z. Rochlin, N. Radinger, and N. S. Kline (1959), "Iproniazid in Depressed and Regressed Patients," in N. S. Kline (1959), *Psychopharmacology Frontiers* (Boston: Little, Brown), pp. 177–179. See also Sandler (1990).

74. N. S. Kline (1961a), "Clinical Experience with Iproniazid (Marsilid)," *Journal of Clinical and Experimental Psychopathology* 19, 2 (suppl. 1):72–78.

75. N. S. Kline (1961b), "Depression: Diagnosis and Treatment," *Medical Clinics in North America* 35:1041–1053; *Time*, September 4, 1963, p. 48, "Medicine: Quick Lift for Depression."

76. Kline (1970).

77. H. M. Salzer and M. L. Lurie (1953), "Anxiety and Depressive States Treated with Isonicotinyl Hydrazide (Isoniazid)" *Archives of Neurology and Psychiatry* 70:317–324.

78. H. M. Salzer and M. L. Lure (1955), "Depressive States Treated with Isonicotinyl Hydrazide (Isoniazid): A Follow-Up Study," *Ohio State Medical Journal* 51:437–441.

79. D. L. Davies and M. Shepherd (1955), "Reserpine in the Treatment of Anxious and Depressed Patients," *Lancet* 2:117–120; Shepherd and Healy (1998).

80. Sandler (1990).

81. F. J. Ayd (1960), "Amitriptyline (Elavil) Therapy for Depressive Reactions," *Psychosomatics* 1:320–325.

82. V. Pedersen, K. Bøgesø, and D. Healy (1998), "Drug-Hunting," in Healy (1998).

83. F. J. Ayd (1961), *Recognizing the Depressed Patient* (New York: Grune and Stratton).

84. See Chapters 4, 6, and 7 for differences between reactive and neurotic depression.

3. Other Things Being Equal

1. R. Kuhn (1990), "Contribution to Discussion on the History of Psychopharmacology at a Cambridge Meeting on the History of Physical Treatments in Psychiatry," *Journal of Psychopharmacology* 4:170.

2. I. Hacking (1975), *The Emergence of Probability* (Cambridge: Cambridge University Press); L. Kruger, L. J. Daston, and M. Heidelberger (1987), *The Probabilistic Revolution*, vol. 1: *Ideas in History* (Cambridge: MIT Press); G. Gigerenzer, Z. Switjtink, T. Porter, L. Daston, J. Beatty, and L. Kruger (1989), *The Empire of Chance* (Cambridge: Cambridge University Press); D. Healy (1990), *The Suspended Revolution* (London: Faber & Faber), chap. 3; A. M. Lilienfeld (1982), "Ceteribus Paribus: The Evolution of Clinical Trial," *Bulletin of the History of Medicine* 56:1–18.

3. Petrarch, cited in Lilienfeld (1982), p. 4.

4. Paracelsus (1979), *Selected Writings*, ed. J. Jacobi (Princeton: Princeton University Press), p. 51.

5. Ibid., p. liii.

6. Ibid., p. 3.

7. Hacking (1975).

8. K. Welman (1992), *La Mettrie: Medicine, Philosophy, and Enlightenment* (Durham: Duke University Press).

9. J. McManners (1981), *Death and the Enlightenment* (Oxford: Oxford University Press).

10. Lilienfeld (1982).

11. Pinel, cited in T. D. Murphy (1981), "Medical Knowledge and Statistical Methods in Early Nineteenth-Century France," *Medical History* 25: 301–319.

12. P. C. A. Louis, cited in Lilienfeld (1982), p. 6.

13. P. C. A. Louis, cited in M. D. Rawlins (1990), "Development of a Rational Practice of Therapeutics," *British Medical Journal* 301:729–733.

14. L. M. Lawson (1849), cited in C. E. Rosenberg (1979), "The Therapeutic Revolution: Medicine, Meaning, and Social Change in Nineteenth-Century America," in M. J. Vogel and C. E. Rosenberg, eds. (1979), *The Therapeutic Revolution* (Philadelphia: University of Pennsylvania Press), p. 20.

15. Double cited in Murphy (1981).

16. Risueno d'Amador cited in Murphy (1981).

17. P. Pichot and D. Healy (1996), "The Discovery of Chlorpromazine and the Place of Psychopharmacology in the History of Psychiatry," in D. Healy, ed. (1996), *The Psychopharmacologists* (London: Chapman & Hall), pp. 1–27.

18. J. P. Bull (1959), "The Historical Development of Clinical Therapeutic Trials," *Journal of Chronic Disease* 10:218–249.

19. Murphy (1981).

20. M. S. Pernick (1985), *A Calculus of Suffering: Pain, Professionalism, and Anesthesia in Nineteenth-Century America* (New York: Columbia University Press).

21. Poisson cited in Murphy (1981).

22. Healy (1990), chap. 3.

23. Lilienfeld (1982).

24. R. A. Fisher (1935), *The Design of Experiments* (Edinburgh: Oliver & Boyd). See also Gigerenzer et al. (1989).

25. G. Gigerenzer (1993), "The Superego, the Ego, and the Id in Statistical Reasoning," in G. Keren and C. Lewis, eds. (1993), *A Handbook for Data Analysis in the Behavioural Sciences* (Hillsdale, N.J.: Erlbaum), pp. 311–339.

26. MRC (1948), "Streptomycin Treatment of Pulmonary Tuberculosis," a Medical Research Council Investigation, *British Medical Journal* 2:769–782.

27. J. O. Cole and D. Healy (1996), "The Evaluation of Psychotropic Drugs," in Healy (1996).

28. M. Shepherd and D. Healy (1998), "Psychopharmacology: From Specific to Non-Specific," in D. Healy, ed. (1998), *The Psychopharmacologists,* vol. 2 (London: Chapman & Hall); H. Beecher (1955), "The Powerful Placebo," *Journal of the American Medical Association* 159:1602–1606; L. Lasagna and D. Healy (1998), "Back to the Future: Evaluation and Drug Development, 1954–1996," in Healy (1998).

29. H. Gold, N. T. Kwit, and H. Otto (1937), "The Xanthines in the Treatment of Cardiac Pain," *Journal of the American Medical Association* 108:

2173–2179; H. Gold (1989), letter, *Pharmacologist* 31:235. Gold is probably the person who coined the term "double-blind," which began to appear in his articles around 1950. He was never given much recognition for his role in these developments. He had a flourishing private practice, and this often militates against recognition of "academic" contributions—for a variety of reasons.

30. H. E. Lehmann (1993), "Before They Called It Psychopharmacology," *Neuropsychopharmacology* 8:291–303.

31. W. L. Rees (1949), "Electronarcosis in the Treatment of Schizophrenia," *Journal of Mental Science* 95:625–637.

32. W. L. Rees and G. N. King (1951), "Desoxycortisone Acetate and Ascorbic Acid Treatment of Schizophrenia," *Journal of Mental Science* 97:376–380; W. L. Rees and G. N. King (1952), "Cortisone in the Treatment of Schizophrenia," *Journal of Mental Science* 98:408–413.

33. J. Elkes and C. Elkes (1954), "Effect of Chlorpromazine on the Behaviour of Chronically Over-Active Psychotic Patients," *British Medical Journal* 2:560–565; J. Elkes and D. Healy (1998), "Towards Footings in a New Science: Psychopharmacology, Receptors, and the Pharmacy Within," in Healy (1998).

34. W. L. Rees (1956a), "A Controlled Study of the Value of Chlorpromazine in the Treatment of Anxiety Tension States," *L'Eencehale* 4:547–549.

35. D. L. Davies and M. Shepherd (1955), "Reserpine in the Treatment of Anxious Patients," *Lancet* 2:117–120; Shepherd and Healy (1998).

36. M. Schou, N. Juel-Nielsen, E. Stromgren, and H. Voldby (1954), "The Treatment of Manic-Psychoses by the Administration of Lithium Salts," *Journal of Neurology and Psychiatry* 17:250–260; M. Schou and D. Healy (1998), "Lithium," in Healy (1998).

37. J. Elkes (1974), in J. P. Swazey (1974), *Chlorpromazine: A Study in Therapeutic Innovation* (Cambridge: MIT Press); J. Elkes (1995), "Psychopharmacology: Finding One's Way," *Neuropsychopharmacology* 12:93–111.

38. Editorial (1954), "Chlorpromazine," *British Medical Journal* 2:581–582.

39. W. L. Rees (1956b), "A Controlled Trial of Chlorpromazine in the Management of Patients Suffering from Asthma Associated with Anxiety and Tension Symptoms," extract from *L'Encephale* 4:555–561; W. L. Rees and D. Healy (1997), "The Role of Clinical Trials in the Development of Psychopharmacology," *History of Psychiatry* 7.

40. Elkes, surprisingly, was not present at this meeting, but Mayer-Gross, also at Birmingham, mentioned the Elkes study there.

41. W. L. Rees and S. Benaim (1960), "The Evaluation of Iproniazid (Marsilid) in the Treatment of Depression," *Journal of Mental Science* 106:193–202.

42. See also M. Shepherd (1981), "Reserpine as a Tranquilliser," in M. Shepherd (1981), *Psychotropic Drugs in Psychiatry* (New York: Jason Aronson), pp. 55–66.

43. See Chapter 5.

44. D. C. Wallace (1955), "Treatment of Hypertension: Hypotensive Drugs and Mental Changes," *Lancet* 2: 116–117.

45. Shepherd and Healy (1998).

46. Schou et al. (1954).

47. M. Schou (1992), "Phases in the Development of Lithium Treatment in Psychiatry," in F. Samson and G. Adelman, *The Neurosciences: Paths of Discovery,* vol. 2 (Boston: Birkhauser), pp. 148–166.

48. H. Bourne (1953), "The Insulin Myth," *Lancet* 2:964–968.

49. H. Pullar-Strecker, R. Davies, J. Gibson, F. B. Charatan, R. Sandison, W. Sargant, R. A. Hunter, W. L. Rees, and W. Mayer-Gross (1953), "The Insulin Myth," *Lancet* 2:1047–1048, 1094–1095, 1151–1152; H. Bourne (1953), "The Insulin Myth," *Lancet* 2:1259.

50. B. Ackner, A. Harris, and A. J. Oldham (1957), "Insulin Treatment of Schizophrenia," *Lancet* 607–611.

51. M. Shepherd (1990), "The 'Neuroleptics' and the Oedipus Effect," *Journal of Psychopharmacology* 4:131–135.

52. L. Kalinowsky (1970), in F. Ayd and B. Blackwell, eds. (1970), *Discoveries in Biological Psychiatry* (Philadelphia: Lippincott).

53. Rees and Healy (1996).

54. Pichot and Healy (1996).

55. J. O. Cole and R. W. Gerard (1959), *Psychopharmacology: Problems in Evaluation,* Publication 583 (Washington, D.C.: National Academy of Sciences/National Research Council); E. Shorter (1996), *A History of Psychiatry: From the Asylums to Prozac* (New York: Wiley), chap 8.

56. Shepherd and Healy (1998).

57. S. Kety (1982), "Seymour Kety," in M. Shepherd, ed. (1982), *Psychiatrists on Psychiatry* (Cambridge: Cambridge University Press), pp. 83–97.

58. Elkes (1959), "Report on Preliminary Screening," in Cole and Gerard (1959), p. 604.

59. Kety (1959), "Summary of the Conference," in Cole and Gerard (1959), pp. 647–652.

60. J. O. Cole, S. C. Goldberg, and G. L. Klerman (1964), "Phenothiazine Treatment in Acute Schizophrenia," *Archives of General Psychiatry* 10:246–261.

61. Kline (1959), "Scales and Checklists," in Cole and Gerard (1959), p. 475; Lasagna and Healy (1998).

62. M. E. Hamilton (1960), "A Rating Scale for Depression," *Journal of Neurology, Neurosurgery, and Psychiatry* 23:56–62.

63. M. E. Hamilton (1972), "Rating Scales in Depression," in P. Kielholz, ed. (1972), *Depressive Illness: Diagnosis, Assessment, Treatment* (Berne: Hans Huber), pp. 100–108.

64. Ibid.

65. A. Broadhurst and D. Healy (1996), "Before and After Imipramine," in Healy (1996), pp. 111–134.

66. A. T. Beck, C. H. Ward, M. Mendelson, J. Mock, and S. Erbaugh (1961), "An Inventory for Measuring Depression," *Archives of General Psychiatry* 4:561.

67. Hamilton (1972).

68. See L. Lasagna (1989), "Congress, the FDA, and New Drug Development: Before and after 1962," *Perspectives in Biology and Medicine* 32:322–343.

69. The recommendation that placebo-controlled RCTs were the best method of assessment and the adoption of this position in the 1962 amendments probably owed most to Louis Lasagna and Walter Modell.

70. S. Brandon (1990), "Ethics of Placebo-Controlled Trials," *Human Psychopharmacology* 5:176–177; T. J. Fahy (1990), "Ethics of Placebo-Controlled Trials," *Human Psychopharmacology* 5:177–178.

71. P. Leber (1996), "The Role of the Regulator in Psychopharmacology," in D. Healy and D. Doogan, eds. (1996), *Psychotropic Drug Development: Social, Economic, and Pharmacological Aspects* (London: Chapman & Hall).

72. In the process this created the conditions for the later success of Prozac by making it the first new antidepressant for a long time in the United States.

73. S. Garattini and D. Healy (1996), "The Role of Independent Science in Psychopharmacology," in Healy (1996), pp. 135–157.

74. C. Dowrick and I. Buchan (1995), "Twelve-Month Outcome of Depression in General Practice: Does Detection and Disclosure Make a Difference? *British Medical Journal* 311:1274–1276; O. Katon, M. Von Korff, E. Lin, E. Walker, G. E. Simon, T. Bush, P. Robinson, and J. Russo (1995), "Collaborative Management to Achieve Treatment Guidelines Impact on Depression in Primary Care," *Journal of the American Medical Association* 273: 1026–1031.

75. It is unlikely that the medical model is the only set of circumstances in which current psychotropic medicines would have appeared efficacious. The fact that an "antidepressant" signal can be detected against the background noise of a variety of depressive and anxiety-related conditions suggests that the signal is quite powerful. An antibiotic, for example, given to a mixed bag of infections, might not appear to have much if any efficacy.

76. What studies there have been in psychiatry have been run by the PRC, MRC, and NIMH, and these will be dissected in the chapters to come. There is a related problem with most studies of both drugs and psychotherapies which it may not be possible to address under the current regulatory framework, and this is that all trials are done by therapists who are believers in or stakeholders in pharmacotherapy.

77. B. Freedman (1992), *Suspended Judgement: AIDs and the Ethics of Clinical Trials,* Learning the Right Lessons in Controlled Clinical Trials, vol. 13, pp. 1–5.

78. K. M. Boyd et al. (1992), "AIDs, Ethics, and Clinical Trials," Institute of Medical Ethics Working Party and the Ethical Implications of AIDS, *British Medical Journal* 305:699–701.

79. Healy (1990), chap. 3.

80. P. M. Elwood (1988), "Outcomes Management: Technology of Patient Experience," *New England Journal of Medicine* 318:1549–1556.

81. A. S. Relman (1988), "Assessment and Accountability: The Third Revolution in Medical Care," *New England Journal of Medicine* 319:1220–1222; A. M.

Epstein (1990), "The Outcomes Movement: Will It Get Us to Where We Want to Go? *New England Journal of Medicine* 323:267–270.

82. S. S. Jicks, A. D. Dean, and H. Jicks (1995), "Antidepressants and Suicide," *British Medical Journal* 311:215–218. It is worth nothing that the data do not seem to point in the direction of the more severely depressed patients having been given the newer drugs.

83. I. Chalmers, K. Dickerson, and T. C. Chalmers (1992), "Getting to Grips with Archie Cochrane's Agenda," *British Medical Journal* 305:786–788

84. T. McKeown (1979), *The Role of Medicine: Dream, Mirage, or Nemesis?* (Oxford: Basil Blackwell).

85. D. Healy (1993), *Images of Trauma: From Hysteria to Post-Traumatic Stress Disorder* (London: Faber & Faber), chap. 10.

86. Public Health Service Agency for Health Care Policy and Research (1993), *Depression in Primary Care: Treatment of Major Depression,* AHCPR Publication no. 93-051 (Rockville, Md.: U.S. Dept of Health and Human Services); American Psychiatric Association (1993), "Practice Guideline for Major Depressive Disorder in Adults," *American Journal of Psychiatry* 150 (suppl. 1):1–26.

87. Katon et al. (1995).

88. The question of reimbursement is at the heart of Chapters 6 and 7.

89. A. B. Hill (1966), "Reflections on the Controlled Trial," *Annals of the Rheumatic Diseases* 25:107–113.

90. Nobody before Gold and Beecher ever thought that nothing might work in its own right.

4. The Trials of Therapeutic Empiricism

1. M. Shepherd (1993), interview, in G. Wilkinson, ed. (1993), *Talking about Psychiatry* (London: Gaskell Press), pp. 230–245.

2. J. Thuillier (1988), "Psychopharmacology and Perspective," in T. A. Ban and H. Hippius (1988), *A Personal Account by the Founders of the Collegium Internationale Neuropharmacologia* (New York: Springer-Verlag), pp. 86–93.

3. J. O. Cole and R. W. Gerard, eds. (1959), *Psychopharmacology: Problems in Evaluation,* Publication 583 (Washington, D.C.: National Academy of Sciences/National Research Council). Cole is not clear on who actually came up with the term; if he did, he suggests it was by accident. See also J. O. Cole and D. Healy (1996), "The Evaluation of Psychotropic Drugs," in D. Healy, ed. (1996), *The Psychopharmacologists* (London: Chapman & Hall), pp. 239–263.

4. Thuillier (1988); H. E. Lehmann (1993), "Before They Called It Psychopharmacology," *Neuropsychopharmacology* 8:291–303; D. I. Macht and C. F. Mora (1921), "Effect of Opium Alkaloids on the Behaviour of Rats in the Circular Maze," *Journal of Pharmacology and Experimental Therapeutics* 16:219–235.

5. R. Boakes (1984), *From Darwin to Behaviour: Psychology and the Mind of Animals* (Cambridge: Cambridge University Press).

6. E. Kraepelin (1987), *Memoirs,* ed. H. Hippius, trans. Cheryl Wooding-Dean (New York: Springer-Verlag).

7. E. Kraepelin (1982), *Ueber die Beinflussing Einfacher Psychischer Vorgange durch Einige Arzneimittel* (Jena: Vorlagg von Gustaff Fischer); D. Healy (1993), "A Hundred Years of Psychopharmacology," *Journal of Psychopharmacology* 7:207–214.

8. K. Jaspers (1963 [1913]), *General Psychopathology,* trans. J. Hoenig and M. W. Hamilton (Manchester: Manchester University Press).

9. G. Claridge and D. Healy (1994), "The Psychopharmacology of Individual Differences," *Human Psychopharmacology* 9:285–298; H. Steinberg and D. Healy (1996), "Bridging the Gap: Psychology, Pharmacology, and After," in Healy (1996), pp. 214–238.

10. J. Moreau (1845), *Du Haschisch et de l'alienation mentale;* A. Carlsson (1990), "Early Psychopharmacology and the Rise of Modern Brain Research," *Journal of Psychopharmacology* 4:120–126.

11. L. K. Altman (1987), *Who Goes First: The Story of Self-Experimentation in Medicine* (New York: Random House).

12. W. James (1986 [1902]), *The Varieties of Religious Experience* (Harmondsworth: Penguin Books).

13. D. Healy (1993), *Images of Trauma: From Hysteria to Post-Traumatic Stress Disorder* (London: Faber & Faber), chap 5.

14. A. Hoffman (1970), "The Discovery of LSD and Subsequent Investigations on Naturally Occurring Hallucinogens," in F. Ayd and B. Blackwell, eds. (1970), *Discoveries in Biological Psychiatry* (Philadelphia: Lippincott), pp. 91–106.

15. H. E. Lehmann (1994), "Review of De Boor's Pharmacopsychology and Psychopathology," in T. A. Ban and H. Hippius, eds. (1994), *Towards CINP* (Brentwood, Tenn.: J. M. Productions).

16. G. Claridge (1969), *Drugs and Human Behaviour* (Harmondsworth: Pelican Books).

17. Claridge and Healy (1994); Healy (1993), chap. 5.

18. D. Healy and F. Watson (1995), "The Use of Healthy Volunteers in Human Psychopharmacology: Problems and Opportunities," in I. Hindmarch and P. D. Stonier, eds. (1995), *Human Psychopharmacology,* vol. 5 (Chichester: John Wiley & Sons), pp. 63–87; J. Stevens (1988), *Storming Heaven: LSD and the American Dream* (London: Heinemann).

19. G. Claridge (1994), "LSD: A Missed Opportunity?" *Human Psychopharmacology* 9:343–351; T. A. Ban and D. Healy (1996), "They Used to Call It Psychiatry," in Healy (1996), pp. 587–620.

20. And these cases may indeed have owed something to iproniazid's co-prescription with barbiturates.

21. N. S. Kline (1970), "Monoamine Oxidase Inhibitors: An Unfinished Picaresque Tale," in Ayd and Blackwell (1970), pp. 194–204.

22. B. Blackwell (1970), "The Process of Discovery," in Ayd and Blackwell (1970), pp. 11–29.

23. M. Lurie (1995), personal communication.

24. C. E. Cole, R. M. Patterson, J. B. Craig, W. E. Thomas, L. P. Ristine, M. Stahly, and B. Pasamanick (1959), "A Controlled Study of Efficacy of Iproniazid in Treatment of Depression," *Archives of General Psychiatry* 1:513–518.

25. Cole and Healy (1996).

26. M. Shepherd and D. Healy (1998), "Psychopharmacotherapy: From Specific to Non-Specific," in D. Healy, ed. (1998), *The Psychopharmacologists,* vol. 2 (London: Chapman & Hall).

27. Ibid.

28. Medical Research Council (1965), "Clinical Trial Treatment of Depressive Illness," *British Medical Journal* 881–886.

29. W. Sargant (1965), correspondence, *British Medical Journal* 1:1495. It was after this trial, in November 1965, that Bradford Hill made his comment about the pendulum coming off its hook if RCTs ever became the only way to assess the efficacy of treatments; see Chapter 3.

30. D. S. Robinson, A. Nies, C. L. Ravaris, and K. R. Lamborn (1973), "The Monoamine Oxidase Inhibitor Phenelzine in the Treatment of Depressive Anxiety States," *Archives of General Psychiatry* 29:407–413.

31. C. M. B. Pare, W. L. Rees, and M. J. Sainsbury (1962), "Differentiation of Two Genetically Specific Types of Depression by the Response to Antidepressants," *Lancet* 2: 1340–1343.

32. C. N. B. Pare (1985), "The Present Status of Monoamine Oxidase Inhibitors," *British Journal of Psychiatry* 146:576–584.

33. J. P. Johnston (1968), "Some Observations upon a New Form of MAO in Brain Tissue," *Biochemical Pharmacology* 17:1285–1297.

34. P. Waldmeier and D. Healy (1996), "From Mental Illness to Neurodegeneration," in Healy (1996), pp. 565–586.

35. Ibid.

36. A. Delini-Stula and D. Healy (1996), "The Changing Face of Psychotropic Drug Development," in Healy (1996), pp. 425–440.

37. M. Da Prada, R. Kettler, H. H. Keller, M. Cesura, J. G. Richards, J. Saura Marti, E. D. Muggli-Maniglio, P. C. Wyss, E. Kyburz, and R. Imhof (1990), "From Moclobemide to Ro-10-6327 and Ro-41-1049: The Development of a New Class of Reversible, Selective MAO-A and MAO-B inhibitors," *Journal of Neural Transmission* 29 (suppl.):279–292.

38. F. N. Johnson (1989), *The History of Lithium Therapy* (London: Macmillan Press).

39. J. Cade (1970), *The Story of Lithium,* in Ayd and Blackwell (1970), pp. 218–229; M. Schou and D. Healy (1998), "Lithium," in Healy (1998).

40. J. Cade (1949), "Lithium Salts in the Treatment of Psychotic Excitement," *Medical Journal of Australia* 36:349–352.

41. M. Schou, N. Juel-Nielsen, E. Stromgren, and H. Voldby (1954), "The Treatment of Manic Psychoses by the Administration of Lithium Salts," *Journal of Neurology, Neurosurgery, and Psychiatry* 17:250–260; M. Schou (1992), "Phases in the Development of Lithium Treatment in Psychiatry," in F. Sam-

son and G. Adelman, eds. (1992), *The Neurosciences: Paths of Discovery*, vol. 2 (Boston: Birkhauser), pp. 148–166; Schou and Healy (1998).

42. S. Gershon and A. Yuwiler (1960), "Lithium Iron: A Specific Pharmacological Approach to the Treatment of Mania," *Journal of Neuropsychiatry* 1:229–241.

43. R. Maggs (1963), "The Treatment of Manic Illness with Lithium Carbonate," *British Journal of Psychiatry* 109:56–65.

44. R. N. Wharton and R. Fieve (1966), "The Use of Lithium in the Affective Psychoses," *American Journal of Psychiatry* 123:706–712.

45. Johnson (1989).

46. J. Angst and D. Healy (1996), "The Myths of Psychopharmacology," in Healy (1996), pp. 287–307.

47. Shepherd and Healy (1998).

48. A. A. Gattozzi (1970), *Lithium in the Treatment of Mood Disorders*, National Clearing House for Mental Health Information, Publication no. 5033.

49. B. Blackwell and M. Shepherd (1968), "Prophylactic Lithium: Another Therapeutic Myth? An Examination of the Evidence to Date," *Lancet* 1:968–971; B. Blackwell (1969), "Lithium: Prophylactic or Panacea?" *Medical Counterpoint*, pp. 52–59; M. H. Lader (1968), "Prophylactic Lithium?" *Lancet* 2:103; B. M. Saran (1969), "Lithium," *Lancet* 1:1208–1209.

50. A. Coppen and D. Healy (1996), "Biological Psychiatry in Britain," in Healy (1996), pp. 265–286.

51. M. Shepherd (1970), "Critical Review of Clinical Drug Trials in Depression in the 1970's," *Excerpta Medica International Congress Series* 239:2–11; M. Shepherd (interview 1993) in Wilkinson (1993).

52. M. Shepherd (1974), "Discussion," in *Psihofarmakologija 3: Proceedings of the 3rd Yugoslav Psychopharmacological Symposium, Opatija 1973, Medicinska Noklada*, pp. 329–330.

53. B. Blackwell (1969), *Journal of the American Medical Association* 125:1131; N. S. Kline (1969), *Journal of American Medical Association* 125:1131–1132.

54. Cole and Healy (1996).

55. M. Sandler and D. Healy (1994), "The Place of Chemical Pathology in the Development of Psychopharmacology," *Journal of Psychopharmacology* 8:124–133. See also Coppen and Healy (1996).

56. P. C. Baastrup and M. Schou (1968), "Prophylactic Lithium," *Lancet*, pp. 1419–1422.

57. J. Angst and P. Weis (1967), "Periodicity of Depressive Psychoses," in A. A. Brill, J. Cole, P. Deniker, H. Hippius, and P. B. Bradley, eds. (1967), *Neuropsychopharmacology: Excerpta Medical Foundation ISC 129, Amsterdam*, pp. 703–710.

58. A. Coppen, R. Noguera, J. Bailey, B. H. Burns, M. S. Swami, E. H. Hare, R. Gardner, and R. Maggs (1971), "Prophylactic Lithium in Affective Disorders: Controlled Trial," *Lancet* 2:275–279.

59. Coppen and Healy (1996).

60. R. S. Mindham, C. Howland, and M. Shepherd (1972), "Continuation Therapy with Tricyclic Antidepressants in Depressive Illness," *Lancet*, pp. 854–855.

61. A. I. M. Glen, A. L. Johnson, and M. Shepherd (1984), "Continuation Therapy with Lithium and Amitriptyline in Unipolar Depressive Illness: A Randomised Double Blind Controlled Trial," *Psychological Medicine* 14:37–50.

62. J. Moncrieff (1995), "Lithium Revisited," *British Journal of Psychiatry* 167:569–573; G. M. Goodwin (1995), "Lithium Revisited: A Reply," *British Journal of Psychiatry* 167:573–574.

63. W. Schindler (1961), "5-H-dibenzazepines," U.S. patent 2,948,718, *Chemical Abstracts* 55:1671; see also R. A. Maxwell and S. B. Eckhardt (1990), "Carbamazepine," in R. A. Maxwell and S. B. Eckhardt, eds. (1990), *Drug Discovery* (Clifton, N.J.: Humana Press), pp. 193–206.

64. D. J. Kupfer and E. Frank (1992), "The Minimum Length of Treatment for Recovery," in S. A. Montgomery and F. Rouillon, eds. (1992), *Long-Term Treatment of Depression* (Chichester: John Wiley & Sons), pp. 197–228; E. Frank, S. Johnson, and D. J. Kupfer (1992), "Psychological Treatments and Prevention of Relapse," in Montgomery and Rouillon (1992), pp. 197–228.

65. E. Frank, D. J. Kupfer, J. N. Perel, C. Comes, D. B. Jarrett, A. G. Mallinger, M. E. Thase, A. B. McEachran, and V. J. Grochocinski (1990), "Three-Year Outcomes for Maintenance Therapies in Recurrent Depression," *Archives of General Psychiatry* 47:1093–1099.

66. J. Braithwaite (1986), *Corporate Crime in the Pharmaceutical Industry* (London: Routledge and Kegan Paul).

67. Ibid.

68. M. Bury (1986), "Caveat Venditor: Social Dimensions of a Medical Controversy," in D. Healy and D. Doogan, eds. (1986), *Psychotropic Drug Development: Social, Economic, and Pharmacological Aspects* (London: Chapman & Hall), pp. 41–57.

69. J. Jonas (1992), "Idiosyncratic Side Effects of Short Half Life Benzodiazepine Hypnotics: Fact or Fancy?" *Human Psychopharmacology* 7:205–216.

70. Macht and Mora (1921).

71. J. P. Swazey (1974), *Chlorpromazine: A Study in Therapeutic Innovation* (Cambridge: MIT Press).

72. J. V. Brady (1993), "The Origin and Development of Behavioural Pharmacology," *European Behavioural Pharmacology Society Newsletter* 7:2–11; J. V. Brady (1959) "Comparative Psychopharmacology," in Cole and Gerard (1959), pp. 46–63; B. F. Skinner (1959), "Animal Research in the Pharmacotherapy of Mental Disease," in Cole and Gerard (1959), pp. 224–235; H. Hunt (1959), "Effect of Drugs on Emotional Responses and Abnormal Behavior in Animals," in Cole and Gerard (1959), pp. 268–283.

73. Steinberg and Healy (1996).

74. Delini-Stula and Healy (1996).

75. M. N. G. Dukes and B. Swartz (1988), *Responsibility for Drug-Induced Injury* (Amsterdam: Elsevier Press).

76. D. Healy (1994), "The Fluoxetine and Suicide Controversy," *CNS Drugs* 1:223–231.
77. H. Hippius and D. Healy (1996), "The Founding of the CINP and the Discovery of Clozapine," in Healy (1996), pp. 187–213.
78. H. Meltzer and D. Healy (1996), "A Career in Biological Psychiatry," in Healy (1996), pp. 483–507; D. Healy (1993), "Psychopharmacology and the Ethics of Resource Allocation," *British Journal of Psychiatry* 162:23–29.
79. R. Pinder (1987), "The Benefits and Risks of Antidepressant Drugs," *Human Psychopharmacology* 3:73–86.
80. Cole and Healy (1996).
81. Hippius and Healy (1996).
82. M. Wilhelm (1972), "The Chemistry of Polycyclic Psycho-active Drugs: Serendipity or Systematic Investigation?" in P. Kielholz, ed. (1972), *Depressive Illness: Diagnosis, Assessment, Treatment* (Berne: Hans Huber), pp. 129–139; R. Kuhn and D. Healy (1998), "From Imipramine to Levoprotiline," in Healy (1998).
83. R. Kuhn (1972), "Clinical Experiences with a New Antidepressant," in Kielholz (1972), pp. 195–208.
84. Ibid.
85. J. Welner (1972), "A Multi-National, Multi-Centre, Double-Blind Trial of a New Antidepressant," in Kielholz (1972), pp. 209–221.
86. A. Delini-Stula (1972), "The Pharmacology of Ludiomil," in Kielholz (1972), pp. 113–128; Waldmeier and Healy (1996).
87. Delini-Stula and Healy (1996); Waldmeier and Healy (1996).
88. Coppen and Healy (1996).

5. A Pleasing Look of Truth

1. L. J. Rather (1965), *Mind and Body in 18th-Century Medicine: A Study Based on Jerome Gaub's De Regimine Mentis* (London: Wellcome Historical Medical Library), p. 17.
2. See also Chapter 7.
3. H. Van Praag and D. Healy (1996), "Psychiatry and the March of Folly," in D. Healy, ed. (1996a), *The Psychopharmacologists* (London: Chapman & Hall), pp. 353–379; G. Curzon and D. Healy (1998), "From Neurochemistry to Neuroscience," in D. Healy, ed. (1998), *The Psychopharmacologists,* vol. 2 (London: Chapman & Hall).
4. E. Clarke and L. S. Jacyna (1987), *Nineteenth-Century Origins of Neuroscientific Concepts* (Berkeley: University of California Press).
5. Ibid. See also J. P. Changeux (1985), *Neuronal Man: The Biology of Mind* (Oxford: Oxford University Press).
6. D. Healy (1993), *Images of Trauma: From Hysteria to Post-Traumatic Stress Disorder* (London: Faber & Faber), chap. 1.
7. K. Wellman (1991), *LaMettrie* (Durham: Duke University Press).
8. See Chapter 7.
9. W. W. Grey (1961), *The Living Brain* (Harmondsworth: Penguin Books).

10. Clarke and Jacyna (1987).

11. Z. M. Bacq (1983), "Chemical Transmission of Nerve Impulses," in M. J. Parnham and J. Bruinvels, eds. (1983), *Discoveries in Pharmacology*, vol. 1: *Psycho- and Neuropharmacology* (Amsterdam: Elsevier), pp. 49–103; D. Healy (1996b), "The History of British Psychopharmacology," in G. E. Berrios and H. Freeman, eds. (1996), *One Hundred and Fifty Years of British Psychiatry*, vol. 2 (London: Athlone Press).

12. A. Carlsson and D. Healy (1996), "The Rise of Neuropsychopharmacology: Impact on Basic and Clinical Neuroscience," in Healy (1996a), pp. 51–80; A. Carlsson (1987), "Perspectives on the Discovery of Central Monoaminergic Neurotransmission," *Annual Review: Neuroscience* 10:19–40.

13. M. Vialli and V. Erpsamer (1933), "Cellule enterocromaffini e cellule basigranulose acidofile nei vertebrati Zischr," *Zellforsch U. Mikr. Anat.* 19:743.

14. M. Rapoport, A. Green, and I. H. Page (1949), "Purification of the Substance Which Is Responsible for Vasoconstrictor Activity of Serum," *Federation Proceedings* 6:184.

15. I. H. Page (1970), "Neurochemistry as I Have Known It," in F. Ayd and B. Blackwell, eds. (1970), *Discoveries in Biological Psychiatry* (Philadelphia: Lippincott), pp. 53–58.

16. S. Garattini and D. Healy (1996), "The Role of Independent Science in Psychopharmacology," in Healy (1996a), pp. 135–157.

17. D. W. Woolley and E. Shaw (1954), "A Biochemical and Pharmacological Suggestion about Certain Mental Disorders," *Science* 119:587–588; J. A. Gaddum (1953), "Antagonism between Lysergic Acid-dieythlamide and 5-hydroxytryptamine," *Journal of Physiology* 121:15.

18. A. Carlsson (1990), "Early Psychopharmacology: The Rise of Modern Brain Research," *Journal of Psychopharmacology* 4:120–127; Carlsson and Healy (1996), pp. 51–80; H. Blaschko (1957), "Metabolism of Mediated Substances," in S. Garattini and V. Ghetti, eds. (1957), *Psychotropic Drugs* (Amsterdam: Elsevier), pp. 3–9.

19. Noradrenaline, adrenaline, and dopamine are catecholamines; 5HT is an indoleamine. Both catecholamines and indoleamines are sometimes referred to as monoamines.

20. J. R. Vane, G. E. Wolstenholme, and M. O'Connor (1960), *Adrenergic Mechanisms*, Ciba Foundation Symposium (London: J. & A. Churchill).

21. R. Kanigel (1983), *Apprentice to Genius: The Making of a Scientific Dynasty* (Baltimore: Johns Hopkins Press).

22. J. Axelrod and D. Healy (1996), "The Discovery of Amine Reuptake," in Healy (1996a), pp. 29–49; Carlsson and Healy (1996).

23. A. Pletscher, P. A. Shore, and B. B. Brodie (1955), "Serotonin Release as a Possible Mechanism of Reserpine Action," *Science* 122:374–374.

24. Carlsson and Healy (1996).

25. M. Sandler and D. Healy (1994), "The Place of Chemical Pathology in the Development of Psychopharmacology," *Journal of Psychopharmacology* 8:124–133; Kanigel (1993).

26. E. Shorter (1987), *The Health Century* (New York: Doubleday).

27. Kanigel (1993).
28. C. M. B. Pare and M. Sandler (1959), "A Clinical and Biochemical Study of Iproniazid in the Treatment of Depression," *Journal of Neurology, Neurosurgery, and Psychiatry* 22:274–251.
29. Carlsson and Healy (1996); T. L. Sourkes and S. Gauthier (1983), "Levodopa and Dopamine Agonists in the Treatment of Parkinson's Disease," in Parnham and Bruinvels (1983), pp. 249–268.
30. D. L. Davies and M. Shepherd (1955), "Reserpine in the Treatment of Anxious Patients," *Lancet* 2:117–120; D. Healy (1987), "The Structure of Psychopharmacological Revolutions," *Psychiatric Developments* 4:349–376.
31. A. Todrick (1991), "Imipramine and 5HT Reuptake Inhibition," *Journal of Psychopharmacology* 5:263–267.
32. Axelrod and Healy (1996); Kanigel (1993).
33. J. Axelrod (1972), "Brain Amines and Their Impact in Psychiatry," *Seminars in Psychiatry* 4:199–210; J. Axelrod, R. Whitby, and G. Hertting (1961), "Effect of Psychotropic Drugs on the Uptake of Tritiated Noradrenaline by Tissues," *Science* 133:383–384.
34. J. Axelrod and J. K. Inscoe (1963), "The Uptake and Binding of Circulating Serotonin in the Effect of Drugs," *Journal of Pharmacology and Experimental Therapeutics* 141:161–165.
35. A. Broadhurst and D. Healy (1996), "Before and After Imipramine," in Healy (1996a), pp. 111–134.
36. N. S. Kline, G. Simpson, and B. B. Brodie (1962), "The Clinical Application of Desmethyl-imipramine: A New Type of Antidepressant Drug," *International Journal of Neuropharmacology* 1:55–60; F. Sulser and R. Mishra (1983), "The Discovery of Tricyclic Antidepressants and Their Action," in Parnham and Bruinvels (1983), pp. 233–247.
37. J. J. Schildkraut (1965), "The Catecholamine Hypothesis of Affective Disorders: A Review of Supporting Evidence," *American Journal of Psychiatry* 122:509–522.
38. W. Bunney and J. Davis (1965), "Norepinephrine in Depressive Reactions," *Archives of General Psychiatry* 13:483–494.
39. A. Coppen (1987), "The Biochemistry of Affective Disorders," *British Journal of Psychiatry* 113:1237–1264.
40. A. Coppen (1972), "Indoleamines and Affective Disorders," *Journal of Psychiatric Research* 9:163–171; A. Coppen and D. Healy (1996), "Biological Psychiatry in Britain," in Healy (1996a), pp. 265–286.
41. Healy (1987).
42. J. Angst (1985), "Switch from Depression to Mania: A Record Study over Decades between 1920 and 1982," *Psychopathology* 18:140–154.
43. J. Maas, S. Koslow, M. Katz, C. Bowden, R. Gibbons, P. Stokes, E. Robins, and J. Davis (1984), "Pre-Treatment Neurotransmitter Levels and Response to Tricyclic Antidepressant Drugs," *American Journal of Psychiatry* 141:1159–1171. See also Healy (1987).
44. A. J. Clark (1933), *The Mode of Action of Drugs on Cells* (London: E. Arnold & Co.).

45. J. M. Stadel and R. J. Lefkowitz (1991), "Beta-Adrenergic Receptors: Identification and Characterization by Radio-ligand Binding Studies," in J. P. Perkins, ed. (1991), *The Beta-Adrenergic Receptors* (Clifton, N.J.: Humana Press), p. 140.

46. G. W. Ashcroft, D. Eccleston, L. G. Murray, A. I. M. Glen, T. V. Crawford, I. A. Pullar, P. J. Shield, D. S. Walter, I. M. Blackburn, J. Connechan, and M. Lonigan (1972), "Modified Amine Hypothesis for the Aetiology of Affective Illness," *Lancet* 2:573–577.

47. J. Vetulani, R. Stawarz, J. Pingell, and F. Sulser (1975), "A Possible Common Mechanism of Action of Antidepressant Drugs," *N. S. Archives of Pharmacology* 243:109–114.

48. H. Y. Meltzer and S. M. Stahl (1976), "The Dopamine Hypothesis of Schizophrenia: A Review," *Schizophrenia Bulletin* 2:19–76.

49. Carlsson and Healy (1996).

50. D. Healy (1997), *Psychiatric Drugs Explained,* 2nd ed. (London: Mosby Yearbooks), chaps. 25, 26.

51. Coppen and Healy (1996); Van Praag and Healy (1996).

52. Carlsson and Healy (1996).

53. Belgian Patent no. 781.105, March 23, 1972.

54. *Acta Psychiatrica Scandinavica* 63 (suppl. 290)(1981).

55. P. Waldmeier and D. Healy (1996), "From Mental Illness to Neurodegeneration," in Healy (1996a), pp. 565–586; A. Delini-Stula and D. Healy (1996), "The Changing Face of Psychotropic Drug Development," in Healy (1996a), pp. 425–440.

56. J. Toumisto and E. Tukianinen (1976), "Decreased Uptake of 5-hydroxytryptamine in Blood Platelets from Patients with Endogenous Depression," *Psychopharmacology* 65:141–147; D. Healy and B. E. Leonard (1987), "Monoamine Reuptake in Depression: Kinetics and Dynamics," *Journal of Affective Disorders* 12:91–105.

57. D. Healy and E. S. Paykel (1992), "Neurochemical Correlates of Affective Processes," in A. K. Ashbury, G. M. McKhann, and W. I. McDonald, eds. (1992), *Diseases of the Nervous System* (New York: Harcourt Brace and Jovanovitch), pp. 815–830.

58. Healy (1987).

59. D. T. Wong, J. S. Horng, F. P. Bymaster, K. L. Hauser, and B. B. Molloy (1974), "A Selective Inhibitor of Serotonin Uptake: Lilly 110140, 3-(p-trifluoromethylphenoxy)-N-methyl-3-phenylpropylamine," *Life Sciences* 15: 471–479.

60. P. Kramer (1993), *Listening to Prozac* (New York: Viking Press); D. T. Wong, F. P. Bymaster, and E. Engleman (1995), "Prozac (Fluoxetine, Lilly 110140), the First Selective Serotonin Uptake Inhibitor and an Antidepressant Drug: Twenty Years since Its First Publication," *Life Sciences* 57: 411–441.

61. Coppen and Healy (1996).

62. Wong, Bymaster, and Engleman (1995).

63. Carlsson and Healy (1996).

64. Kramer (1993).
65. H. J. Eysenck (1947), *Dimensions of Personality* (London: Routledge and Keegan Paul).
66. H. J. Eysenck (1957), "Drugs in Personality," *Journal of Mental Science* 103:119–131; H. J. Eysenck (1961), "Psychosis, Drive, and Inhibition: A Theoretical and Experimental Account," *American Journal of Psychiatry* 118:198–204.
67. C. Shagass and J. Naiman (1956), "The Sedation Threshold as an Objective Index of Manifest Anxiety in Psycho-Neurosis," *Journal of Psycho-somatic Research* 1:49–57.
68. G. Claridge and D. Healy (1994), "The Psychopharmacology of Individual Differences," *Human Psychopharmacology* 9:285–298.
69. M. Zuckerman (1991), *Psychobiology of Personality* (Cambridge: Cambridge University Press); S. E. Hampson (1982), *The Construction of Personality* (London: Routledge & Kegan Paul).
70. G. S. Claridge (1972), "The Schizophrenias as Nervous Types," *British Journal of Psychiatry* 121:1–17.
71. G. S. Claridge (1987), "The Schizophrenias as Nervous Types, Revisited," *British Journal of Psychiatry* 151:735–743.
72. C. R. Cloninger, D. M. Svrakic, and T. R. Przybeck (1993), "A Psychobiological Model of Temperament and Character," *Archives of General Psychiatry* 50:975–990; H. M. Van Praag (1993), *"Make-Believes" in Psychiatry or The Perils of Progress* (New York: Brunner-Mazel).
73. Van Praag and Healy (1996).
74. T. A. Ban and D. Healy (1996), "They Used to Call It Psychiatry," in Healy (1996a), pp. 587–620.
75. Delini-Stula and Healy (1996); Waldmeier and Healy (1996).
76. Van Praag and Healy (1996).

6. The Luke Effect

1. T. Merton (1965), *The Way of Chuang Tzu* (London: Unwin), pp. 67–68.
2. R. W. Merton (1968), "The Matthew Effect in Science," *Science* 159:59–63.
3. D. Healy (1990), "The Psychopharmacological Era: Notes towards a History," *Journal of Psychopharmacology* 4:152–167.
4. P. Breggin (1991), *Toxic Psychiatry* (New York: St. Martin's Press).
5. Rhône-Poulenc and Sandoz, however, flirted with antidepressant indications for their antipsychotics—Rhône-Poulenc with levomepromazine and Sandoz more successfully with thioridazine (Mellaril).
6. D. F. Klein and M. Fink (1962a), "Psychiatric Reaction Patterns to Imipramine," *American Journal of Psychiatry* 119:432–438.
7. H. Hippius and D. Healy (1996), "The Foundation of the CINP and the Discovery of Clozapine," in D. Healy, ed. (1996a), *The Psychopharmacologists* (London: Chapman & Hall), pp. 187–213.
8. J. F. Ravn (1970), "The History of Thioxanthenes," in F. Ayd and B. Blackwell, eds. (1970), *Discoveries in Biological Psychiatry* (Philadelphia: Lippincott), pp. 180–193.

9. I. M. Sonne (1966), "Flupenthixol in the Treatment of Depressive States," *Nord Psychiat Tidssk* 20:322–324; P. J. Reiter (1969), "On Flupenthixol: An Antidepressant of a New Chemical Group," *British Journal of Psychiatry* 115:1399–1402; V. Predescu, T. Ciurezu, G. Timofte, and I. Roman (1973), "Symptomatic Relief with Flupenthixol of the Anxious-Algetic-Depressive Syndrome Complex in Neurotic States," *Acta Psychiatrica Scandinavia* 49:16–27; V. Pedersen, K. Bøgesø, and D. Healy (1998), "Drug-Hunting," in D. Healy, ed. (1998), *The Psychopharmacologists,* vol. 2 (London: Chapman & Hall).

10. F. Frolund (1974), "Treatment of Depression in General Practice: A Controlled Trial of Flupenthixol," *Current Medical Research and Opinion* 2:78–89; J. P. R. Young, W. C. Hughes, and M. H. Lader (1976), "A Controlled Comparison of Flupenthixol and Amitriptyline in Depressed Outpatients," *British Medical Journal* 1:1116–1118; D. A. W. Johnson (1979), "A Double-Blind Comparison of Flupenthixol, Nortriptyline, and Diazepam in Neurotic Depression," *Acta Psychiatrica Scandinavia* 59:1–8.

11. M. N. Robertson and M. R. Trimble (1982), "Major Tranquillisers Used as Antidepressants," *Journal of Affective Disorders* 4:173–193.

12. A version of the flupenthixol story was replayed more recently with the marketing of Amoxapine (Asendis). See Healy (1990).

13. Garattini (1996), "Experimental and Clinical Activity of Antidepressant Drugs," in D. Healy and D. Doogan, eds. (1996), *Psychotropic Drug Development: Social, Economic, and Pharmacological Aspects* (London: Chapman & Hall), pp. 1–13; S. Garattini and D. Healy (1996), "The Role of Independent Science in Psychopharmacology," in Healy (1996a), pp. 135–157.

14. C. Fear, H. Sharp, and D. Healy (1996), "Cognitive Processes in Delusional Disorders," *British Journal of Psychiatry* 166:61–67.

15. Medical Research Council Trial (1965), "Clinical Trial of the Treatment of Depressive Illness," *British Medical Journal* 1:881–886.

16. E. D. West and P. J. Dally (1959), "Effects of Iproniazid in Depressive Syndromes," *British Medical Journal* 1:1491–1499.

17. J. Pollitt and J. Young (1971), "Anxiety State or Masked Depression? A Study Based on the Action of Monoamine Oxidase Inhibitors," *British Journal of Psychiatry* 119:143–149.

18. D. Kelly, W. Guirguis, E. Frommer, N. Mitchell-Heggs, and W. Sargant (1970), "Treatment of Phobic States with Antidepressants: A Retrospective Study of 246 Patients," *British Journal of Psychiatry* 116:387–398.

19. I. M. Marks (1970), "The Classification of Phobic Disorders," *British Journal of Psychiatry* 116:377–386.

20. I. M. Marks (forthcoming), "Discussion," *Journal of Psychopharmacology* 11 (suppl.); I. M. Marks and D. Healy (1998), "Marketing the Evidence," in Healy (1998).

21. M. R. Liebowitz, J. M. Gorman, A. J. Fyer, and D. F. Klein (1985), "Social Phobia: A Review of a Neglected Disorder," *Archives of General Psychiatry* 42:729–736.

22. American Psychiatric Association (1980/1987), *Diagnostic and Statistical Manual of Mental Disorders,* 3rd ed. (rev.) (Washington, D.C.).

23. T. A. Widiger (1992), "Generalised Social Phobia vs. Avoidant Personality Disorder: A Commentary on Three Studies," *Journal of Abnormal Psychology* 101:340–343.

24. S. M. Turner, D. C. Beidel, and R. M. Townsley (1990), "Social Phobia: Relationships to Shyness," *Behavioural Research and Therapy* 28:497–505.

25. D. Healy (1995), "Social Phobia in Primary Care," *Primary Care Psychiatry* 1:31–38; D. Healy (1996c), "Psychopharmacology in the New Medical State," in Healy and Doogan (1996), pp. 14–39; J. R. T. Davidson (1991), "Social Phobia in Review," *Journal of Clinical Psychiatry* 52 (suppl. 11):3–4; R. C. Kessler, D. K. McGonagle, S. Zhao, C. B. Nelson, M. Hughes, S. Eshleman, H. U. Wittchen, and K. S. Kendler (1994), "Life Time and Twelve Month Prevalence of DSM IIIR Psychiatric Disorders in the United States: Results from the National Co-Morbidity Survey," *Archives of General Psychiatry* 51:8–19; J. R. T. Davidson, D. C. Hughes, L. K. George, and D. G. Biazer (1994), "The Boundary of Social Phobia: Exploring the Threshold," *Archives of General Psychiatry* 51:975–983; J. Ross (1991), "Social Phobia: The Anxiety Disorders," *Journal of Clinical Psychiatry* 52 (suppl. 11):43–47.

26. M. R. Liebowitz, F. R. Schneier, E. R. Hollander, L. A. Welkowitz, J. B. Saoud, J. Feerick, R. Campeas, B. A. Fallon, L. Street, and A. Gitow (1991), "Treatment of Social Phobia with Drugs Other than Benzodiazepines," *Journal of Clinical Psychiatry* 52 (suppl. 11):10–15.

27. M. Versiani, A. E. Nardi, F. D. Mundin, A. B. Alves, M. R. Liebowitz, and R. Amrein (1992), "Pharmacotherapy of Social Phobia: The Controlled Study with Moclobemide and Phenelzine," *British Journal of Psychiatry* 161:353–360.

28. Y. Kasahara (1987), "Social Phobia in Japan," in *Social Phobia in Japan and Korea: Proceedings of the First Cultural Psychiatry Symposium between Japan and Korea* (Seoul: East Asian Academy of Cultural Psychiatry), pp. 3–15. See also Healy (1995).

29. D. F. Klein and M. Fink (1962b), "Reaction Patterns to Psychotropic Drugs and the Discovery of Panic Disorder," in Healy (1996a), pp. 329–352.

30. D. F. Klein and M. Fink (1962a); D. F. Klein and M. Fink (1962c), "Behavioural Reaction Patterns with Phenothiazine," *Archives of General Psychiatry* 7:449–459.

31. M. Roth (1960), "The Phobic Anxiety De-personalisation Syndrome and Some General Aetiological Problems in Psychiatry," *Journal of Neuropsychiatry* 1:293–300.

32. D. V. Sheahan, J. Ballenger, and G. Jacobsen (1980), "Treatment of Endogenous Anxiety with Phobic, Hysterical, and Hypochondriacal Symptoms," *Archives of General Psychiatry* 37:51–57.

33. D. V. Sheahan, interview, November 1994.

34. Marks and Healy (1998); M. Weissman and D. Healy (1998), in Healy (1998).

35. Breggin (1991).

36. G. L. Klerman (1988), "Overview of the Cross-National Collaborative Panic Study," *Archives of General Psychiatry* 45:407–412.

37. I. M. Marks, A. De Albuquerque, J. Cotteaux, V. Gentil, J. Greist, I. Hand, R. J. Liberman, J. S. Relvas, A. Tobean, P. Tyrer, and H. U. Wittchen (1989), "The Efficacy of Alprazolam in Panic Disorder and Agoraphobia: A Critique of Recent Reports," *Archives of General Psychiatry* 46:670–672.

38. G. L. Klerman, J. C. Ballenger, G. D. Burrows, R. L. Dupont, R. Notes, J. C. Pecknold, Q. A. Rifkind, R. T. Ruben, and R. P. Swinson (1989), "In Reply to Marks et al.," *Archives of General Psychiatry* 46:670–672. See also Marks and Healy (1998).

39. G. L. Klerman et al. (1992), "Drug Treatment of Panic Disorder: Comparative Efficacy of Alprazolam, Imipramine, and Placebo. Cross National Collaborative Panic Study Second Phase Investigators," *British Journal of Psychiatry* 160:191–202.

40. I. M. Marks, J. Griest, M. Basoglu, H. Noshirvani, and G. O'Sullivan (1992), "Comment on the Second Phase of the Cross National Collaborative Panic Study," *British Journal of Psychiatry* 160:202–205.

41. G. L. Klerman (1992), "Drug Treatment of Panic Disorder: Reply to Comment by Marks and Associates," *British Journal of Psychiatry* 161:465–471.

42. I. M. Marks, R. P. Swinson, M. Basoglu, K. Cuch, H. Noshirvani, G. O'Sullivan, E. T. Lelliott, M. Kirby, G. MacNamee, S. Sengun, and K. Wickwire (1993), "Alprazolam and Exposure Alone and Combined in Panic Disorder with Agoraphobia: A Controlled Study in London and Toronto," *British Journal of Psychiatry* 162:776–787.

43. D. A. Spiegel, M. Roth, M. Weissman, P. Lavori, J. Gorman, J. Rush, and J. Ballenger, "Comment on the London/Toronto Study of Alprazolam in Exposure in Panic Disorder with Agoraphobia," *British Journal of Psychiatry* 162:788–789.

44. I. M. Marks, R. P. Swinson, M. Basoglu, H. Noshirvani, K. Cuch, G. O'Sullivan, and E. T. Lelliott (1993), "Reply to Comment on London/Toronto Study," *British Journal of Psychiatry* 162:790–794; I. M. Marks, M. Basoglu, H. Noshirvani, J. Greist, R. P. Swinson, and G. O'Sullivan (1993), "Drug Treatment of Panic Disorder: Further Comment," *British Journal of Psychiatry* 162:795–796.

45. *Journal of Psychiatric Research* (1990), 24 (suppl. 1).

46. Richard Swinson, discussion at Clarke Institute, October 1995.

47. Marks and Healy (1998).

48. Marks, Swinson, Basoglu, Cuch, Noshirvani, O'Sullivan, Lelliott, Kirby, MacNamee, Sengun, and Wickwire (1993).

49. D. Healy (1990), "The Psychopharmacological Era: Notes toward a History," *Journal of Psychopharmacology* 4:152–167.

50. G. Beaumont and D. Healy (1993), "The Place of Clomipramine in the Development of Psychopharmacology," *Journal of Psychopharmacology* 7:383–393.

51. J. Guyotat, M. Favre-Tissot, and M. Marie-Gardine (1968), "A Clinical Trial with a New Antidepressant G34586," in *Congress du Psychiatrie du Neurologie, Dijon 1967* (Paris: Masson).

52. E. F. Cordoba and J. J. Lopez-Ibor (1967), "Monochlorimipramine in Psychiatric Patients Resistant to Other Forms of Treatment," *Actas Luso-españolas Neurol Psiquiat* 26:119–147.

53. W. K. Marshall and V. Micev (1973), "Chlomipramine in the Treatment of Obsessional Illnesses in Phobic Anxiety States," *Journal of International Medical Research* 1:403–412; N. Capstick (1973), "The Graylingwell Study," *Journal of International Medical Research* 1:392–396; C. J. Walter (1973), "Clinical Impressions and Treatment of Obsessional States with Intravenous Chlomipramine," *Journal of International Medical Research* 1:413–416; D. Waxman (1973), "A General Practitioner Investigation on the Use of Chlomipramine in Obsessional and Phobic Disorders," *Journal of International Medical Research* 1:417–420; P. H. Rack (1973), "Chlomipramine in the Treatment of Obsessional States with Special Reference to the Latent Obsessional Inventory," *Journal of International Medical Research* 1:397–402.

54. Beaumont and Healy (1993).

55. I. M. Marks, R. S. Stern, D, Mawson, J. Cobb, and R. MacDonald (1980), "Chlomipramine and Exposure for Obsessive Compulsive Rituals," *British Journal of Psychiatry* 136:1–25.

56. S. A. Montgomery (1980), "Clomipramine in Obsessional Neurosis: A Placebo Controlled Trial," *Pharmaceutical Medicine* 1:189–192.

57. I. M. Marks and M. Basoglu (1989), "Obsessive Compulsive Rituals," *British Journal of Psychiatry* 154:650–658.

58. I. M. Marks and G. O'Sullivan (1988), "Drugs and Psychological Treatments for Agoraphobia/Panic and Obsessive Compulsive Disorders: A Review," *British Journal of Psychiatry* 153:650–658.

59. T. L. Perse, J. H. Greist, J. W. Jefferson, R. Rosenfeld, and R. Dar (1987), "Fluvoxamine Treatment of Obsessive Compulsive Disorder," *American Journal of Psychiatry* 144:1543–1548.

60. W. K. Goodman, L. H. Price, S. A. Rasmussen, P. L. Delgado, G. R. Henninger, and D. S. Chamey (1989), "Efficacy of Fluvoxamine in Obsessive Compulsive Disorder," *Archives of General Psychiatry* 46:36–44.

61. J. L. Rapoport (1989), *The Boy Who Couldn't Stop Washing* (New York: E. P. Dutton).

62. C. Fear, H. Sharp, and D. Healy (1995), "Obsessive-Compulsive and Delusional Disorders: Notes on Their History, Nosology, and Interface," *Journal of Serotonin Research* 1 (suppl. 1):1–18.

63. I. M. Marks (1989), "The Gap between Research and Policy in Mental Health Care," *Journal of the Royal Society of Medicine* 82:514–517.

64. D. Healy (1993), *Images of Trauma: From Hysteria to Post-Traumatic Stress Disorder* (London: Faber & Faber), chap. 10.

65. Marks and Healy (1998).

66. G. Beaumont (1973), "Sexual Side-effects of Clomipramine," *Journal of International Medical Research* 1:469–472. H. Eaton (1973), "Chlorimipramine in the Treatment of Premature Ejaculation," *Journal of International Medical Research* 1:432–434.

67. J. D. Couper-Smartt and R. Rodham (1973), "A Technique for Surveying Side Effects of Tricyclic Drugs with Reference to Reported Sexual Side Effects," *Journal of International Medical Research* 1:473–476.

68. R. E. Goodman (1977), "The Management of Premature Ejaculation," *Journal of International Medical Research* 5 (suppl. 1):78–81; R. E. Goodman (1980), "An Assessment of Chlomipramine in the Treatment of Premature Ejaculation," *Journal of International Medical Research* 8 (suppl. 3):53–59.

69. E.-E. Baulieu (1991), *The Abortion Pill* (London: Random Century).

70. D. Healy (1997), *Psychiatric Drugs Explained,* 2nd ed. (London: Mosby Yearbooks), chaps. 25, 26.

71. H. Saul (1993), "Fat as a Pharmaceutical Issue," *New Scientist* 1883:28–31.

72. W. M. Carley (1995), "Papers Detail Ortho's Retin-A Deception," *Wall Street Journal,* March 1.

73. G. Beaumont (1969), "Discussion," in G. Beaumont, ed. (1969), *Chlorimipramine,* Symposium Report (Stratford-upon-Avon).

74. J. Kroll (1993), *PTSD/Borderlines in Therapy: Finding the Balance* (New York: Norton), chap. 8.

75. J. K. Galbraith (1967), *The New Industrial State* (London: Hamish Hamilton). See also Healy (1996c).

76. F. J. Ayd and D. Healy (1996), "The Discovery of Antidepressants," in Healy (1996a), pp. 81–110.

77. L. A. Bero, A. Galbraith, and D. Rennie (1992), "The Publication of Sponsored Symposiums in Medical Journals," *New England Journal of Medicine* 327:1135–1140.

78. T. A. Sheldon and G. D. Smith (1993), "Consensus Conferences as Drug Promotion," *Lancet,* pp. 100–102.

7. From Oedipus to Osheroff

1. Initially called the Psychopharmacology Service Center; see J. O. Cole and D. Healy (1996), "The Evaluation of Psychotropic Drugs," in D. Healy, ed. (1996a), *The Psychopharmacologists* (London: Chapman & Hall), pp. 239–263.

2. J. O. Cole, S. C. Goldberg, and G. L. Klerman (1964), "Phenothiazine Treatment in Acute Schizophrenia," *Archives of General Psychiatry* 10:246–261.

3. D. J. Armor and G. L. Klerman (1968), "Psychiatric Treatment Orientations and Professional Ideology," *Journal of Health and Social Behaviour* 9:243–255.

4. M. Shepherd (1961), "The Present Status of Psychopharmacology," *Journal of Chronic Disease* 13:289–292.

5. A. Lewis (1959), "Response to H. Brill," in P. Bradley, P. Deniker, and C. Radouco-Thomas, eds. (1959), *Neuropsychopharmacology* (Amsterdam:

Elsevier), p. 211; M. Gelder (1991), "Adolf Meyer and His Influence on British Psychiatry," in G. E. Berrios and H. Freeman, eds. (1991), *One Hundred and Fifty Years of British Psychiatry*, vol. 1 (London: Athlone Press), pp. 419–435.

6. D. Healy (1996b), "The History of British Psychopharmacology," in G. E. Berrios and H. Freeman, eds. (1996), *One Hundred and Fifty Years of British Psychiatry*, vol. 2 (London: Athlone Press).

7. A. Coppen and D. Healy (1996), "Biological Psychiatry in Britain," in Healy (1996a), pp. 265–286.

8. Cole and Healy (1996).

9. Lewis (1959).

10. M. Shepherd (1990), "The 'Neuroleptics' and the Oedipus Effect," *Journal of Psychopharmacology* 4:131–135.

11. J. P. Swazey (1974), *Chlorpromazine: A Study in Therapeutic Innovation* (Cambridge: MIT Press); E. Shorter (1996), *A History of Psychiatry: From the Asylums to Prozac* (New York: Wiley); D. Tantam (1991), "The Anti-Psychiatry Movement," in Berrios and Freeman (1991), pp. 333–349.

12. H. Van Praag and D. Healy (1996), "Psychiatry and the March of Folly," in Healy (1996a).

13. Shorter (1996).

14. A. Scull (1995), "Psychiatrists and Historical Facts," *History of Psychiatry* 6:225–242.

15. G. N. Grob (1991), *From Asylum to Community: Mental Health Policy in Modern America* (Princeton: Princeton University Press).

16. W. Reich (1981), "American Psycho-Ideology," *New York Times*, January 11; rpt. *Psychiatric Bulletin*, 1982, p. 43.

17. D. F. Klein and D. Healy (1996), "Reaction Patterns to Psychotropic Drugs and the Discovery of Panic Disorder," in Healy (1996a).

18. A. Scull (1994), "Somatic Treatments and the Historiography of Psychiatry," *History of Psychiatry* 5:1–11; H. Merskey (1994), "Somatic Treatments: Ignorance and the Historiography of Psychiatry," *History of Psychiatry* 5:387–391; J. L. Cranmer (1994), "Britain in the Fifties: Leucotomy and Open Doors," *History of Psychiatry* 5:393–396.

19. H. E. Lehmann and D. Healy (1996), "Psychopharmacotherapy," in Healy (1996a), pp. 159–186.

20. E. S. Valenstein (1986), *Great and Desperate Cures* (New York: Basic Books). Although see D. Crossley (1994), "The Introduction of Leucotomy: A British Case History," *History of Psychiatry* 5:553–564.

21. J. T. Braslow (1995), "Effect of Therapeutic Innovation on Perception of Disease and the Doctor-Patient Relationship: A History of General Paralysis of the Insane and Malaria Fever Therapy, 1910–1950," *American Journal of Psychiatry* 152:660–665.

22. T. A. Ban and D. Healy (1996), "They Used to Call It Psychiatry," in Healy (1996a), pp. 587–620; A. Collins (1985), *In the Sleep Room* (Toronto: Lester and Orpen Dennys); G. Simpson and D. Healy (1998), "Early Clinical Psy-

chopharmacology," in D. Healy, ed. (1998), *The Pharmacologists,* vol. 2 (London: Chapman & Hall).

23. M. S. Pernick (1985), *A Calculus of Suffering: Pain, Professionalism, and Anesthesia in Nineteenth-Century America* (New York: Columbia University Press).

24. J. M. Masson (1989), *Against Therapy* (London: Collins).

25. M. Shepherd and D. Healy (1998), "Psychopharmacotherapy: From Specific to Non-Specific," in Healy (1998).

26. Pernick (1985).

27. A. Huxley (1951), *Brave New World* (Harmondsworth: Penguin Books).

28. M. C. Smith (1991), *A Social History of the Minor Tranquillisers: The Quest for Small Comfort in the Age of Anxiety* (New York: Haworth Press).

29. S. Yolles cited in Smith (1991), p. 179.

30. Smith (1991).

31. D. Manheimer, S. Davidson, M. Balter, G. Mellinger, I. Cisin, and H. Parry (1973), "Popular Attitudes and Beliefs about Tranquillizers," *American Journal of Psychiatry* 130:1246–1253.

32. M. Bury and J. Gabe (1990a), "Hooked? Media Responses to Tranquilliser Dependence," in P. Abbott and G. Payne, eds. (1990), *New Direction in the Sociology of Health* (Basingstoke: Falmer, Brass), pp. 87–103; M. Bury and J. Gabe (1990b), "A Sociological View of Tranquilliser Dependence: Challenges and Responses," in I. Hindmarch, G. Beaumont, S. Brandon, and B. E. Leonard, *Benzodiazepines: Current Concepts* (New York: John Wiley & Sons); M. Bury and J. Gabe (1991), "Tranquillisers and Health Care in Crisis," *Social Science and Medicine* 32:449–454; M. Bury and J. Gabe (1996), "Halcion Nights: A Sociological Account of a Medical Controversy," *Sociology* 30:447–469; M. Bury (1996), "Caveat Venditor: Social Dimensions of a Medical Controversy," in D. Healy and D. Doogan, eds. (1996), *Psychotropic Drug Development: Social, Economic, and Pharmacological Aspects* (London: Chapman & Hall), pp. 41–57.

33. M. Balter, J. Levine, and D. Mannheimer (1974), "Cross-National Study of the Extent of Anti-Anxiety/Sedative Drug Use," *New England Journal of Medicine* 290:769–774.

34. M. Pflanz, H. Basler, and D. Schwoon (1977), "Use of Tranquillising Drugs by a Middle-Aged Population in a West German City," *Journal of Health and Social Behaviour* 18:194–205.

35. Shorter (1996), chap 8.

36. Smith (1991).

37. Shepherd and Healy (1998); M. Shepherd, B. Cooper, A. C. Brown, and G. Kalton (1966), *Psychiatric Illness in General Practice* (London: Oxford University Press).

38. D. Goldberg and P. Huxley (1980), *Mental Illness in the Community: The Pathways to Psychiatric Care* (London: Tavistock Press); C. V. R. Blacker and A. Clare (1987), "Depressive Disorder in Primary Care," *British Journal of Psychiatry* 150:737–751.

39. E. S. Paykel and R. Priest (1992), "Recognition and Treatment of Depression in General Practice: Consensus Statement," *British Medical Journal* 305:1198–1202.

40. The recognition of depression in primary care is quite a different phenomenon from the turn to community psychiatry, which preceded the introduction of the psychotropic drugs and which was clearly not something that was fostered by the pharmaceutical industry, as Gerald Grob (1994), *The Mad among Us* (Cambridge: Harvard University Press), makes clear.

41. P. Breggin (1991), *Toxic Psychiatry* (New York: St. Martin's Press); P. Breggin (1994), *Talking Back to Prozac* (New York: St. Martin's Press).

42. E. Shorter (1992), *From Paralysis to Fatigue: A History of Psychosomatic Illness in the Modern Era* (New York: Free Press); E. Shorter (1994), *From the Mind into the Body: The Cultural Origins of Psychosomatic Syndromes* (New York: Free Press).

43. P. Duncan-Jones, D. M. Fergusson, J. Ormel, and L. J. Horwood (1990), "A Model of Stability and Change in Minor Psychiatric Symptoms: Results from Three Longitudinal Studies," *Psychological Medicine*, Monograph Supplement 18; J. Ormel (1995), "Epidemiology of Anxiety and Depression in Primary Care," *Primary Care Psychiatry* 1 (suppl. 1):3–8.

44. G. N. Grob (1991), "Origins of DSM-1: A Study in Appearance and Reality," *American Journal of Psychiatry* 148:421–431.

45. R. E. Kendell et al. (1971), "Diagnostic Criteria of American and British Psychiatrists," *Archives of General Psychiatry* 25:123–130.

46. R. Bayer and R. L. Spitzer (1985), "Neurosis, Psychodynamics, and DSM-111," *Archives of General Psychiatry* 42:187–196.

47. R. M. Hirschfeld (1994), "Diagnosis and Classification in Psychiatry: Gerald Klerman's Contribution," *Harvard Review of Psychiatry* 1:306–309; M. Weissman and D. Healy (1998), "Gerald Klerman and Psychopharmacotherapy," in Healy (1998).

48. R. Blashfield (1982), "Feighner et al.: Invisible Colleges and the Matthew Effect," *Schizophrenia Bulletin* 8:1–12 (with commentary by Guze, Strauss, Katz, and Kendell). See also Shorter (1996).

49. M. M. Katz and G. L. Klerman (1979), "Introduction: Overview of the Clinical Studies Program (Psychobiology of Depression Study)," *American Journal of Psychiatry* 136:49–51.

50. R. L. Spitzer, J. Endicott, and E. Robins (1975), *Research Diagnostic Criteria (RDC) for a Selected Group of Functional Disorders* (New York: New York State Department of Mental Hygiene, Biometrics Branch).

51. Ibid.

52. Bayer and Spitzer (1985); M. Wilson (1993), "DSM-III and the Transformation of American Psychiatry: A History," *American Journal of Psychiatry* 150:399–410; G. L. Klerman (1984), "The Advantages of DSM-III," *American Journal of Psychiatry* 141:539–542.

53. G. E. Vaillant (1984), "The Disadvantages of DSM-III Outweigh Its Advantages," *American Journal of Psychiatry* 141:542–545. It should be noted that

a number of the neo-Kraepelinians, such as George Winokur, leaned toward a therapeutic nihilism rather than enthusiasm for pharmacotherapy.

54. R. Michels (1984), "First Rebuttal," *American Journal of Psychiatry* 141:548–560.

55. Vaillant (1984).

56. Bayer and Spitzer (1985).

57. G. L. Klerman, J. Endicott, R. Spitzer, and R. M. Hirschfeld (1979), "Neurotic Depression," *American Journal of Psychiatry* 136:57–61.

58. Bayer and Spitzer (1985); Wilson (1993).

59. Klerman (1984). The sentence omitted from the middle of the quotation is "one of my premises is that professional groups, such as medical specialities, are able to determine their own destiny to a greater extent than most other occupational groups"—a statement of almost greater interest than anything in the rest of the debate.

60. G. L. Klerman (1990), "Administrative Obstacles to Innovations in the U.S. at the Federal Level," in I. Marks and R. Scott, eds. (1990), *Mental Health Care Delivery* (Cambridge: Cambridge University Press), pp. 179–188.

61. H. Pardes (1990), "The Demise of a Major Innovation: Carter's 1980 Community Mental Health Systems Act in Reagan's Hands," in Marks and Scott (1990), pp. 189–200; D. A. Regier, M. J. Kyers, M. Kramer, L. N. Robins, D. G. Blazer, R. L. Hough, W. W. Eaton, and B. Z. Locke (1984), "The NIMH Epidemiologic Catchment Area Program," *Archives of General Psychiatry* 41:934–941.

62. H. J. Eysenck (1952), "The Effects of Psychotherapy: An Evaluation," *Journal of Consulting and Clinical Psychology* 16:319–324.

63. Shorter (1996).

64. A. T. Beck (1967), *Depression: Causes and Treatment* (Philadelphia: University of Pennsylvania Press); Beck (1976), *Cognitive Therapy and the Emotional Disorders* (New York: International Universities Press).

65. J. M. G. Williams (1992), *The Psychological Management of Depression* (Beckenham: Croon & Helm).

66. E. S. Paykel, J. Myers, M. Dienelt, G. L. Klerman, J. J. Lindenthal, and M. Pepper (1969), "Life Events and Depression," *Archives of General Psychiatry* 21:753–760.

67. E. S. Paykel, interview on the origins of interpersonal therapy, July 18, 1995; Weissman and Healy (1998).

68. G. L. Klerman, M. M. Weissman, B. Rounsaville, and E. S. Chevron (1984), *Interpersonal Therapy of Depression* (New York: Basic Books).

69. I. Elkin, M. B. Parloff, S. W. Hadley, and J. H. Autry (1985), "NIMH Treatment of Depression Collaborative Research Program: Background and Research Plan," *Archives of General Psychiatry* 42:305–316; I. Elkin, P. A. Pilkonis, J. P. Docherty, and S. M. Sotsky (1985), "Conceptual and Methodological Issues in Comparative Studies of Psychotherapy and Pharmacotherapy, 1: Active Ingredients and Mechanism of Change," *American Journal of Psychiatry* 145:909–917; 2: Nature and Timing of Treatment Effects," *American Journal of Psychiatry* 145:1070–1076.

70. I. Elkin, M. T. Shea, J. T. Watkins, S. D. Imber, S. M. Sotsky, J. F. Collins, D. R. Glass, P. A. Pilkonis, W. R. Leber, J. P. Docherty, S. J. Fiester, and M. B. Parloff (1989), "NIMH Treatment of Depression Collaborative Research Program: General Effectiveness of Treatments," *Archives of General Psychiatry* 46:971–982; S. M. Sotsky, D. R. Glass, M. T. Shea, P. A. Pilkonis, J. F. Collins, I. Elkin, J. T. Watkins, S. D. Imber, W. R. Leber, J. Moyer, and M. E. Oliveri (1991), "Patient Predictors of Response to Psychotherapy and Pharmacotherapy: Findings of the NIMH Treatment of Depression Collaborative Research Program," *American Journal of Psychiatry* 148:997–1008; M. T. Shea, I. Elkin, S. D. Imber, S. M. Sotsky, J. T. Watkins, J. F. Collins, P. A. Pilkonis, E. Beckham, D. R. Glass, R. T. Dolan, and M. B. Parloff (1992), "Course of Depressive Symptoms over Follow Up: Findings from the NIMH Treatment of Depression Collaborative Research Program," *Archives of General Psychiatry* 49:762–787.

71. D. X. Freedman (1989), "Editorial Note (Especially for the Media)," *Archives of General Psychiatry* 46:983.

72. D. F. Klein (1990), "NIMH Collaborative Research on Treatment of Depression," *Archives of General Psychiatry* 47:682–683; I. Elkin, M. T. Shea, J. F. Collins, C. J. Clett, S. D. Imber, S. M. Sotsky, J. T. Watkins, and M. B. Parloff (1990), "Reply to Klein," *Archives of General Psychiatry* 47:684–685; R. M. A. Hirschfeld (1990), "Reply to Klein," *Archives of General Psychiatry* 47:685–686; G. L. Klerman (1990), "Reply to Klein," *Archives of General Psychiatry* 47:686–688; H. A. Rashkin (1990), "Déjà Vu in Depression Research," *Archives of General Psychiatry* 47:688.

73. N. S. Jacobson, K. S. Dobson, P. A. Truax, M. E. Addis, K. Koerner, J. K. Gollan, E. Gortner, and S. E. Prince (1996), "A Component Analysis of Cognitive Behavioral Treatment for Depression," *Journal of Consulting & Clinical Psychology* 64:295–304.

74. Both drug and talking therapies tend to take one or the other side of the brain-mind split, and this is aggravated by the professional interests that then accrue to particular treatment approaches. There is an extensive literature and considerable recent evidence to the effect that exercise, for instance, may have substantial effects on mental state. In the hunt for specificity, however, we seem to have lost our ability to manipulate nonspecific factors such as these.

75. C. Spanier, E. Frank, A. B. McEachran, V. I. Grochocinski, and D. J. Kupfer (1996), "The Prophylaxis of Depressive Episodes in Recurrent Depression Following Discontinuation of Drug Therapy: Integrating Psychological and Biological Factors," *Psychological Medicine* 26:461–475.

76. Pharmacotherapy can be ridiculously inexpensive. I recently calculated the drug budget for a year for a sixty-bed general psychiatric unit that had 812 separate admissions that year; the total spent on all psychotropic medications, major and minor tranquilizers, hypnotics, anticholinergics, and antidepressants came to $30,000 for the year. The money to be made is clearly in sales to individuals in the community. See F. Johnstone et al., *British Journal of Psychiatry* 167:112–113.

77. Kline, still active in the field, died on February 11, 1983, at the age of sixty-nine, while undergoing heart surgery for an accidentally discovered aortic aneurysm

78. Shorter (1996).

79. G. L. Klerman (1990), "The Psychiatric Patient's Right to Effective Treatment: Implications of Osheroff vs. Chestnut Lodge," *American Journal of Psychiatry* 147:409–418.

80. Chestnut Lodge was, it can be noted, maintaining a pre-Esquirol (or post-Kramer) focus on the larger personality aspects of the case.

81. G. L. Klerman (1983), "The Efficacy of Psychotherapy as the Basis for Public Policy," *American Psychologist* 38:929–934

82. A. A. Stone (1990), "Law, the Science, and Psychiatric Malpractice: A Response to Klerman's Indictment of Psycho-analytic Psychiatry," *American Journal of Psychiatry* 147:419–427.

83. G. L. Klerman (1991), "The Osheroff Debate: Finale," *American Journal of Psychiatry* 148:387–388; A. A. Stone (1991), "Dr. Stone Replies," *American Journal of Psychiatry* 148:388–390.

84. H. J. Eysenck (1994), "The Outcome Problem in Psychotherapy: What Have We Learned?" *Behavior Research and Therapy* 32:477–495.

85. Klerman died on April 3, 1992.

86. G. L. Klerman (1992), "Drug Treatment of Panic Disorder," *British Journal of Psychiatry* 161:470.

87. Marks and Scott (1990), p. 202; I. M. Marks and D. Healy (1998), "Marketing the Evidence," in Healy (1998).

88. Shepherd and Healy (1998).

89. M. Shepherd (1993), "The Placebo: From Specificity to the Non-specific and Back," *Psychological Medicine* 23:569–578.

90. This focus is unlike that of cognitive therapy.

91. M. Shepherd (1995), "Two Faces of Emil Kraepelin," *British Journal of Psychiatry* 166:174–183.

92. Klerman (1983).

93. Data derived from ongoing studies involving P. Michael, M. Jones, D. Hirst, T. McMonagle, M. Savage, and D. Healy (see Chapter 1). I have no reason to believe that these figures are not representative of what was happening elsewhere in Western psychiatry.

94. Shepherd and Healy (1998).

95. B. Williams (1996), "The Role of the Person in People-Centred Mental Health Care," Ph.D. thesis, University of Wales.

96. A. Kleinman (1988), *Rethinking Psychiatry: From Cultural Category to Personal Experience* (New York: Free Press); A. Kleinman (1988), *The Illness Narratives: Suffering, Healing, and the Human Condition* (New York: Basic Books).

97. D. Healy (1990), *The Suspended Revolution* (London: Faber & Faber), chap. 1.

98. S. J. Jachuk, H. Brierley, S. Jachuk, and P. M. Willcox (1982), "The Effect of Hypotensive Drugs on the Quality of Life," *Journal of the Royal College of General Practitioners* 32:103–105.

99. Despite the neo-Kraepelinian revolution, opinion leaders in psychiatry were still arguing in 1997 that the profession faced disaster if it did not stop offering to solve social ills and pull back to a biomedical focus (T. Detre and M. C. McDonald [1997], "Managed Care and the Future of Psychiatry," *Archives of General Psychiatry* 54:201–204). It is worth noting that the co-morbid occurrence of a number of different disorders in one individual is at odds with Kraepelin's notion of *Krankheitseinheit*, according to which a variety of clinical presentations should be explicable by an appeal to one disease entity.

Postscript

1. Cosmetic surgery occupies something of a similar role today. In general there is an increasingly ambiguous interface between therapeutics and the possibilities offered by medical engineering. The antidepressant story has implications for the use of agents such as HRT (hormone replacement therapy) or growth hormone and for "conditions" which have a significant social construction, of which hyperactivity and the addictions are perhaps two obvious examples within the mental health arena at present.

2. The new generation of antipsychotic drugs—olanzapine, sertindole, ziprasidone, quetiapine, and others—do not at present appear to be a radical departure from what has already been available.

3. T. Ustun and N. Sartorius (1995), *Mental Illness in General Health Care* (Chichester: John Wiley & Sons).

4. D. Healy (1987), "Rhythm and Blues," *Psychopharmacology* 93:271–285; C. L. Ehlers, E. Frank, and D. J. Kupfer (1988), "Social Zeitgebers and Biological Rhythms: A Unified Approach to Understanding," *Archives of General Psychiatry* 45:948–952; D. Healy and J. M. Waterhouse (1991), "Reactive Rhythms and Endogenous Clocks: A Shift Work Model of Affective Disorders," *Psychological Medicine* 21:557–564; D. Healy and J. M. Waterhouse (1995), "The Circadian System and the Therapeutics of the Affective Disorders," *Pharmacology and Therapeutics* 65:241–263; R. Adeniran, D. Healy, H. M. Sharp, J. M. G. Williams, D. S. Minors, and J. M. Waterhouse (1996), "Interpersonal Sensitivity, Circadian Rhythm Disruption and Depressive Symptoms," *Psychological Medicine* 26:1211–1221; E. Frank, D. J. Kupfer, C. L. Ehlers, T. H. Monk, C. Cornes, S. Carter, and D. Frankel (1995), "Interpersonal and Social Rhythm Therapy for Bipolar Disorder: Integrating Interpersonal and Behavioral Approaches, *Behavior Therapist* 17:143–149.

5. A. Kleinman (1988), *Rethinking Psychiatry: From Cultural Category to Personal Experience* (New York: Free Press).

6. D. Healy (1993), "Involving Users in Mental Health Services in the Era of the Word-Processor and the Database," in C. Crosby and M. Barry, eds. (1993),

Community Care: Evaluation of the Provision of Mental Health Services (Aldershot: Avebury Press), pp. 209–231; K. Fitzgerald, B. Williams, and D. Healy (1996), "Shared Care?" *Journal of Mental Health* 6:37–46.

7. D. Healy (1997), *Psychiatric Drugs Explained* (London: Mosby Yearbooks).

8. The increasing availability of melatonin and melatonin analogues may make a difference here.

9. C. M. Vize and R. Priest (1993), "Defeat Depression Campaign: Attitudes toward Depression," *Psychiatric Bulletin* 17:573–574.

10. R. Carson (1962), *Silent Spring* (Boston: Houghton Mifflin). The ironies abound: Galenism was the height of orthodoxy and involved a strict adherence to a theoretical model; neo-Galenism is heretical and eschews theory. Humoralism in the Galenic sense does not seem retrievable, as it required a certain uniformity of culture, where our societies seem irredeemably pluralistic.

11. This is not an argument against regulation per se but a questionning of one component of current regulations. It would also be possible to question another—the emphasis on developing therapies for specific disease indications rather than for broader indications such as anxiolysis or anti-inflammatory effects. Some regulation and especially evaluation of effectiveness for the indications claimed seems absolutely necessary.

12. Mahlon Johnson, a neuropathologist, was infected with HIV in 1992 while doing a postmortem. He treated himself with a radical combination of drugs. In 1996 he got $1.3 million from Bantam for his story and over $1 million from Twentieth-Century Fox for the film rights to *Working on a Miracle*.

13. See J. A. Trostle (1986), "Medical Compliance as an Ideology," *Social Science and Medicine* 27:1299–1308.

14. E. Shorter (1985), *Bedside Manners: The Troubled History of Doctors and Patients* (New York: Viking Press).

Index